Graduate Texts in Physics

Graduate Texts in Physics

Graduate Texts in Physics publishes core learning/teaching material for graduate- and advanced-level undergraduate courses on topics of current and emerging fields within physics, both pure and applied. These textbooks serve students at the MS- or PhD-level and their instructors as comprehensive sources of principles, definitions, derivations, experiments and applications (as relevant) for their mastery and teaching, respectively. International in scope and relevance, the textbooks correspond to course syllabi sufficiently to serve as required reading. Their didactic style, comprehensiveness and coverage of fundamental material also make them suitable as introductions or references for scientists entering, or requiring timely knowledge of, a research field.

Series Editors

Professor Richard Needs
Cavendish Laboratory
JJ Thomson Avenue
Cambridge CB3 0HE
UK
rn11@cam.ac.uk

Professor William T. Rhodes
Department of Computer and Electrical Engineering and Computer Science
Imaging Science and Technology Center
Florida Atlantic University
777 Glades Road SE, Room 456
Boca Raton, FL 33431
USA
wrhodes@fau.edu

Professor Susan Scott
Department of Quantum Science
Australian National University
Canberra ACT 0200, Australia
susan.scott@anu.edu.au

Professor H. Eugene Stanley
Center for Polymer Studies Department of Physics
Boston University
590 Commonwealth Avenue, Room 204B
Boston, MA 02215
USA
hes@bu.edu

Professor Martin Stutzmann
Technische Universität München
Am Coulombwall
Garching 85747, Germany
stutz@wsi.tu-muenchen.de

Alexandre Zagoskin

Quantum Theory
of Many-Body Systems

Techniques and Applications

Second Edition

Alexandre Zagoskin
Department of Physics
Loughborough University
Leicestershire
UK

ISSN 1868-4513   ISSN 1868-4521   (electronic)
ISBN 978-3-319-37429-1   ISBN 978-3-319-07049-0   (eBook)
DOI 10.1007/978-3-319-07049-0
Springer Cham Heidelberg New York Dordrecht London

Printed on acid-free paper

Springer is part of Springer Science+Business Media (www.springer.com)

To my parents

Preface to the Second Edition

Over the last 15 years, there has been a considerable amount of advancements in condensed matter physics: graphene, pnictide superconductors, and topological insulators, to name just a few. The understanding, and to a large degree the very discovery, of these new phenomena required the use of advanced theoretical tools. The knowledge of the basic methods of quantum many-body theory thus becomes more important than ever for each student in the field.

Some of the most challenging current problems stem from the spectacular progress in quantum engineering and quantum computing, more specifically, in developing solid-state based—mostly superconducting—quantum bits and qubit arrays. During this short period, we arrived from the first experimental demonstration of coherent quantum tunnelling in single qubits (which are, after all, quite macroscopic objects) to precise manipulation of quantum state of several qubits, their quantum entanglement over macroscopic distances and, recently, signatures of quantum coherent behaviour in devices comprising hundreds of qubits. The difficulty is that it is impossible to directly simulate such large, partially coherent, essentially nonequilibrium quantum systems, due to the sheer volume of computation—which was the motivation behind quantum computing in the first place. It would seem that one needs a quantum computer in order to make a quantum computer! The hope is that appropriate generalizations of the methods of nonequilibrium many-body theory would provide good enough approximations and keep the research going until the time when (and if) the task can be handed to quantum computers themselves.

Given the above considerations, I did not feel the need to change the scope or the approach of the book. I have, though, added a new chapter, in order to introduce bosonization and elements of conformal field theory. These are beautiful and powerful ideas, especially useful when dealing with low-dimensional systems with interactions, and belong to the essential condensed matter theory toolkit. I have also corrected some typos—hopefully introducing fewer new ones in the process.

In addition to those of my teachers and colleagues, whom I had the opportunity to thank in the preface to the first edition, I would like to express my gratitude to Profs. A. N. Omelyanchouk, F. V. Kusmartsev, Jeff Young, and Franco Nori, and to all my colleagues at the University of British Columbia, D-Wave Systems Inc., RIKEN, and Loughborough University, with whom I had the pleasure and honour to collaborate during this time. My special thanks to Dr. Uki Kabasawa, who translated the first edition of this book to the Japanese, and whose questions and helpful remarks contributed to improving the book you hold.

Loughborough, UK   Alexandre Zagoskin

Preface to the First Edition

This book grew out of lectures that I gave in the framework of a graduate course in quantum theory of many-body systems at the Applied Physics Department of Chalmers University of Technology and Göteborg University (Göteborg, Sweden) in 1992–1995. Its purpose is to give a compact and self-contained account of basic ideas and techniques of the theory from the "condensed matter" point of view. The book is addressed to graduate students with knowledge of standard quantum mechanics and statistical physics. (Hopefully, physicists working in other fields may also find it useful.)

The approach is—quite traditionally—based on a quasiparticle description of many-body systems and its mathematical apparatus—the method of Green's functions. In particular, I tried to bring together all the main versions of diagram techniques for normal and superconducting systems, in and out of equilibrium (i.e., zero-temperature, Matsubara, Keldysh, and Nambu–Gor'kov formalisms) and present them in just enough detail to enable the reader to follow the original papers or more comprehensive monographs, or to apply the techniques to his own problems. Many examples are drawn from mesoscopic physics—a rapidly developing chapter of condensed matter theory and experiment, which deals with macroscopic systems small enough to preserve quantum coherence throughout their volume; this seems to me a natural ground to discuss quantum theory of many-body systems.

The plan of the book is as follows.

In Chapter 1, after a semi-qualitative discussion of the quasiparticle concept, Green's function is introduced in the case of one-body quantum theory, using Feynman path integrals. Then its relation to the S-operator is established, and the general perturbation theory is developed based on operator formalism. Finally, the second quantization method is introduced.

Chapter 2 contains the usual zero-temperature formalism, beginning with the definition, properties, and physical meaning of Green's function in the many-body system, and then building up the diagram technique of the perturbation theory.

In Chapter 3, I present equilibrium Green's functions at finite temperature, and then the Matsubara formalism. Their applications are discussed in relation to linear response theory. Then Keldysh technique is introduced as a means to handle essentially nonequilibrium situations, illustrated by an example of quantum conductivity of a point contact. This gives me an opportunity to discuss both Landauer and tunneling Hamiltonian approaches to transport in mesoscopic systems.

Finally, Chapter 4 is devoted to applications of the theory to the superconductors. Here the Nambu–Gor'kov technique is used to describe superconducting phase transition, elementary excitations, and current-carrying state of a superconductor. Special attention is paid to the Andreev reflection and to transport in mesoscopic superconductor–normal metal–superconductor (SNS) Josephson junctions.

Each chapter is followed by a set of problems. Their solution will help the reader to obtain a better feeling for how the formalism works.

I did not intend to provide a complete bibliography, which would be far beyond the scope of this book. The original papers are cited when the results they contain are either recent or not widely known in the context, and in a few cases where interesting results would require too lengthy a derivation to be presented in full detail (those sections are marked by a star*). For references on more traditional material, I have referred the reader to existing monographs or reviews.

For a course in quantum many-body theory based on this book, I would suggest the following tentative schedule[1]:

Lecture 1 (Sect. 1.1); Lecture 2 (Sect. 1.2.1); Lecture 3 (Sect. 1.2.2, 1.2.3); Lecture 4 (Sect. 1.3); Lecture 5 (Sect. 1.4); Lecture 6 (Sect. 2.1.1); Lecture 7 (Sect. 2.1.2); Lecture 8 (Sect. 2.1.3, 2.1.4); Lecture 9 (Sect. 2.2.1, 2.2.2); Lecture 10 (Sect. 2.2.3); Lectures 11–12 (Sect. 2.2.4); Lecture 13 (Sect. 3.1); Lecture 14 (Sect. 3.2); Lecture 15 (Sect. 3.3); Lecture 16 (Sect. 3.4); Lecture 17 (Sect. 3.5); Lecture 18 (Sect. 3.6); Lecture 19 (Sect. 3.7); Lecture 20 (Sect. 4.1); Lecture 21 (Sect. 4.2); Lecture 22 (Sect. 4.3.1, 4.3.2); Lecture 23 (Sect. 4.3.3, 4.3.4); Lecture 24 (Sect. 4.4.1, 4.4.2); Lectures 25–26 (Sect. 4.4.3–5); Lecture 27 (Sect. 4.5.1); Lecture 28 (Sect. 4.5.2–4); Lecture 29 (Sect. 4.6).

Acknowledgments

I am deeply grateful to Professor R. Shekhter, collaboration with whom in preparing and giving the course on quantum theory of many-body system significantly influenced this book.

I wish to express my sincere thanks to the Institute for Low Temperature Physics and Engineering (Kharkov, Ukraine) and Professor I. O. Kulik, who first taught me what condensed matter theory is about; to the Applied Physics Department of Chalmers University of Technology and Göteborg University (Göteborg, Sweden) and Professor M. Jonson, and to the Physics and Astronomy

---

[1] Based on a "two hours" (90 min) lecture length.

Department of the University of British Columbia (Vancouver, Canada) and Professor I. Affleck for support and encouragement.

I thank Drs. S. Gao, S. Rashkeev, P. Hessling, R. Gatt, and Y. Andersson for many helpful comments and discussions.

Last, but not least, I am grateful to my wife Irina for her unwavering support and for actually starting this project by the comment, "Well, if you are spending this much time on preparing these handouts, you should rather be writing a book," and to my daughter Ekaterina for her critical appreciation of the illustrations.

Vancouver, British Columbia   Alexandre Zagoskin

Contents

Chapter 1
Basic Concepts

> When asked to calculate the stability of a dinner table with four
> legs, a theorist rather quickly produces the results for tables
> with one leg and with an infinite number of legs. He spends the
> rest of his life in futile attempts to solve the problem for a table
> with an arbitrary number of legs.
>
> A popular wisdom.
> From the book "Physicists keep joking."

Abstract Basic ideas of quantum many-body theory. Qualitative picture of quasi-particles. Thomas-Fermi screening. Plasmons. Propagators and path integrals in a one-body quantum theory. Aharonov-Bohm effect. Perturbation theory for a propagator. Second quantization and field operators.

1.1 Introduction: Whys and Hows of Quantum Many-Body Theory

Technically speaking, physics deals only with one-body and many-body problems (because the two-body problem reduces to the one-body case, and the three-body problem does not, and is already insolvable, (Fig. 1.1)). Still, what an average physicist thinks of as "many" in this context is probably something of the order of $10^{19} - 10^{23}$, the number of particles in a cubic centimeter of a gas or a solid, respectively. When you have this many particles on your hands, you need a many-body theory. At these densities, the particles will spend enough time at several de Broglie wavelengths from each other, and therefore we need a quantum many-body theory. (A good thing too: What we really should not mess with is the classical chaos!)

The real reason why you want to deal with such a large collection of particles in the first place, instead of quietly discussing a helium atom, is of course that 10^{23} is much closer to infinity. The epigraph, or intuition, or both, tell us that the infinite

A. Zagoskin, *Quantum Theory of Many-Body Systems*,
Graduate Texts in Physics, DOI: 10.1007/978-3-319-07049-0_1,
© Springer International Publishing Switzerland 2014

Fig. 1.1 One, two, many …

number of particles (or legs) is almost as easy to handle as one, and much, much easier than, say, 3, 4, or 7.

The basic idea of the approach is that instead of following a large number of strongly interacting real particles, we should try to get away with considering a relatively small number of weakly interacting *quasiparticles*, or *elementary excitations*.

An elementary excitation is what its name implies: something that appears in the system after it has suffered an external perturbation, and to which the reaction of the system to this perturbation can be almost completely ascribed—like a ripple on the surface of a pond, only in quantum theory those ripples will be quantized. In a crystal lattice, such quantized ripples are *phonons*, sound quanta, which carry both energy and *quasi*momentum, and only weakly interact with each other and, e.g., electrons. Strike a solid, or heat it, and you excite (that is, generate) a whole bunch of phonons, which will carry away the energy and momentum of your influence (Fig. 1.1).

Phonons form a rather dilute Bose gas, and therefore are much easier to deal with, than the actual particles—atoms or ions—that constitute the lattice. The phonons are called quasiparticles not only because they don't exist outside the lattice; they also have finite lifetime, unlike the stable "proper" particles. A key point here is that the quasiparticles must be stable enough: if they decay faster than they can be created, the whole description loses sense.

Let us now consider a system of interacting electrons in a metal lattice (which we will describe by the standard "jellium" model of uniformly distributed positive charge, neutralizing the total charge of free electrons). Here we have real particles, which interact through strong Coulomb forces, which moreover have an infinite radius (because they decay only as $1/r^2$). For a given electron, we must thus take into account influences of all the other electrons. Therefore, nothing actually depends on the details of behavior of any of those electrons! We can safely replace their action by some average field, depending on averaged electronic density $n(\mathbf{r})$, thus arriving at the *mean field approximation* (MFA). We immediately use it to calculate the screening of Coulomb interaction, to see that not only particles, but interactions as well, are changed in the many-body systems.

1.1.1 Screening of Coulomb Potential in Metal

Suppose we place an external charge Q in the system. It will create a potential $\Phi(\mathbf{r})$, which will change the initial uniform distribution of electronic density,

$$n = \frac{p_F^3}{3\pi^2\hbar^3}. \tag{1.1}$$

Here $p_F = \sqrt{2m\mu}$ is the Fermi momentum, and we have used the well-known relation between p_F and density of the electron gas. Of course, if the electronic density becomes coordinate dependent, so is the Fermi momentum: $p_F \to p_F(n(\mathbf{r}))$. In equilibrium, the electrochemical potential of the electrons must be constant, that is,

$$\mu = \frac{p_F^2(n(\mathbf{r}))}{2m} + e\Phi(\mathbf{r}) = \text{const}, \tag{1.2}$$

and we easily find that

$$n(\mathbf{r}) = \frac{(2m(\mu - e\Phi(\mathbf{r})))^{3/2}}{3\pi^2\hbar^3}. \tag{1.3}$$

If there is no external potential, we return to the unperturbed case (1.1).

Now let us employ the electrostatics. The potential must satisfy Poisson's equation,

$$\nabla^2\Phi(\mathbf{r}) = 4\pi\rho \equiv 4\pi e\Delta n,$$

where ρ is the charge density induced on the neutral background by the probe charge, and $\Delta n(\mathbf{r}) = n(\mathbf{r}) - n$ is the change in electronic density. (The positive "jellium" neutralized the negative charge of the electrons, remember? Besides, we assume that it has unit dielectric permeability, $\varepsilon = 1$.) Therefore, we can write

$$\nabla^2\Phi(\mathbf{r}) = 4\pi e\left[\frac{(2m(\mu - e\Phi(\mathbf{r})))^{3/2} - (2m\mu)^{3/2}}{3\pi^2\hbar^3}\right]. \tag{1.4}$$

This is the *Thomas–Fermi equation*, first obtained in the theory of electron density distribution in atoms.

Generally, this nonlinear equation can be solved only numerically. If, though, we assume that $e\Phi$ is much smaller than the Fermi energy, μ, and expand the right-hand side of (1.4) in powers of Φ to the lowest order—that is, taking

$$\Delta n(\mathbf{r}) = -\frac{3}{2}\frac{e\Phi(\mathbf{r})n}{\mu}, \tag{1.5}$$

we obtain a linear equation,

$$\nabla^2\Phi(\mathbf{r}) = \frac{1}{\lambda_{TF}^2}\Phi(\mathbf{r}). \tag{1.6}$$

Here λ_{TF} is the *Thomas–Fermi screening length*,

$$\lambda_{TF} = \frac{\mu^{1/2}}{\sqrt{6\pi}en^{1/2}} = \frac{\pi^{1/6}}{2\cdot 3^{1/6}}\frac{\hbar}{em^{1/2}}n^{-1/6}. \tag{1.7}$$

To find the physical meaning of λ_{TF}, let us solve (1.6) for $\Phi(\mathbf{r})$, imposing the condition that at small distances $\Phi(\mathbf{r}) \approx Q/r$. This is reasonable, because close enough to the probe charge—at $r \ll n^{-1/3}$—there will be, on average, no electrons around, and we should observe the same potential as in vacuum. (Of course, the constant potential of the "jellium" does not matter.) Therefore, we obtain the following result:

$$\Phi(\mathbf{r}) = \frac{Q}{r}\exp[-r/\lambda_{TF}]. \tag{1.8}$$

As you see, the Coulomb potential is modified, and it now exponentially decays at a distance of order λ_{TF} from the source—it was screened.[1]

Physically, when a positive charge is brought into our system, the electrons will be attracted to it, forming a negatively charged cloud. Formula (1.8) tells us that the total charge of this cloud is equal and opposite to the external charge, so that for the electrons at $r \gg \lambda_{TF}$ it is completely compensated. The size of the cloud is determined by the interplay between the Coulomb attraction of electrons to the charge and their Coulomb repulsion from each other. The latter we took into account in our approximate formula (1.5). In the case of negative external charge, the electrons will be repulsed, laying bare the positive charge of the "jellium" background, which again will compensate for it. In either case, the charge is being surrounded by the charged cloud of equal and opposite sign. This holds, of course, for every single electron in the system (Figs. 1.2, 1.3).

Now we can refine the criterion for the applicability of our "averaging" approach. Previously, we thought that since the Coulomb interaction reaches to infinity, the number of electrons acting on any single electron is always very large—the same as the number of electrons in the whole system, and their action could be replaced by

---

[1] The dependence (1.8) is often called the *Yukawa potential*, though in the context of screening of the Coulomb potential in (classical) plasma it was derived by Debye and Hückel, with a different screening length, $\lambda_D \sim \sqrt{k_B T}/(ne)$. (The difference is due to the use of the Boltzmann instead of the Fermi distribution in a nondegenerate gas.)

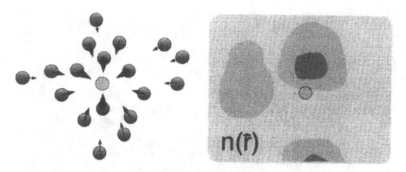

Fig. 1.2 Forces acting on an electron and mean field approximation

the action of some average charge density, ne. Now we see that the actual number of electrons influencing any single electron is only of order $n\lambda_{TF}^3$. Then our considerations are exact as long as

$$n\lambda_{TF}^3 \gg 1. \tag{1.9}$$

Since

$$n^{1/3}\lambda_{TF} \propto \sqrt{\frac{\mu}{n^{1/3}e^2}} \alpha \sqrt{\frac{p_F^2 \cdot h}{m \cdot p_F \cdot e^2}} \propto \sqrt{\frac{\hbar v_F}{e^2}},$$

the above criterion reduces to

$$\frac{\hbar v_F}{e^2} \gg 1. \tag{1.10}$$

Recalling that $e^2/(\hbar c) \approx 1/137$, we find on the left-hand side $137 v_F/c$, a ratio of order one.

This is a nasty surprise, of the sort that are abundant in the many-body theory. It indicates that even if our approximation gives a qualitatively correct answer (which it does), we must be ready to take into account the effects of deviations from the "mean field" picture, and preferably in a systematic way. (A pleasant surprise, which also occasionally can be encountered here, is that mean field approximation often—though not always—gives excellent results even outside its domain of applicability.)

One all-important qualitative conclusion that we can make. based on our results is that in metal, an electron is surrounded by the screening cloud of other electrons. Any force applied to it will have to accelerate the whole cloud. Therefore, the electron will behave as if it had a larger effective mass, m^*, than in vacuum! (We have not yet considered its interactions with the crystal lattice itself.) Instead of being a point particle, it acquires a finite size, the size of the cloud. We can thus call this complex entity "electron + cloud" a *quasi* particle. As they say, an electron is *dressed*. (Logically, the "lone" electron is called a *bare* particle.)

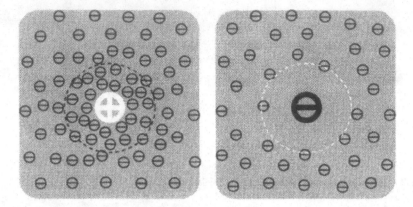

Fig. 1.3 Screening of external charge in metal

The quasi electrons are what we will see when probing the system. Pleasantly, the problem of considering these quasielectrons, interacting through some sort of short-range potential (even if it is not exactly the Yukawa potential we derived), is much simpler than the initial problem dealing with electrons strongly interacting with infinite-range Coulomb forces. We need only accurately find effective masses and potentials. This "only" is actually the very subject of the many-body theory!

1.1.2 Time-Dependent Effects: Plasmons

To better understand possible underwater rocks, let us recollect that so far we have considered only the static case, which is all right for a probe charge in rest, but we implied that this picture should hold for one in motion. One can guess that if a screening cloud is likely to follow a slow electron, it might lose the fast one: after all, the cloud formation takes some time.

To this end, let us single out an electron (number zero) with coordinate $\mathbf{r}(t)$ and write for it the classical equation of motion:

$$m\ddot{\mathbf{r}}(t) = e\mathbf{E}(\mathbf{r}(t)),$$

where the electric field $\mathbf{E}$ arises due to local deviation of the electronic density from its equilibrium value and satisfies the Maxwell equation

$$\nabla_{\mathbf{r}} \cdot \mathbf{E}(\mathbf{r}(t)) = 4\pi e \left[\sum_{i \neq 0} \delta(\mathbf{r} - \mathbf{r}(t)) - n \right].$$

Here we have used for the moment the exact electronic density at the point $\mathbf{r}$, $\sum_{i \neq 0} \delta(\mathbf{r}_i - \mathbf{r})$ (summation is taken over all other electrons). If now write

$\mathbf{r}(t) = \mathbf{r}_0 + \Delta\mathbf{r}(t)$ and take into account that the unperturbed density of electrons can be written as $n = n(\mathbf{r}_0) = \sum_{i \neq 0} \delta(\mathbf{r}_i - \mathbf{r}_0)$, we can write a linearized equation at $\mathbf{r}_0$:

$$\nabla_{\mathbf{r}_0} \cdot \mathbf{E}(\mathbf{r}_0) = -4\pi e \Delta\mathbf{r}(t) \cdot \nabla_{\mathbf{r}_0} \sum_{i \neq 0} \delta(\mathbf{r}_i - \mathbf{r}_0) = -4\pi e \nabla_{\mathbf{r}_0} \cdot \Delta\mathbf{r}(t) \sum_{i \neq 0} \delta(\mathbf{r}_i - \mathbf{r}_0).$$

(1.11)

Since if $\Delta\mathbf{r}(t) = 0$ there will be no field, we can write simply

$$\mathbf{E}(\mathbf{r}_0) = -4\pi e \Delta\mathbf{r}(t) n.$$

Substituting this into the equation of motion, performing Fourier transformation over frequencies, and inserting the expression obtained for $\Delta\mathbf{r}$ back into the Maxwell equation, we find

$$\Delta\mathbf{r}(\omega) = -\frac{e\mathbf{E}(\omega)}{m\omega^2};$$

(1.12)

$$\mathbf{E}(\omega) = \frac{4\pi e^2 n}{m\omega^2}\mathbf{E}(\omega).$$

(1.13)

The result is consistent if the frequency

$$\omega = \omega_p \equiv \sqrt{\frac{4\pi e^2 n}{m}},$$

(1.14)

the *plasma frequency*. This is the frequency of small oscillations of a uniform electron gas. The period of plasma oscillations gives the characteristic time of any charge redistribution in the metal.

Now we see that the screening cloud will be able to follow the electron only as long as its velocity

$$v \ll \lambda_{\mathrm{TF}}\omega_p.$$

(1.15)

Otherwise, the surrounding electrons simply will not have time to react! (Figs. 1.4, 1.5).

The quanta of plasma oscillations are called *plasmons*. They can propagate across the system and are created whenever the charge neutrality of the metal is disturbed. This is yet another example of quasiparticles. The screening of Coulomb potential, e.g., and dressing of bare electrons in the metal can be directly described in terms of plasmons.

We have already mentioned phonons. Interactions of the electrons with the crystal lattice can lead to what can be described as a phonon cloud around an electron, forming a *polaron*. Since the characteristic phonon frequency (Debye frequency, ω_D) is much lower than ω_p in metals, there electrons always leave the phonon cloud behind. Nevertheless, such a cloud can be run into by another electron; as a result, an effective electron–electron interaction arises, which can lead to such a spectacular phenomenon as superconductivity.

Fig. 1.4 **a** Bare particle. **b** Interaction

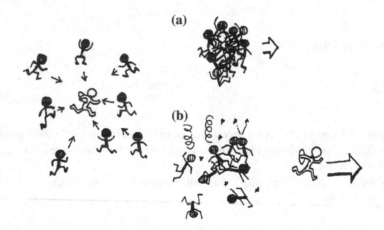

Fig. 1.5 Particles in the field. **a** $v \ll \lambda\omega$; **b** $v \gg \lambda\omega$

In short, the many-body systems seem to yield to a quasiparticle approach. The question is, how to make it work.

1.2 Propagation Function in a One-Body Quantum Theory

1.2.1 Propagator: Definition and Properties

The introduction of quasiparticles would hardly be an improvement if for each new problem we had to invent a completely new method.

Fortunately, a general mathematical apparatus, based on the famous Green's functions and Feynman diagrams, takes care of all details, and thus makes a quasiparticle approach efficient. Actually this apparatus works so excellently that often it is applied to problems that can be more easily solved by other methods. And as with any efficient tool, people tend to forget about its natural limitations. But in itself, the approach is a rare example of mathematical beauty, physical insight, and practical efficiency. It originates from quantum field theory, and therefore is designed to deal with a system of an infinite number of degrees of freedom. This is exactly what the condensed matter theory needs. Moreover, usually in condensed matter problems we are not interested in relativistic covariance—which simplifies the necessary apparatus and makes mastering it easier. On the other side of this is the fact that the enormous wealth of effects in solid-state physics already provides a much wider field for field theorists, than the "standard" field theory, including many effects that either were not, or simply cannot, be observed elsewhere (like $1 + 1$ or $2 + 1$ field theory).

We have advised you against using Green's functions and Feynman diagrams where simpler methods can be applied. Nevertheless, we now start from the case of a *single quantum particle*. The reason is that in this way we will be able to derive all the general expressions that actually do not depend on the number of particles in the system, to see clearly the structure of the theory, and to recognize where the many-particle properties *really* enter the picture.

It is well known that the probability of observing a single quantum particle at point x at time t is determined by the square modulus of the wave function of this particle at this place and time, $|\Psi(x, t)|^2$. In order to find $\Psi(x, t)$ we could, e.g., solve the Schrödinger equation, given initial and boundary conditions. But many properties of the solution can be obtained directly from general principles.

First, the superposition principle. Mathematically, this means that $\Psi(x, t)$ satisfies a linear differential equation. Physically, the wave functions follow the Huygens principle; i.e., each point of the wave front acts as a secondary emitter. Anyway, this allows us to write for the wave function at some time t,

$$\Psi(x, t) = \int dx' K(x, t; x', t')\Psi(x', t'), \qquad t > t'. \tag{1.16}$$

The kernel $K(x, t; x't')$ describes the propagation of the Ψ-wave from (x', t') to (x, t) and therefore is called the *propagation function*, or *propagator*. It is fundamental for all our theory. Note that due to the *causality principle*

$$K(x, t; x', t') = 0, \qquad t < t', \tag{1.17}$$

so that the future does not affect the past. (It would not be this easy if we had to deal with relativistic covariance!)

Let's now suppose that the particle at the initial moment is strictly localized: $\Psi(x', t') = \delta(x' - x_0)$. Then from (1.16)

$$\Psi(x, t) = K(x, t; x't'). \tag{1.18}$$

That is, more specifically, the propagator is the *transition amplitude* of the particle between the points (x', t') and (x, t), and its square modulus gives the transition *probability*.

In (1.16) we did not specify the moment t', except that it must precede the observation moment t. Then, for some $t'' > t'$ we obtain

$$\Psi(x, t) = \int dx'' K(x, t; x'', t'') \Psi(x'', t'')$$

$$= \int dx'' \int dx' K(x, t; x'', t'') K(x'', t''; x', t') \Psi(x', t'). \quad (1.19)$$

Since both expressions must be identical, for $(t > t'' > t')$ we obtain

$$K(xt; x't') = \int dx'' K(xt; x''t'') K(x''t''; x't'). \quad (1.20)$$

This is the *composition property* of the propagator, and we will heavily use it later. Of course, this is a reformulation of the Huygens principle, from the wave point of view. But what does it mean from the particle point of view? If we want to know the probability amplitude for a particle, starting at (x_i, t_i) to reach point x_f at t_f, at *any* intermediate moment t' we must take into account *all* conceivable positions the particle can occupy in order to obtain a proper result. This situation is often illustrated by the famous double-slit experiment (we do not know for certain through which slit the particle passed). It is a close, though a slightly different, situation. There (in the double-slit experiment) we know the relevant region in space over which we should integrate (the slits), but we do not know *when* the particle passes it. Here we know the relevant time, but we have to integrate over *all* the available space. You can ponder how these two situations complement each other (just think about the stationary wave propagation).

In principle, the above picture is *almost* identical to one of Brownian motion of a classical particle. The only difference is that the reasoning is applied rather to probabilities themselves than to their complex amplitudes, and this, as you know, changes a lot.

Returning to the properties of the propagator, we have decided that for negative times it is strictly zero, while for positive times it certainly is not. This might imply the singular behavior for $t - t' = 0$. Indeed, for $t = t'$ we must get an identity:

$$\Psi(x, t) \equiv \int dx' K(xt; x't) \Psi(x', t),$$

so that

$$K(x, t; x', t) = \delta(x - x'). \quad (1.21)$$

We have now received all the information available about the properties of the propagator that could be obtained from the most general principles of quantum mechanics.

To proceed, we need more specific data. One way is to use the Schrödinger equation. The other is to formulate instead some statement from which the Schrödinger equation itself could be derived. We will take both ways, because if we are here most interested in the inner workings of the formalism (we are), it is wise to run it in both directions.

If the wave function obeys the Schrödinger equation,

$$\left[i\hbar\frac{\partial}{\partial t} - \mathcal{H}(x, \partial_x, t) \right] \Psi(x, t) = 0, \tag{1.22}$$

then it follows from (1.16) that for $t > t'$ the propagator satisfies the same equation:

$$\left[i\hbar\frac{\partial}{\partial t} - \mathcal{H}(x, \partial_x, t) \right] K(x, t; x', t') = 0. \tag{1.23}$$

Besides, we have seen that $K(x, t; x', t' = t) = \delta(x - x')$ and $K(x, t; x', t' > t) = 0$. That is, we can write

$$K(x, t; x', t') \equiv \theta(t - t')K(x, t; x', t'), \tag{1.24}$$

where $\theta(t - t')$ is the Heaviside step function. All these properties can be taken care of by the following equation:

$$\left[i\hbar\frac{\partial}{\partial t} - \mathcal{H}(x, \partial_x, t) \right] K(x, t; x', t') = i\hbar\delta(x - x')\delta(t - t'). \tag{1.25}$$

Indeed, for $t > t'$ this reduces to (1.23), while for $t \to t' + 0$ we can keep in the left-hand side of (1.25) only the term with $\partial\theta(t - t')/\partial t$, which matches the right-hand side.

From (1.25) we see that up to the factor of $i/\hbar$, the propagator is *Green's function* of the Schrödinger equation in the mathematical sense. (If $\hat{L}$ is a linear differential operator, then Green's function of the equation $\hat{L}\psi = 0$ is the solution to the equation $\hat{L}G_\psi = -\delta(x - x')$.) Therefore, quantum-mechanical propagators are more often called Green's functions, especially in the many-particle case; since our solution vanishes for $t < t'$, it is called the *retarded* Green's function. We, though, will keep calling the function $K(x, t; x', t')$ a propagator, to stress that we are still working on the one-particle problem.

It is easy to see that for a free particle of mass m (described by the Hamiltonian $(\mathcal{H} = (-\hbar^2/2m)(\partial_x)^2)$) the propagator depends only on differences of its arguments, and the solution to (1.25) is given by

$$K_0(x - x', t - t') = \left(\frac{m}{2\pi i\hbar(t - t)} \right)^{d/2} \exp\left(\frac{im(x - x')^2}{2\hbar(t - t)} \right) \theta(t - t'). \tag{1.26}$$

Here d is the space dimensionality.

In the simplest case of one dimension (generalizations to $d > 1$ are straightforward) formula (1.26) immediately follows after we Fourier transform (1.25):

$$\left(\hbar\omega - \frac{(\hbar k)^2}{2m}\right) K_0(k, \omega) = i\hbar;$$

$$K_0(k, \omega) = \frac{i\hbar}{\hbar\omega - \frac{(\hbar k)^2}{2m}}. \tag{1.27}$$

Now we can find

$$K_0(x, t) = \int\limits_{-\infty}^{\infty} \frac{dk}{2\pi} \int\limits_{-\infty}^{\infty} \frac{d\omega}{2\pi} e^{ikx - i\omega t} K_0(k, \omega).$$

The integral over ω is convenient to take using complex analysis:

$$\oint\limits_C \frac{d\omega}{2\pi} e^{-i\omega t} K_0(k, \omega) = \pm i \sum_{\omega_0} \text{Res}\left[K_0(k, \omega) e^{-i\omega t}\right],$$

where the sum is taken over the residues at all the poles of $K_0(k, \omega)$ as a function of the complex variable ω, and the closed contour C consists of the real axis and an infinitely remote half-circle (we assume that the integral converges). The sign depends on whether the contour is circumscribed in the positive or negative direction.

Since the integrand contains the factor $e^{-i\omega t} = e^{-it\Re\omega + t\Im\omega}$, we must close the contour in the upper half-plane of ω if $t < 0$ and in the lower half-plane if $t > 0$. Then the factor $e^{t\Im\omega}$ ensures exponential decay of the integrand on the half-circle and convergence of the integral (Watson's lemma).

As a matter of fact, the only pole of $K_0(k, \omega)$, at $\omega = \hbar k^2/2m$, lies on the very integration contour, and adding an infinitesimal imaginary part to ω displaces the pole to either the positive or negative imaginary half plane, which will dramatically change the answer (Fig. 1.6).

If, for example, we write $\omega \to \omega + i0$, the pole will shift below the real axis. Then for $t < 0$ the contour does not contain any singularity, and $K_0(x, t)$ will be identically zero. This is exactly what we need: a retarded propagator! On the other hand, for $t > 0$ the contour encloses the pole, yielding $\exp(-i\hbar k^2 t/2m)$. The momentum integral is now straightforward: it has a Gaussian form

$$K_0(x, t) = \theta(t) \int\limits_{-\infty}^{\infty} \frac{dk}{2\pi} e^{ikx - i\frac{\hbar k^2 t}{2m}},$$

and directly yields (1.26).

Note that had we displaced the pole to the upper half-plane (with $\omega \to \omega - i0$), the result would be an *advanced* Green's function, disappearing at all positive times.

Fig. 1.6 Integration contour
in the complex frequency
plane

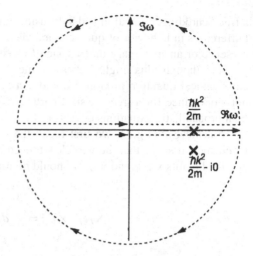

We could leave the pole on the contour, with still another answer. This game of infinitesimals reflects the fact that besides the differential Eq. (1.25), we need initial conditions—e.g., that the solution is a retarded Green's function.

The above result was almost too easy to obtain, and it leaves an unpleasant after-taste of having cheated. Indeed, it was not worth the trouble to introduce the prop-agator "from the most general principles," only to resort finally to the Schrödinger equation. Of course, in the one-particle case propagator may seem to be simply a mathematical tool to solve the wave equation, without important physics involved (as in the many-body case). But as often happens in physics, mathematical reformu-lation here also provides a tool for deeper understanding of fundamentals, which we will see in the next section.

1.2.2 Feynman's Formulation of Quantum Mechanics: Path (Functional) Integrals

To start with, note the striking similarity between the formula (1.26) for the prop-agator and the well-known formula for the probability distribution of the classical Brownian particle. The latter quantity, $P(x, t|x', t')$, gives the conventional proba-bility of finding the particle at x at some time t, *if* at some earlier time t' it was at x' (see, e.g., [3]):

$$P(x, t|x', t') = (4\pi D(t - t'))^{-d/2} \exp\left(-\frac{(x - x')^2}{4D(t - t)}\right)\theta(t - t'). \qquad (1.28)$$

The diffusion coefficient D in the quantum case is replaced by $2m/i\hbar$. From the mathematical point of view, the similarity between (1.26) and (1.28) is due to the fact that both K_0 and P are Green's functions of similar linear differential equations:

a free Schrödinger equation and a diffusion equation, $\partial_t f(x, t) = D(\partial_x)^2 f(x, t)$. Differences in behavior of quantum and classical Brownian particles are due to the presence of an imaginary unit in one of these equations. From the physical point of view, though, this might indicate some deeper link between how we describe classical and quantum motion. But a direct analogy with Brownian motion would not work, since for a free classical particle, we should obtain deterministic, rather than probabilistic, equations of motion.

To achieve this goal, we first recall the extremal action principle of classical mechanics. It states that the particle's trajectory, or path, $\mathbf{x}_{cl}(t)$, between the initial and final points $\mathbf{x}_i$, t_i and $\mathbf{x}_f$, t_f should minimize the *action*

$$S[\mathbf{x}_f t_f, \mathbf{x}_i t_j] = \int_{t_i}^{t_f} dt L(\mathbf{x}, \dot{\mathbf{x}}, t), \tag{1.29}$$

and this is the only admissible—real-path for a classical particle. The action is the so-called *functional* of the trajectory, not a function, since it depends on the behavior of $x(t)$ on the whole interval $[t_i, t_f]$.

Here $L(\mathbf{x}, \dot{\mathbf{x}}, t)$ is the *Lagrange function* of the system: in the simplest case

$$L(\mathbf{x}, \dot{\mathbf{x}}, t) = T(\dot{\mathbf{x}}(t)) - V(\mathbf{x}(t)), \tag{1.30}$$

with $T(\dot{\mathbf{x}}(t))$ and $V(\mathbf{x}(t))$ being the kinetic and potential energy respectively.

As you know, the condition of extremum means that the action is not sensitive to small deviations from the classical (extremal) path. More specifically, if we take, instead of the real path $\mathbf{x}_{cl}(t)$, a trial one, $\mathbf{x}_{tr}(t) = \mathbf{x}_{cl}(t) + \delta\mathbf{x}(t)$, where $\delta\mathbf{x}(t)$ is small, then the the change in the action integral (1.29) will be only of second order in $\delta\mathbf{x}(t)$:

$$\delta S = O(\delta\mathbf{x}(t)^2). \tag{1.31}$$

This condition is employed in derivation of the Lagrange equations of analytical mechanics, but this is not our concern for the moment. For the free particle the action is obtained directly:

$$S_0[\mathbf{x}_f t_f, \mathbf{x}_i t_i] = \int_{t_i}^{t_f} dt \frac{m\dot{\mathbf{x}}^2}{2} = \frac{m}{2} \frac{(\mathbf{x}_f - \mathbf{x}_i)^2}{(t_f - t_i)^2}(t_f - t_i) = \frac{m(\mathbf{x}_f - \mathbf{x}_i)^2}{2(t_f - t_i)}. \tag{1.32}$$

This is—up to a factor of $i/\hbar$—the very expression we have seen in the exponent of the free quantum-mechanical propagator!

The numerical factor here is very important. The role of $\hbar$ is more or less clear: since only dimensionless quantities are allowed in the exponent, and $\hbar$ is the *action quantum*, the ratio $S/\hbar$ must appear in the quantum case. The imaginary unit plays a

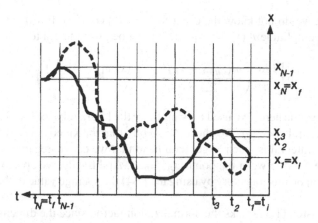

Fig. 1.7 Slicing of classical trajectories

somewhat subtler role: it brings out the interference, which distinguishes propagation of a quantum particle from its classical counterpart. But anyway, we see that the quantum-mechanical propagation is somehow related to the classical action S, or more specifically to $\exp[i/\hbar S]$.

It was the very idea first suggested by Dirac, and then implemented by Feynman, that the *propagation amplitude of a quantum particle between two points is given by a coherent sum of terms* $\exp[(i/h)S[q,\dot{q}]]$ *corresponding to* all *possible* classical *trajectories* $q(t)$ *between these points.* Instead of *propagation amplitude* here you can read propagator, $K(x_f, t_f; x_i, t_i)$, since we have established that they are the same.

What does this give us in the classical limit, when by definition the action $S \gg \hbar$? Then $\exp[i/\hbar S]$ will very quickly oscillate in response to any minute change in $q(t)$. This means that the contributions to the transition amplitude from virtually all trajectories cancel! The only exclusion will be the classical trajectory: by definition, small deviations from it do not change the action, so that the contribution of this trajectory will survive. Now you see why classical particles choose classical paths! (Similar reasoning, long ago, helped to reconcile the wave theory of light with the fact that light usually propagates along straight lines.)

In order to develop the fundamental idea that we have just described, we need some way of counting the trajectories and summing up their contributions. Let us divide the time interval $[t_i, t_f]$ into a large number $(N-1)$ of "slices" each of length $\Delta t = (t_f - t_i)/(N-1)$. The N partition moments are thus $t_1 \equiv t_i, t_2, \ldots, t_N \equiv t_f$. Each classical trajectory thus is sliced into $(N-1)$ pieces (see Fig. 1.7): $[x_1 \equiv x_i, x_2], [x_2, x_3], \ldots, [x_{N-1}, x_N \equiv x_f]$. Now we can use the composition property of the propagator, Eq. (1.20), and obtain the expression

$$K(x_N t_N; x_1 t_1) = \int_{-\infty}^{\infty} dx_{N-1} \int_{-\infty}^{\infty} dx_{N-2} \cdots \int_{-\infty}^{\infty} dx_2$$
$$\times K(x_N t_N; x_{N-1} t_{N-1}) K(x_{N-1} t_{N-1}; x_{N-2} t_{N-2}) \cdots K(x_2 t_2; x_1 t_1). \quad (1.33)$$

Of course, we do not know the exact form of $K(xt; x't')$. But if $\Delta t \to 0$, then the transition amplitude $K(x_{n+1}t_{n+1}; x_n t_n)$ must be proportional to

$$\exp\left[\frac{i\Delta t}{\hbar}\left(\frac{m(x_{n+1} - x_n)^2}{2\Delta t^2} - \frac{V(x_{n+1}) + V(x_n)}{2}\right)\right]. \qquad (1.34)$$

What we have done here? We used for simplicity the Lagrangian of Eq. (1.30) (which is, though, general enough). We chose Δt so tiny that the kinetic energy term in the action on this interval is much larger than $\hbar$, so that we can disregard all except the classical trajectory between the points x_n, x_{n+1}. And finally, we approximated the classical action on this trajectory by using in (1.34) the average value of the Lagrange function.

The expression (1.34) lacks the normalization factor, since the dimensionality of the propagator is inverse volume, L^{-d}. It can be restored from condition (1.21). If we recall one of limit representations of the delta function,

$$\delta(x) = \lim_{\alpha \to 0} \frac{1}{\sqrt{\alpha \pi i}} e^{ix^2/\alpha}, \qquad (1.35)$$

we see that the factor in question will be $(m/(2\pi\hbar i \Delta t))^{d/2}$. (This is the very factor that we obtained for the free propagator from the Schrödinger equation, but here we did *not* exploit this equation at all.)

Now substitute this form of the propagator (for infinitesimal Δt)

$$K(x\Delta t; x'0) = \left(\frac{m}{2\pi h i \Delta t}\right)^{d/2} e^{\left(\frac{i}{\hbar}\frac{m(x-x')^2}{2\Delta t^2} - \frac{V(x)+V(x')}{2}\right)\Delta t^2} \qquad (1.36)$$

into the composition equation, to obtain

$$K(x_N t_N; x_1 t_1) = \lim_{N \to \infty} \int_{-\infty}^{\infty} dx_{N-1} \int_{-\infty}^{\infty} dx_{N-2} \cdots \int_{-\infty}^{\infty} dx_2$$
$$\times \left(\prod_{n=2}^{N}\left(\frac{m}{2\pi\hbar i \Delta t}\right)^{d/2}\right) e^{\frac{i}{\hbar}\sum_{n=2}^{N}\Delta t\left(\frac{m(x_n - x_{n-1})^2}{2\Delta t^2} - \frac{V(x_n)+V(x_{n-1})}{2}\right)}.$$
$$(1.37)$$

As you see, the exponent of this nontrivial construction contains $i/\hbar$ times the Riemannian sum for the integral, giving the classical action along some path, $x(t)$. The limit of the infinite number of consequent integrations over intermediate coordinates, x_j, with the proper normalization factors, is called a *continual, functional*, or simply *path* integral, and is denoted by $\int \mathcal{D}x$. Thus,

$$K(x, t; x', t') = \int_{x'(t')}^{x(t)} \mathcal{D}x \, e^{\frac{i}{\hbar} S[x(t), \dot{x}(t)]}. \qquad (1.38)$$

This also can be written in a more symmetric (and more general) form:

$$K(x,\ t;x',\ t') = \int\limits_{x'(t')}^{x(t)} \mathcal{D}x \int \mathcal{D}\frac{p}{2\pi\hbar}e^{\frac{i}{\hbar}S[p(t),x(t),t]}, \tag{1.39}$$

where

$$S[p(t),\ x(t),\ t] = \int_{t'}^{t} dt[p\dot{x} - H(p(t),\ x(t),\ t)]$$

is the action expressed through the canonical variables;

$$H(p,\ x,\ t) = \dot{x}\partial L/\partial\dot{x} - L(x,\dot{x},\ t)$$

is the *Hamiltonian function* of the particle. Since the above expression explicitly contains H, it proves more useful in applications of path integral methods to the systems with many degrees of freedom. But for our limited goals, it will be enough to demonstrate the equivalence of (1.38) and (1.39), simultaneously explaining the meaning of the symbol $\mathcal{D}p$ (before that the expression (1.39) is, of course, null and void).

The simplest way to do that is to employ the Schrödinger equation for the propagator. Now we do not postulate, but *derive* it from (1.38), thus preserving the consistency of speculation. It is clear that given the form of the propagator for infinitesimal times, (1.36), the integral composition equation can be reduced to a differential one.

Using expression (1.36), we can write (for the one-dimensional case, generalizations are trivial)

$$K(x_N t_{N-1} + \Delta t; x_1 t_1) \approx \int\limits_{-\infty}^{\infty} dx_{N-1}\left(\frac{m}{2\pi\hbar i\,\Delta t}\right)^{1/2}$$

$$\times e^{\frac{i}{\hbar}\left(\frac{m(x_N-x_{N-1})^2}{2\Delta t^2} - \frac{V(x_N)+V(x_{N-1})}{2}\right)\Delta t} K(x_{N-1}t_{N-1}; x_1 t_1)$$

on one hand, and

$$K(x_N t_{N-1} + \Delta t; x_1 t_1) \approx K(x_{N-1}t_{N-1}; x_1 t_1) + \Delta t\frac{\partial}{\partial t_{N-1}}K(x_{N-1}t_{N-1}; x_1 t_1)$$

on the other. Now expand the functions under the integral:

$$e^{\frac{i}{\hbar}\left(\frac{m(x_N-x_{N-1})^2}{2\Delta t^2} - \frac{V(x_N)+V(x_{N-1})}{2}\right)\Delta t} \approx e^{\frac{i}{\hbar}\frac{m(x_N-x_{N-1})^2}{2\Delta t}}\left(1 - \frac{i}{\hbar}V(x_N)\Delta t\right);$$

$$K(x_{N-1}t_{N-1}; x_1 t_1) \approx K(x_N t_{N-1}; x_1 t_1)$$

$$- (x_N - x_{N-1})\frac{\partial}{\partial x_N}K(x_N t_{N-1}; x_1 t_1)$$

$$+ \frac{(x_N - x_{N-1})^2}{2}\frac{\partial^2}{\partial x_N^2}K(x_N t_{N-1}; x_1 t_1).$$

Integrating over x_{N-1} (which is easy, since the integrals are of Gaussian type) and keeping the leading terms in Δt, we obtain:

$$\Delta t \frac{\partial}{\partial t_{N-1}} K(x_{N-1} t_{N-1}; x_1 t_1) = -\frac{i}{\hbar} V(x_N) \Delta t K(x_N t_{N-1}; x_1 t_1)$$

$$+ \frac{i\hbar}{2m} \Delta t \frac{\partial^2}{\partial x_N^2} K(x_N t_{N-1}; x_1 t_1) + o(\Delta t).$$

Dividing by Δt, and in the limit $\Delta t \to 0$, we finally obtain the Eq. (1.25) for the propagator for $t > t'$, thus having demonstrated that the Schrödinger equation follows from the Dirac–Feynman conjecture about the structure of the transition amplitude. (Of course, the opposite is true as well.)

Now let us return to the basic Eq. (1.16), which determines the action of the propagator on the wave function. Using Dirac's "bra" and "ket" notation, in which the wave function $\Psi(x)$ is presented as a scalar product of two abstract vectors in Hilbert space,

$$\Psi(x) \equiv \langle x | \Psi \rangle, \tag{1.40}$$

we can rewrite it as follows:

$$\langle x | \Psi(t) \rangle = \sum_{x'} \langle x | \mathcal{S}(t, t') | x' \rangle \langle x' | \Psi(t') \rangle \equiv \langle x | \mathcal{S}(t, t') | \Psi(t') \rangle. \tag{1.41}$$

We have used the closure relation (completeness) of the quantum states of the particle with definite coordinate (coordinate eigenstates), $|x\rangle$, that is,

$$\sum_{x'} |x'\rangle \langle x'| = \mathcal{I}, \tag{1.42}$$

where $\mathcal{I}$ is the unit operator.

We see that the propagator, $K(xt; x't')$ for $t > t'$, is a matrix element of some time-dependent operator $\mathcal{S}(t, t')$ between the coordinate eigenstates, $K(xt; x't') = \langle x | \mathcal{S}(t, t') | x' \rangle$. The equation for the propagator then can be written in a general form, notwithstanding the basis (representation):

$$i\hbar \frac{\partial}{\partial t} \mathcal{S}(t, t') = \mathcal{H} \mathcal{S}(t, t'), \tag{1.43}$$

and its formal solution is found immediately:

$$\mathcal{S}(t, t') = e^{-\frac{i}{\hbar} \mathcal{H}(t-t')}. \tag{1.44}$$

Since $\langle x | x' \rangle = \delta(x - x')$ (orthonormality condition for eigenstates of coordinate), the above solution indeed satisfies the initial condition for the propagator $K(xt; x't - 0) = \delta(x - x')$.

What is the benefit? It is that now we are not limited to the coordinate representation, and can easily work in, say, momentum space. This is what we actually need to prove (1.39). Besides, we will need the closure relation for the momentum eigenstates $|p\rangle$:

$$\sum_{p'} |p'\rangle\langle p'| = \mathcal{I}. \tag{1.45}$$

Recall that in the coordinate (momentum) representation the coordinate and momentum eigenstates look as follows:

$$\Psi_x(x') \equiv \langle x'|x\rangle = \delta(x' - x); \; \Psi_p(x') \equiv \langle x'|p\rangle = e^{\frac{i}{\hbar}px'}, \tag{1.46}$$

and respectively

$$\tilde{\Psi}_x(p') \equiv \langle p'|x\rangle = e^{-\frac{i}{\hbar}p'x}; \; \tilde{\Psi}_p(p') \equiv \langle p'|p\rangle = \delta(p' - p). \tag{1.47}$$

Now at last we can return to the path-integral calculation of the propagator. In complete agreement with our previous treatment, we slice the time interval $[t_f (= t_N); t_i (= t_1)]$ in tiny bits $\Delta t = (t_f - t_i)/(N - 1)$ and use the composition property;

$$\langle x_N|\mathcal{S}(t_N, t_1)|x_1\rangle = \lim_{N\to\infty} \langle x_N|\mathcal{S}(t_N, t_{N-1})\mathcal{S}(t_{N-1}, t_{N-2})\cdots\mathcal{S}(t_2, t_1)|x_1\rangle.x1 \tag{1.48}$$

The Hamiltonian here is a function of coordinate and momentum *operators*, $\mathcal{H} = \mathcal{H}(\hat{p}, \hat{x})$.

Now we can insert between each of the two propagators in (1.48) the unit operator $\sum_x |x\rangle\langle x| \sum_p |p\rangle\langle p|$. Evidently

$$\langle x_m|e^{-\frac{i}{\hbar}\mathcal{H}(\hat{p},\hat{x})\Delta t}|p_m\rangle\langle p_m|x_{m-1}\rangle \approx \langle x_m|e^{-\frac{i}{\hbar}H(p_m,x_m)\Delta t}|p_m\rangle\langle p_m|x_{m-1}\rangle$$
$$= e^{\frac{i}{\hbar}p_m x_m} e^{-\frac{i}{\hbar}H(p_m,x_m)\Delta t} e^{-\frac{i}{\hbar}p_m x_{m-1}} \tag{1.49}$$

(note that now instead of the (operator) Hamiltonian, we have obtained the classical Hamiltonian function, depending on usual coordinates and momenta). Therefore, Eq. (1.48) is reduced to

$$\langle x_N|\mathcal{S}(t_N, t_1)|x_1\rangle = \lim_{N\to\infty} \int \prod_{n=2}^{N-1} dx_n \int \prod_{n=2}^{N} \frac{dp_n}{2\pi\hbar}$$
$$\times e^{\frac{i}{\hbar}\sum_{n-2}^{N}\Delta t\left[p_n\frac{x_n-x_{n-1}}{\Delta t} - H(p_n,x_n)\right]}. \tag{1.50}$$

We have restored the continuous case notation (i.e., $\sum_x \to \int dx; \sum_p \to \int dp/(2\pi\hbar)$). This is the very path integral in the phase space (that is, over coordinates *and* momenta), the shorthand notation of which was given above by (1.39). Keep in mind that here we did not include the normalization factors $\left[\frac{m}{2\pi\hbar i\,\Delta t}\right]^{1/2}$ in definition of $\mathcal{D}x$. Actually they will be given by integrations over momenta, and there is no general convention whether such factors should be written explicitly or not.

As you see, the expression (1.50) contains $(N-1)$ momenta and N coordinates, but there are $N-2$ integrations over coordinates and $N-1$ over momenta. As a result, we have two "loose" coordinates, initial and final ones, as it should be for the propagator in coordinate representation. But nothing prevents us from calculating a different matrix element of S, say, $\langle p_f|S(t_f,\ t_i)|p_i\rangle$. Evidently, this should be the propagator in momentum representation, giving the probability amplitude for the particle to change its momentum from p_i to p_f. You can easily demonstrate that the corresponding path integral can be written as (see Problem 1.1)

$$K(p,\ t;\ p',\ t') = \int\limits_{p'(t')}^{p(t)} \mathcal{D}\frac{p}{2\pi\hbar} \int \mathcal{D}x e^{\frac{i}{\hbar}S[p(t),x(t),t]}. \tag{1.51}$$

Thus, path integrations generally do not commute!

The last thing we should do now is to demonstrate that (1.50) yields the initial expression (1.38), i.e. the path integral in the configuration space. This will complete our argument. To do this, let us take the Hamiltonian function in the form $H(p,\ x) = \frac{p^2}{2m} + V(x)$. In (1.50) we can then easily integrate out the momenta, since the corresponding integrals are Gaussian,

$$\int \frac{dp_n}{2\pi\hbar} e^{\frac{i}{\hbar}\left[p_n\frac{x_n-x_{n-1}}{\Delta t} - \frac{p_n^2}{2m}\right]\Delta t} = \left[\frac{m}{2\pi\hbar i\,\Delta t}\right]^{1/2} e^{i\frac{m(x_n-x_{n-1})^2}{2\hbar\,\Delta t}}, \tag{1.52}$$

and we are back to the initial formula (1.38).

1.2.3 Quantum Transport in Mesoscopic Rings: Path Integral Description

The path integral description as we have introduced it is nice and clear when we deal with a single particle. Then it seems inapplicable to the problems of condensed matter, with giant numbers of particles involved: we have to develop a more subtle approach, equivalent to the technique of Green's functions, etc.

Nevertheless, there exists a class of solid systems where the single particle approach holds and gives sensible results, namely, the *mesoscopic systems* (see, e.g., [5]). These are the systems of intermediate size, i.e., macroscopic but small enough ($\leq 10^{-4}$cm). In these systems quantum interference is very important, since at low enough temperatures (<1 K) the phase coherence length of quasiparticles ("electrons") exceeds the size of the system. This means that the electrons preserve their "individuality" when passing through the system.

Since the wave function of the quantum particle depends on its energy as $e^{-iEt/\hbar}$, any inelastic interaction spoils the phase coherence. Then the condition

$$l_\phi \approx l_i > L \tag{1.53}$$

must hold. Here l_ϕ is the phase coherence length, l_i is the inelastic scattering length, L is the size of the system. The above condition can be satisfied in experiment, due to the fact we have discussed above: that in the condensed matter we can deal with weakly interacting quasiparticles instead of strongly interacting real particles.

Because the inelastic scattering length of the quasielectron exceeds the size of the mesoscopic system, we can regard it as a single particle in the external potential field and apply to it the path integral formalism in the simplest possible version.

1.2.3.1 Aharonov–Bohm Effect in Normal Metal Rings

Imagine a metal ring threaded by a solenoid with a magnetic flux $\Phi = \int d\mathbf{S} \cdot \mathbf{B} = \oint_C d\mathbf{x} \cdot \mathbf{A}$, where C is any contour encircling the solenoid. There is no magnetic field in the bulk.

The conductivity between points A and B is related to the probability for an electron to travel from A to B, given by a square modulus of amplitude

$$\langle Bt_B | At_A \rangle = \int\limits_A^B \mathcal{D}\mathbf{x} e^{\frac{i}{\hbar} \int\limits_{t_A}^{t_B} dt L(\mathbf{x},\dot{\mathbf{x}},t)}. \tag{1.54}$$

(We have for the sake of brevity denoted the transition amplitude—propagator—between the points $\mathbf{x}_A$, t_A and $\mathbf{x}_B$, t_B simply by $\langle Bt_B | At_A \rangle$; in the next section we will see that this is not only a shorthand.)

The Lagrange function of the electron in the magnetic field is given by a Legendre transformation:

$$L(\mathbf{x}, \dot{\mathbf{x}}) = \mathbf{P} \cdot \dot{\mathbf{x}} - H(\mathbf{P}, \mathbf{x}); \tag{1.55}$$

$$H(\mathbf{P}, \mathbf{x}) = \frac{(\mathbf{P} - \frac{e}{c}\mathbf{A})^2}{2m^*} + V(\mathbf{x}). \tag{1.56}$$

Here $V(\mathbf{x})$ is a static random potential, $\mathbf{A}$ is the vector potential, and $\mathbf{P} = m^*\dot{\mathbf{x}} + \frac{e}{c}\mathbf{A}$ is the canonical momentum.

Performing the transformation (1.55), we find that the Lagrange function is related to one *without* the magnetic field, L_0, by

$$L(\mathbf{x}, \dot{\mathbf{x}}) = \frac{e}{c}\mathbf{A} \cdot \dot{\mathbf{x}} + L_0(\mathbf{x}, \dot{\mathbf{x}}). \tag{1.57}$$

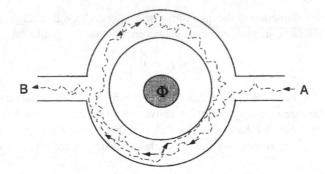

Fig. 1.8 $hc/2e$ oscillations in a mesoscopic ring

Therefore, the transition amplitude is

$$\langle Bt_B|At_A\rangle = \int_A^B \mathcal{D}\mathbf{x}\, e^{\frac{ie}{\hbar c}\int_{t_A}^{t_B} dt\,\mathbf{A}\cdot\dot{\mathbf{x}}}\cdot e^{\frac{i}{\hbar}\int_{t_A}^{t_B} dt\,L_0(\mathbf{x},\dot{\mathbf{x}},t)} \tag{1.58}$$

$$= \int_A^B \mathcal{D}\mathbf{x}\, e^{\frac{ie}{\hbar c}\int_{t_A}^{t_B} dt\,\mathbf{A}\cdot\dot{\mathbf{x}}}\, e^{\frac{i}{\hbar}S_0[\mathbf{x},\dot{\mathbf{x}}]}.$$

There exists a special class of trajectories that loop around the hole and have a self-intersection (see Fig. 1.8). Each of them has a counterpart with an opposite direction of motion around the hole. Each pair of thus conjugated trajectories has the same value of the $\exp(\frac{i}{\hbar}S_0[\mathbf{x},\dot{\mathbf{x}}])$ factor (since without the magnetic field the motion is reversible), while the rest of the expression gives

$$\frac{ie}{\hbar c}\int_{t_A}^{t_B} dt\,\mathbf{A}\cdot\dot{\mathbf{x}} = \pm\frac{ie}{\hbar c}\oint \mathbf{A}\cdot d\mathbf{x} = \pm\frac{ie}{\hbar c}\int \mathrm{rot}\mathbf{A}\cdot d\mathbf{S} = \pm\frac{ie}{\hbar c}\Phi. \tag{1.59}$$

Then the following term in the transition amplitude arises:

$$\langle Bt_B|At_A\rangle_\odot = \left(e^{\frac{ie}{\hbar c}\Phi} + e^{-\frac{ie}{\hbar c}\Phi}\right)\int_\odot \mathcal{D}\mathbf{x}\, e^{\frac{i}{\hbar}S_0[\mathbf{x},\dot{\mathbf{x}}]} \equiv 2F_\odot\cos\frac{e\Phi}{\hbar c}. \tag{1.60}$$

The transition *probability* is then

$$|\langle B|A\rangle|^2 = |\langle B|A\rangle_\odot|^2 + |\langle B|A\rangle_{\mathrm{other}}|^2 + 2\Re\langle B|A\rangle_\odot\langle B|A\rangle_{\mathrm{other}}^*. \tag{1.61}$$

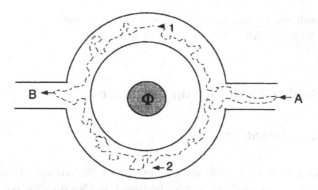

Fig. 1.9 hc/e oscillations in a mesoscopic ring

The third term vanishes due to phase randomness; the second term does not contain any pronounced Φ-dependence. But the first one[2] is *periodic in Φ with a period equal to the superconducting flux quantum,* $\Phi_0 = hc/2e$:

$$|\langle B|A\rangle_{\circlearrowright}|^2 = 2|F_{\circlearrowright}|^2 \left(1 + \cos 2\pi \frac{\Phi}{\Phi_0}\right). \tag{1.62}$$

The doubling of the period is, of course, not due to the Cooper pairing and double electric *charge,* but due to the simple fact that the transition amplitude contains the difference between the contributions of particles that encircle the hole clockwise and counterclockwise, thus doubling the *path.*

Another type of oscillation originates from a different class of trajectories (see Fig. 1.9), that run from A to B on the different sides of the hole. Each pair of trajectories from this class produces in the transition *probability* the term

$$2\Re e^{\frac{ie}{\hbar c}(\int_1 \mathbf{A}\cdot d\mathbf{x} - \int_2 \mathbf{A}d\mathbf{x})e^{\frac{i}{\hbar}(\int_1 L_0(\mathbf{x},\dot{\mathbf{x}}) - \int_2 L_0(\mathbf{x},\dot{\mathbf{x}}))}} \tag{1.63}$$

$$= 2\Re e^{\frac{ie}{\hbar c}\oint_1 \mathbf{A}\cdot d\mathbf{x}} e^{i\chi_{12}} = 2\cos\left(2\pi\frac{\Phi}{2\Phi_0} + \chi_{12}\right).$$

These oscillations have a doubled period, $2\Phi_0 = hc/e$, but they include a random phase, χ_{12}. Therefore, they are sensitive to the number of possible trajectories of this class, and quickly vanish when it grows. For example, in the metal rings both hc/e and $hc/2e$ oscillations were observed, while in the metal cylinders only the

---

[2] It is not so easy to calculate the prefactor $F_{\circlearrowright}$; but it is not difficult to show that it is small only as a power of the parameter λ_F/L.

latter exist, while the former are averaged to zero. (A cylindrical conductor can be regarded as a huge number of rings stacked together.)

1.3 Perturbation Theory for the Propagator

1.3.1 General Formalism

Though we always can write a path-integral formula (1.38), an explicit expression for the propagator in the general case cannot be found, neither directly, nor by solving the Schrödinger equation (1.25). Apart from exactly solvable cases (which are as beautiful as rare—and worse still, usually known for years), the only regular way to deal with a problem is to use some sort of perturbation theory.

Fortunately, the propagator formalism is uniquely suited for the task.

We will work again in Dirac's notation. Let us start with the *Schrödinger representation*, where, as you know from quantum mechanics, the operators of observables are time independent (except possible *explicit* time dependence), while the state vectors (wave functions) evolve according to the Schrödinger equation:

$$i\hbar \frac{\partial}{\partial t} |\Phi(t)\rangle_S = \mathcal{H} |\Phi(t)\rangle_S. \tag{1.64}$$

If the Hamiltonian is time independent, the formal solution to this is given by

$$|\Phi(t)\rangle_S = e^{-\frac{i}{\hbar}\mathcal{H}t} |\Phi(0)\rangle_S. \tag{1.65}$$

We have operated with the Hamiltonian as if it were a number; of course, the operator exponent is meaningful only as a power series,

$$e^{-\frac{i}{\hbar}\mathcal{H}t} = \mathcal{I} - \frac{i}{\hbar}\mathcal{H}t + \frac{1}{2}\left(-\frac{i}{\hbar}\mathcal{H}t\right)^2 + \cdots, \tag{1.66}$$

where $\mathcal{I}$ is the unit operator, and the justification of (1.65) is in the fact that we can rewrite the Schrödinger equation as

$$|\Phi(t)\rangle_S = |\Phi(0)\rangle_S - \frac{i}{\hbar}\int_0^t \mathcal{H}|\Phi(t')\rangle_S dt' \tag{1.67}$$

and then iterate it, which will give us the series for $\exp(-i/\hbar\mathcal{H}t)$.

The operator

$$\mathcal{U}(t) = e^{-\frac{i}{\hbar}\mathcal{H}t} \tag{1.68}$$

is for an obvious reason called the *evolution operator*. Written in this form, it satisfies the Schrödinger equation with time-independent Hamiltonian. What if $\mathcal{H}$ is time dependent? For a usual number, the solution would be

$$e^{-\frac{i}{\hbar} \int\limits_0^t dt' \mathcal{H}(t')},$$

but here we are dealing with operators. There is no reason to believe that at different moments of time, t_1 and t_2, $\mathcal{H}(t_1)$ and $\mathcal{H}(t_1)$ commute, and the above expression will be invalid. But we can still iterate the Schrödinger equation,

$$\mathcal{U}(t) = \mathcal{I} + \left(-\frac{i}{\hbar}\right) \int\limits_0^t dt' \mathcal{H}(t'), \tag{1.69}$$

to yield

$$\mathcal{U}(t) = \mathcal{I} + \left(-\frac{i}{\hbar}\right) \int\limits_0^t dt_1' \mathcal{H}(t_1') + \left(-\frac{i}{\hbar}\right)^2 \int\limits_0^t dt_1' \int\limits_0^{t_1'} dt_2' \mathcal{H}(t_1')\mathcal{H}(t_2')$$

$$+ \left(-\frac{i}{\hbar}\right)^3 \int\limits_0^t dt_1' \int\limits_0^{t_1'} dt_2' \int\limits_0^{t_2'} dt_3' \mathcal{H}(t_1')\mathcal{H}(t_2')\mathcal{H}(t_3') + \cdots \tag{1.70}$$

In the above expression the operators are *time-ordered* (or *chronologically ordered*), that is, the operator of larger time argument always stands to the left. We can introduce the *time-ordering operator* $\mathcal{T}$, whose action on any set of time-dependent operators is exactly this:

$$\mathcal{T}[\mathcal{A}(t_A)\mathcal{B}(t_B)\mathcal{C}(t_C)\cdots] = \begin{cases} \mathcal{A}(t_A)\mathcal{B}(t_B)\mathcal{C}(t_C)\cdots \text{ if } t_A > t_B > t_C \cdots \\ \mathcal{B}(t_B)\mathcal{A}(t_A)\mathcal{C}(t_C)\cdots \text{ if } t_B > t_A > t_C \cdots \\ \mathcal{A}(t_A)\mathcal{C}(t_C)\mathcal{B}(t_B)\cdots \text{ if } t_A > t_C > t_B \cdots \\ \cdots \end{cases} \tag{1.71}$$

This operator allows us to present the series (1.70) in an elegant form:

$$\mathcal{U}(t) = \mathcal{T} e^{-\frac{i}{\hbar} \int\limits_0^t d\tau \mathcal{H}(\tau)}. \tag{1.72}$$

Indeed, let us expand the exponent and take the nth term,

$$\frac{1}{n!} \mathcal{T}[\int\limits_0^t d\tau_1 \int\limits_0^t d\tau_2 \cdots \int\limits_0^t d\tau_n \mathcal{H}(\tau_1)\mathcal{H}(\tau_2) \cdots \mathcal{H}(\tau_n)].$$

The n-dimensional integral is taken over the region $\{0 \leq \tau_1 \leq t; 0 \leq \tau_2 \leq t; \ldots; 0 \leq \tau_n \leq t\}$. We can take a part of this region, where, e.g., $\tau_1 \geq \tau_2 \geq \cdots \geq \tau_n$. The corresponding integral coincides with the nth term in the expansion (1.70), if we forget about the $1/n!$ factor. But the integration variables are dummy, and can be rearranged in exactly $n!$ ways, giving the same result. (The time-ordering operator will ensure that the operators are always in proper order.) Therefore, we can simply multiply the result by $n!$, thus proving the validity of (1.72).

We can sum up the important properties of the evolution operator, proceeding from its definition:

$$i\hbar \frac{\partial}{\partial t} \mathcal{U}(t) = \mathcal{H}(t)\mathcal{U}(t); \tag{1.73}$$

$$\mathcal{U}^\dagger(t) = \mathcal{U}^{-1}(t); \tag{1.74}$$

$$\mathcal{U}(0) = \mathcal{I}. \tag{1.75}$$

The second line contains the all-important *unitarity condition*, which physically means that probability is not getting lost when the quantum state evolves – if we start with one particle, we will not end up with 1/4 (or 22/7). Indeed, the norm of the state vector, related to the probability,

$$\|\Phi(t)\| \equiv \sqrt{\langle \Phi(t)|\Phi(t)\rangle} = \sqrt{\langle \Phi(0)|\mathcal{U}^\dagger(t)\mathcal{U}(t)|\Phi(0)\rangle}$$
$$= \sqrt{\langle \Phi(0)|\Phi(0)\rangle} = \text{const}$$

is conserved.

Of course, there is nothing special in the moment $t = 0$, and we can follow the evolution of the quantum state from any point: evidently, for any t, t',

$$|\Phi(t)\rangle_S = \mathcal{U}(t)\mathcal{U}^\dagger(t')|\Phi(t')\rangle_S \equiv S(t, t')|\Phi(t')\rangle_S, \tag{1.76}$$

where the *S-operator* is defined by

$$S(t, t') \equiv \mathcal{U}(t)\mathcal{U}^\dagger(t'). \tag{1.77}$$

Now we can, for example, express the wave function of the particle in coordinate space at time t via its value at *some* previous time t', by

$$\Psi(x, t) = \langle x|\Phi(t)\rangle_S = \langle x|S(t, t')|\Phi(t')\rangle_S$$
$$= \int dx' \langle x| S(t, t')|x'\rangle\langle x'||\Phi(t')\rangle_S$$
$$= \int dx' \langle x|S(t, t')|x'\rangle \Psi(x', t'). \tag{1.78}$$

Now we see that it is the very operator $\mathcal{S}$, related to the propagator, that we have previously introduced (see Eq. 1.41): for $t > t'$,

$$K(x, t; x', t') = \langle x|\mathcal{S}(t, t')|x' \rangle.$$

This operator (sometimes called the *S-matrix*) has the following properties:

$$i\hbar\frac{\partial}{\partial t}\mathcal{S}(t, t') = \mathcal{H}\mathcal{S}(t, t'); \tag{1.79}$$

$$\mathcal{S}(t, t) = \mathcal{I}; \tag{1.80}$$

$$\mathcal{S}^{\dagger}(t, t') = \mathcal{S}^{-1}(t, t') = \mathcal{S}(t', t); \tag{1.81}$$

$$\mathcal{S}(t, t'')\mathcal{S}(t'', t') = \mathcal{S}(t, t'); \tag{1.82}$$

$$\text{for } t > t' \, \mathcal{S}(t, t') = Te^{-\frac{i}{\hbar}\int_{t'}^{t}d\tau\mathcal{H}(\tau)}$$

$$= \left(e^{-\frac{i}{\hbar}\mathcal{H}(t-t')} \text{ if } \mathcal{H} \neq \mathcal{H}(t)\right). \tag{1.83}$$

Equation (1.81) is the unitarity condition. Equation (1.82) follows directly from the definition of $\mathcal{S}$ and the unitarity of the evolution operator, but it is the very composition property that we introduced for the propagator in the beginning (see Eq. 1.20).

The last line follows from (1.72). This is an elegant formula, but not very practical: the Hamiltonian of the system is "of order one," and the expansion would converge very slowly, or simply diverge! Fortunately, in most cases the Hamiltonian can be split into two parts: the *unperturbed*, time-independent Hamiltonian (for which we presumably know the solution) and a small, possibly time-dependent perturbation:

$$\mathcal{H}(t) = \mathcal{H}_0 + \mathcal{W}(t).$$

The goal is to present the solution for $\mathcal{H}$ as one for $\mathcal{H}_0$ plus corrections in powers of the small perturbation. The latter series will hopefully be rapidly convergent.

Until now we have worked in the Schrödinger representation, i.e., the state vectors were time dependent (governed by $\mathcal{U}(t)$), while the operators were constant (if not explicitly time dependent). The opposite picture is provided by the Heisenberg representation. It can be arrived at by the canonical transformation, using the evolution operators:

$$|\Phi\rangle_H = \mathcal{U}^{\dagger}(t)|\Phi(t)\rangle_S \equiv |\Phi(0)\rangle_S;$$
$$\mathcal{A}_H(t) = \mathcal{U}^{\dagger}(t)\mathcal{A}_S\mathcal{U}(t).$$

Now, the operators evolve over time, while state vectors do not. The operators satisfy the *Heisenberg equations of motion*:

$$i\hbar\frac{d}{dt}\mathcal{A}_H(t) = [\mathcal{A}_H(t), \mathcal{H}_H(t)] + i\hbar\frac{\partial}{\partial t}\mathcal{A}_H(t). \tag{1.84}$$

The above equation follows immediately from the definition of $\mathcal{A}_H$ and properties of the evolution operator. Here the partial derivative deals with explicit time dependence of the operator (say, due to changing of external conditions); the Hamiltonian, if time dependent, should be taken in Heisenberg representation as well, $\mathcal{H}_H(t) = \mathcal{U}^\dagger(t)\mathcal{H}(t)\mathcal{U}(t)$. The above equation follows immediately from the definition of $\mathcal{A}_\mathcal{H}$ and properties of the evolution operator.

For our goals it is more convenient to employ an intermediate, *interaction representation*, first suggested by Dirac. In this representation both operators and state vectors are time dependent, but the evolution of operators is governed by the unperturbed Hamiltonian (we will not use the index I to label operators and state vectors in interaction representation, since this will be our working representation):

$$\mathcal{A}(t) = e^{i\mathcal{H}_0 t/\hbar}\mathcal{A}_S e^{-i\mathcal{H}_0 t/\hbar}; \tag{1.85}$$

$$i\hbar\frac{d}{dt}\mathcal{A}(t) = [\mathcal{A}(t),\ \mathcal{H}_0] + i\hbar\frac{\partial}{\partial t}A(t). \tag{1.86}$$

The last line is exactly the Heisenberg equation for an operator in an unperturbed system. (Notice that since $\mathcal{H}_0$ is time independent, it is the same in the Schrödinger and interaction representations.)

The state vectors in interaction representation undergo the corresponding canonical transformation,

$$|\Phi(t)\rangle = e^{i\mathcal{H}_0 t/\hbar}|\Phi(t)\rangle_S, \tag{1.87}$$

and obey the equation

$$i\hbar\frac{\partial}{\partial t}|\Phi(t)\rangle = \mathcal{W}(t)|\Phi(t)\rangle. \tag{1.88}$$

(Here $\mathcal{W}(t)$ is also in interaction representation, $\mathcal{W}(t) = e^{i\mathcal{H}_0 t/\hbar}\mathcal{W}e^{-i\mathcal{H}_0 t/\hbar}$, as you can see when deriving this formula from the original Schrödinger equation. The trick is that due to unitarity you can insert the operator $\mathcal{U}(t)\mathcal{U}^\dagger(t) \equiv \mathcal{I}$ wherever it is needed.) The benefit of this representation is, therefore, that the state vectors are affected only by the perturbation. Now we can almost literally repeat all the calculations from the beginning of this section. For example, iterating (1.88), we find the solution

$$|\Phi(t)\rangle = \mathcal{T}e^{-\frac{i}{\hbar}\int_0^t d\tau \mathcal{W}(\tau)}|\Phi(0)\rangle, \tag{1.89}$$

with the same time-ordering operator.

Now we can find the explicit expression for the S-operator in interaction representation. The easiest way is to introduce an auxiliary operator $\mathcal{O}(t,\ t') \equiv \exp$

$(i'\mathcal{H}_0 t/\hbar)\mathcal{S}_S(t,\ t')$, which, as is easy to see, satisfies the same equation as the state operator:

$$i\hbar\frac{\partial}{\partial t}\mathcal{O}(t,\ t') = \mathcal{W}(t)\mathcal{O}(t,\ t').$$

Then, of course,

$$\mathcal{O}(t,\ t') = \mathcal{T}e^{-\frac{i}{\hbar}\int_{t'}^{t}d\tau\mathcal{W}(\tau)}\mathcal{O}(t',\ t'),$$

so that the S-operator itself can be written as follows (for $t > t'$);

$$\mathcal{S}_S(t,\ t') = e^{(-i\mathcal{H}_0 t/\hbar)}\mathcal{T}e^{-\frac{i}{\hbar}\int_{t'}^{t}d\tau\mathcal{W}(\tau)}e^{(i\mathcal{H}_0 t'/\hbar)}. \tag{1.90}$$

This is the so-called *Dyson's expansion* for the S-operator in Schrödinger representation. Transforming the $\mathcal{U}$-operators according to (1.85), we get that in interaction representation the S-operator takes the simple form

$$S(t,\ t') = e^{(i\mathcal{H}_0 t/\hbar)}\mathcal{S}_S(t,\ t')e^{(-i\mathcal{H}_0 t'/\hbar)}$$

$$= \mathcal{T}e^{-\frac{i}{\hbar}\int_{t'}^{t}d\tau\mathcal{W}(\tau)} \quad (t > t'). \tag{1.91}$$

It looks as if it depends only on the perturbation! (Of course, the unperturbed Hamiltonian is hidden in $\mathcal{W}(\tau)$; but we presumably know how everything behaves without perturbation.)

We have expressed the propagator as a matrix element of the S-operator. There is another expression, sometimes useful in the many-body case. In order to arrive at it, let us return to Heisenberg representation with its time-dependent operators. The coordinate operator $\mathcal{X}_H(t)$ will be time dependent as well. In Schrödinger representation this operator had time-independent eigenstates, which constituted the complete basis of the Hilbert space:

$$\mathcal{X}_S|x\rangle = x|x\rangle; \tag{1.92}$$

$$\sum_x |x\rangle\langle x| = \mathcal{I}. \tag{1.93}$$

(In coordinate representation they are simply delta functions, in momentum representation, plane waves.) Let us now introduce the set of *instantaneous* eigenstates of the coordinate operator in Heisenberg representation, $\{|xt\rangle\}$:

$$\mathcal{X}_H(t)|xt\rangle = x|xt\rangle. \tag{1.94}$$

Since $\chi_H(t)|xt\rangle = \mathcal{U}^\dagger(t)\mathcal{X}_S\mathcal{U}(t)|xt\rangle$, then $\mathcal{U}(t)|xt\rangle = |x\rangle$, and we see that the time evolution of these states is governed by $\mathcal{U}^\dagger(t)$ instead of $\mathcal{U}(t)$, and they still constitute a complete basis at any moment t;

$$|xt\rangle = \mathcal{U}^\dagger(t)|x\rangle; \tag{1.95}$$

$$\sum_x |xt\rangle\langle xt| = \sum_x \mathcal{U}^\dagger(t)|x\rangle\langle x|\mathcal{U}(t)$$

$$= \mathcal{U}^\dagger(t)\mathcal{U}(t)$$

$$= \mathcal{I}. \tag{1.96}$$

Now we can rewrite the propagator in coordinate space simply as an overlap of two states from this basis (cf. Eq. (1.54) of the previous section!):

$$K(x, t; x', t') = \langle x|\mathcal{S}(t, t')|x\rangle$$

$$\equiv \langle x|\mathcal{U}(t)\mathcal{U}^\dagger(t')|x\rangle = \langle xt|x't'\rangle. \tag{1.97}$$

From this expression straightforwardly follows an expression for the propagator in the momentum space:

$$K(p, t; p', t') = \int dx \int dx' \langle p|x\rangle \langle xt\ |x't'\rangle\langle x'|p'\rangle$$

$$= \int dx \int dx' \langle pt|xt\rangle\langle xt|x't'\rangle\langle x't'|p't'\rangle \tag{1.98}$$

$$= \langle pt|p't'\rangle$$

The states $\{\langle pt\rangle\}$ are, of course, instantaneous eigenstates of the momentum operator in Heisenberg representation, $\mathcal{P}_H(t)$.

1.3.2 An Example: Potential Scattering

The above formulae are very general: actually they are applicable to any quantum system, notwithstanding the number of particles and type of interaction. This was one reason why we went to such lengths to derive them: we will use them later throughout this book.

Now let us apply them to our initial case of one, structureless, quantum particle. Now we have a single option for the perturbation operator, a scalar external potential, so that its coordinate matrix element is

$$\langle x|\mathcal{W}(t)|x'\rangle = V(x, t)\delta(x - x'). \tag{1.99}$$

Fig. 1.10 Feynman diagram for the potential scattering

Table 1.1 Feynman rules for a particle in the external potential field

xt	x't'	$K(\mathbf{x},\ t;\mathbf{x}',\ t')$	Propagator
xt	x't'	$K_0(\mathbf{x},\ t;\mathbf{x}',\ t')$	Free (unperturbed) propagator
		$= \dfrac{m^{3/2}\exp\left(\frac{im(\mathbf{x}-\mathbf{x}')^2}{2\hbar(t-t')}\right)}{(2\pi\ i\hbar(t-t'))^{3/2}}\theta(t-t')$	
xt		$-i\ V(\mathbf{x},t)/\hbar$	External potential (in interaction representation)

The integration over all intermediate coordinates and times is implied

Table 1.2 Feynman rules for a particle in the external potential field (momentum representation)

pE	p'E'	$K(\mathbf{p},\ E;\mathbf{p}',\ E')$	Propagator
pE	p'E'	$K_0(\mathbf{p},\ E;\mathbf{p}',\ E')$	Free (unperturbed)
		$= (2\pi\hbar)^4\dfrac{i\hbar\delta(\mathbf{p}-\mathbf{p}')\delta(E-E')}{E-p^2/2m+i0}$	propagator
pE		$-i\ V(\mathbf{p},\ E)/\hbar$	Fourier transform of the external potential

The integration over all intermediate momenta and energies is implied, taking into account energy/momentum conservation in every vertex

It is natural then to work in coordinate representation; taking a corresponding matrix element of (1.90) we obtain the *perturbation expansion for the propagator*:

$$
\begin{aligned}
K(x,\ t;x',\ t') = {} & K_0(x,t;x',t') \\
& + \int dx''dt''K_0(x,\ t;x'',\ t'')\left(-\frac{i}{\hbar}\right) \\
& \times V(x'',\ t'')K_0(x'',\ t'';x',\ t') + \cdots.
\end{aligned}
\tag{1.100}
$$

This expression is presented graphically in Fig. 1.10, the elements of which are explained in Table 1.1.

Of course, our discourse is not limited to the coordinate representation; as a matter of fact, it is more often than not easier to use momentum representation. The Feynman rules for the momentum representation can be found in Table 1.2 (see Problem 1.2).

The graph in Fig. 1.10 is the simplest example of a *Feynman diagram*. In this case its use seems superfluous, because of the simple structure of the perturbation involved. In the many-body case, though, the structure of the terms entering the perturbation series is much more complicated, and the graphs provide great help in comprehending their structure and making physically consistent approximations. The graph under consideration, e.g., suggests to us a clear picture of a quantum particle repeatedly scattered by an external potential, but propagating freely between the scattering acts. It will be useful to look into how (and whether) this intuitive picture fits into a path-integral description of the behavior of the quantum particle. We shall see that this very result can indeed be easily derived directly from formula (1.37) for the propagator, in a slightly changed form;

$$
K(x_N t_N; x_1 t_1) = \lim_{N \to \infty} \int_{-\infty}^{\infty} dx_{N-1} \int_{-\infty}^{\infty} dx_{N-2} \cdots \int_{-\infty}^{\infty} dx_2
$$

$$
\times \left(\prod_{n=2}^{N} \left(\frac{m}{2\pi \hbar i \, \Delta t} \right)^{d/2} \right) e^{\frac{i}{\hbar} \sum_{n-2}^{N} \Delta t \frac{m(x_n - x_{n-1})^2}{2\Delta t^2}} e^{-\frac{i}{\hbar} \sum_{k-2}^{N} \Delta t V(x_k, t_k)}.
$$

All we need to do is expand the exponents containing potential and rearrange this expression as a power series over the external potential, V.

The zero-order term is, evidently, unperturbed propagator, $K_0(x_N t_N; x_1 t_1)$. The first-order term is

$$
K_1(x_N t_N; x_1 t_1)
$$

$$
= \lim_{N \to \infty} \int_{-\infty}^{\infty} dx_{N-1} \int_{-\infty}^{\infty} dx_{N-2} \cdots \int_{-\infty}^{\infty} dx_2 \left(\prod_{n=2}^{N} \left(\frac{m}{2\pi \hbar i \, \Delta t} \right)^{d/2} \right)
$$

$$
\times e^{\frac{i}{\hbar} \sum_{n-2}^{N} \Delta t \frac{m(x_n - x_{n-1})^2}{2\Delta t}} \left\{ -\frac{i}{\hbar} \sum_{k=2}^{N} \Delta t V(x_k, t_k) \right\}.
$$

We can rewrite the last expression as

$$
K_1(x_N t_N; x_1 t_1)
$$

$$
= \lim_{N \to \infty} \left\{ -\frac{i}{\hbar} \sum_{k=2}^{N} \Delta t \int_{-\infty}^{\infty} \cdots \int dx_{N-1} dx_{N-2} \cdots dx_2 \right.
$$

$$
\times \left(\prod_{n=2}^{N} \left(\frac{m}{2\pi \hbar i \, \Delta t} \right)^{d/2} \right) e^{\frac{i}{\hbar} \sum_{n-2}^{N} \Delta t \frac{m(x_n - x_{n-1})^2}{2\Delta t^2}} V(x_k, t_k) \right\}
$$

$$
= \lim_{N \to \infty} \left\{ -\frac{i}{\hbar} \sum_{k=2}^{N} \Delta t \int_{-\infty}^{\infty} dx_k \left[\int_{-\infty}^{\infty} \cdots \int dx_{k-1} dx_{k-2} \cdots dx_2 \right. \right.
$$

$$\times \left(\prod_{n=2}^{k} \left(\frac{m}{2\pi\hbar i \, \Delta t} \right)^{d/2} \right) e^{\frac{i}{\hbar} \sum_{n-2}^{k} \Delta t \frac{m(x_n - x_{n-1})^2}{2\Delta t^2}} \right] V(x_k, t_k)$$

$$\times \left[\int_{-\infty}^{\infty} \cdots \int dx_N dx_{N-1} \cdots dx_{k+1} \left(\prod_{n=k+1}^{N} \left(\frac{m}{2\pi\hbar i \, \Delta t} \right)^{d/2} \right) \right.$$

$$\times \left. e^{\frac{i}{\hbar} \sum_{n-k+1}^{N} \Delta t \frac{m(x_n - x_{n-1})^2}{2\Delta t^2}} \right] \Bigg\}.$$

Now we see that

$$K_1(x_N t_N; x_1 t_1) = \int_{t_1}^{t_N} dt \int_{-\infty}^{\infty} dx \, K_0(x_N t_N; xt) V(x, t) K_0(xt; x_1 t_1)$$

$$\equiv \int_{-\infty}^{\infty} dt \int_{-\infty}^{\infty} dx \, K_0(x_N t_N; xt) V(x, t) K_0(xt; x_1 t_1)$$

(the last transformation has taken into account that for $t < t_1$ or $t > t_N$ the integrand is identically zero). It is clear from our derivation that indeed, in the path integral picture we can regard the effects of external potential as a result of multiple scatterings of an otherwise free particle.

The next terms of the expansion can be derived in the same way. Factors $1/n!$ in the expansion of the exponent will be canceled because we will have exactly $n!$ ways to relabel the points x_k, t_k where the scattering occurs.

1.4 Second Quantization

1.4.1 Description of Large Collections of Identical Particles: Fock's Space

By now, we have successfully handled single quantum particle. We have written the propagator as a path integral, obtained the perturbation series in the presence of external scattering potential, and explained the magnetoconductance oscillations in mesoscopic systems. A clear overkill: it would suffice to employ the Schrödinger equation and standard methods from the theory of partial differential equations. But thus we have prepared ourselves to the formidable task of dealing with macroscopic collectives of quantum particles.

To write explicitly the wave function for such a system, even in the simplest case of a collective of $N \approx 10^{23}$ identical particles, is as impossible as to determine the momentary position and velocity of each particle in its classical counterpart.

In classical statistical mechanics, though, the problem is avoided by introducing the *distribution functions*, giving the *probabilities* of, e.g., finding a particle with velocity v_1 at point r_1, and another with v_2 at r_2, and so on. In other words (and with other normalization constant), the distribution functions give information on how many particles occupy each cubicle of the phase space.

Of course, there is a whole hierarchy of them, including one, two,...N-particle distributions functions, and together they contain exactly the same amount of information as the record of velocities and positions of all particles in the system. The *enormous* advantage is that we usually need only the few first functions of this hierarchy. Hopefully, such an approach will pay off for the quantum particles as well.

The difference is that in quantum statistics we have to operate somehow with the *wave function of the system as a whole*, $\Phi(\xi_1, \xi_2, \ldots, \xi_i, \ldots, \xi_j, \ldots, \xi_N)$, and have trouble in trying to "glue" characteristics to individual particles per se. The more so, identical quantum particles cannot be distinguished in principle. This quantum-mechanical *principle of indistinguishability of identical particles* imposes the following property on the wave function. If we exchange two particles, the wave function can only acquire a phase factor (since its modulus, which determines the observable effects of such interchange, should stay the same):

$$\Phi(\xi_1, \xi_2, \ldots, \xi_i, \ldots, \xi_j, \ldots) = e^{i\chi}\Phi(\xi_1, \xi_2, \ldots, \xi_j, \ldots, \xi_i, \ldots, \xi_N). \quad (1.101)$$

After making the second permutation of the same particles, we have

$$\Phi(\xi_1, \xi_2, \ldots, \xi_i, \ldots, \xi_j, \ldots) = e^{2i\chi}\Phi(\xi_1, \xi_2, \ldots, \xi_i, \ldots, \xi_j, \ldots, \xi_N), \quad (1.102)$$

so that $e^{i\chi} = \pm 1$, and we are left with two choices

$$\Phi(\xi_1, \xi_2, \ldots, \xi_i, \ldots, \xi_j, \ldots)$$
$$= \begin{cases} +\Phi(\xi_1, \xi_2, \ldots, \xi_i, \ldots, \xi_j, \ldots) \text{ (Bose–Einstein statistics)} \\ -\Phi(\xi_1, \xi_2, \ldots, \xi_i, \ldots, \xi_j, \ldots) \text{ (Fermi–Dirac statistics)} \end{cases} \quad (1.103)$$

The N-particle wave function $\Phi(\xi_1, \xi_2, \ldots)$ can be expanded over a complete set of functions, which are provided, e.g., by the eigenfunctions of the one-particle Hamiltonian, $\mathcal{H}_1$:

$$\Phi(\xi_1, \xi_2, \ldots, \xi_N) = \sum C_{p1,p2,\ldots,pN}\phi_{p1}(\xi_1)\phi_{p2}(\xi_2)\cdots\phi_{pN}(\xi_N);$$
$$\mathcal{H}_1\phi_j(\xi) = \varepsilon_j\phi_j(\xi).$$

Here ξ_j denotes the coordinates and spin of the jth particle, and p_j labels the *one-particle state*. Very often (but not always) one chooses for $\phi_j(\xi)$ plane waves, $\phi_j(\xi) \propto \exp(i(\mathbf{p}_j\mathbf{x}_j - \varepsilon_j t)/\hbar)$. This is an excellent choice when dealing with a uniform infinite system. Otherwise, it is more convenient to use a different complete set of one-particle functions that would explicitly express the symmetry of the problem or nontrivial boundary conditions.

The condition of (anti)symmetry (1.103) means that we can use only properly symmetrized products of one-particle functions. For bosons we have thus

$$\Phi_B^{N_1,N_2,\cdots}(\xi_1, \xi_2, \ldots, \xi_N) \equiv |N_1, N_2, \ldots\rangle_{(B)}$$

$$= \sqrt{\frac{N_1!N_2!\cdots}{N!}} \sum \phi_{p1}(\xi_1)\phi_{p2}(\xi_2)\cdots\phi_{pN}(\xi_N). \quad (1.104)$$

Here the nonnegative number N_i shows how many times the ith one-particle eigenfunction ϕ_i enters the product ($N_1+N_2+\cdots = N$, the number of particles in the system). It is called the *occupation number* of the ith one-particle state. Summation is extended over all distinguishable permutations of indices $\{p1, p2, \ldots, pN\}$. Notice that since all the N_j are nonnegative and add up to N, the sequence $N_0, N_1, \ldots$ always contains a rightmost nonzero term, $N_{j_{max}}$, followed by zeros to infinity.

For fermions we use *Slater's determinants*

$$\Phi_F^{N_1,N_2,\cdots}(\xi_1, \xi_2, \cdots, \xi_N) \equiv |N_1, N_2, \ldots\rangle_{(F)}$$

$$= \frac{1}{\sqrt{N!}} \begin{vmatrix} \phi_{p1}(\xi_1) & \phi_{p1}(\xi_2) & \cdots & \phi_{p1}(\xi_N) \\ \phi_{p2}(\xi_1) & \phi_{p2}(\xi_2) & \cdots & \phi_{p2}(\xi_N) \\ \cdots & \cdots & \cdots\cdots \\ \phi_{pN}(\xi_1) & \phi_{pN}(\xi_2) & \cdots & \phi_{pN}(\xi_N) \end{vmatrix}. \quad (1.105)$$

The properties of determinants guarantee the necessary antisymmetry of the wave function. Indeed, a transmutation of two particles in this case corresponds to transmutation of two columns in the determinant, which by definition changes its sign. Then, if two columns are equivalent, the determinant equals zero. Physically this means that two fermions cannot occupy the same quantum state (*Pauli principle*).

Given the basis of one-particle states, any N-particle wave function is completely defined by the set of occupation numbers and can be written as $|N_1, N_2, \rangle_{(B,F)}$. The set of states $|N_j\rangle$ will now provide us a basis for the Hilbert space of N-particle states (for Bose or Fermi system).

Do we really need now a condition $\sum N_j = N$, which seems rather awkward? After all, as often as not in a solid-state problem the system under consideration can exchange particles with the exterior. On the other hand, (quasi)particles (like phonons) can be created and annihilated, so that their total number will fluctuate. Can't we simply give the occupation numbers arbitrary values?

As a matter of fact, no. If we had no limitations on N_j, the set $|N_j\rangle$ would be *non-denumerably infinite*; each element would be more like a dot on the real axis line, than an integer, and a space spanned by such a set would possess unpleasant mathematical properties and be very unlike a "N-is-very-big-but-finite-dimensional" vector space, which we are accustomed to. Luckily, not all "infinite" states are bad, only "actually infinite" ones, like $|1, 1, 1, 1, 1, 1, \ldots\rangle$. On the other hand, if we keep the condition $\sum N_j = N$ but allow N to be arbitrarily large, the problem is solved. The set of states $|N_j\rangle$ satisfying that condition is called the [0]-set and is *denumerably infinite*. Indeed, since for each state $\sum N_j < \infty$, there is some $N_{j_{max}}$, and the product

$(\sum N_j)_{j_{\max}}$ is a finite nonnegative integer, say M. Moreover, there is only a finite number of states $|N_j\rangle$ with the same value of M. Therefore we can count all states in the [0]-set (first n_0 states for $M = 0$, then n_1 states for $M = 1$, and so on ad infinitum). This exactly means that there are "as many" states in the [0]-set as integer numbers; i.e., the set is denumerably infinite.

The Hilbert space spanned by a [0]-set is called *Fock's space*, and it is in Fock's space that *second-quantized operators* act. The state vectors here, as we have said, are defined by the corresponding set of the *occupation numbers*, and the second quantized operators change these numbers. Thus, any operator can be represented by some combination of basic *creation/annihilation* operators, which act as follows Annihilation operator(we will establish the proper factors a little later):

Annihilation operator:

$$c_j|\ldots, N_j, \ldots\rangle \propto |\ldots, N_j - 1, \ldots\rangle. \tag{1.106}$$

Creation operator:

$$c_j^\dagger|\ldots, N_j, \ldots\rangle \propto |\ldots, N_j + 1, \ldots\rangle. \tag{1.107}$$

Evidently, any element of the [0]-set can be obtained by the repeated action of creation operators on the *vacuum state* (state with no particles) $|0\rangle = |0, 0, 0, 0, \ldots\rangle$:

$$|N_1, N_2, \ldots, N_j, \ldots\rangle \propto \left(c_1^\dagger\right)^{N_1} \left(c_2^\dagger\right)^{N_2} \cdots \left(c_j^\dagger\right)^{N_j} \cdots |0\rangle. \tag{1.108}$$

The vacuum state is *annihilated* by *any annihilation operator*:

$$c|0\rangle = 0. \tag{1.109}$$

What is extremely important to keep in mind is that while in the representation of second quantization we explicitly deal with occupation numbers only, our calculations make sense only as long as we can point out the correct (consistent with the system's properties) one-particle basic functions. This may also be stated as a problem of choice of the vacuum state, for starting from a wrong vacuum we never build the proper Fock space. (It turns out that sentences like "I know that you don't have a dog, but I must know what sort of dog you don't have" are quite reasonable when you deal with a vacuum state of a many-body system.) We will see a good example of this problem later, in superconductivity.

1.4.2 Bosons

Let us begin with a one-particle operator

$$\mathcal{F}_1 = \sum_j f_1(\xi_j). \tag{1.110}$$

Here $f_1(\xi_j)$ is an operator, acting on a one-particle state of the system $\phi(\xi_j)$, and summation is taken over all particles. An example is provided by the kinetic energy operator,

$$\mathcal{K} = \sum_i \left(-\frac{\hbar^2}{2m}\right) \nabla_j^2. \tag{1.111}$$

Let us take a matrix element of $\mathcal{F}_1$ between two N-particle Bose states, $\langle \Phi_{B'}|\mathcal{F}_1|\Phi_B \rangle$. Since in order to make a number (matrix element) of an operator $f_1(\xi_j)$ we need two one-particle wave functions, and since any two such functions $\phi_i(\xi)$, $\phi_j(\xi)$ are orthogonal for $i \neq j$, then there will be only two sorts of nonzero matrix elements of the operator $\mathcal{F}_1$: (a) diagonal, and (b) between the states $|\Phi_{B'}\rangle$, $|\Phi_B\rangle$, which differ by *one* particle, which from some (initial) state $\phi_i(\xi)$ was transferred to another (final) state $\phi_f(\xi) : |\Phi_B\rangle = |\ldots, N_i, \ldots, N_f - 1, \ldots\rangle$, and $|\Phi_{B'}\rangle = |\ldots, N_i - 1, \ldots, N_f, \ldots\rangle$.

The diagonal matrix element is

$$\langle \Phi_B|\mathcal{J}_1|\Phi_B \rangle = \sum_a \left(\frac{N_1! \cdots N_a! \cdots}{N!}\right) \int \int \cdots \int d\xi_1 d\xi_1 \cdots d\xi_N$$

$$\times \sum_{\rho, \rho'} \mathcal{P}_s[\phi_{\rho_1'}^*(\xi_1)\phi_{\rho_2'}^*(\xi_2) \cdots \phi_{\rho_N'}^*](\xi_N) f_1(\xi_a) \mathcal{P}_s[\phi_{\rho_1}(\xi_1)\psi_{\rho_2}(\xi_2) \cdots \phi_{p_N}](\xi_N).$$

Here $\mathcal{P}_s[\ldots]$ denotes symmetrization of indices. Due to the fact that sets of indices p_i and p_i' coincide, $p_i = p_i'$. Let us say that the particle that is affected by the operator is in the state p_u. After we take the (diagonal) matrix element of $f_1(\xi_a)$, $\langle p_a|f_1(\xi_a)|p_a \rangle = \int d\xi_a \phi_i^*(\xi_a) f_1(\xi_a)\phi_i(\xi_a)$, and calculated the other integrals (which are all equal to one due to orthonormality), we can symmetrically rearrange the rest of the occupied states in $(N-1)!$ ways. This must be divided by $N_1!, N_2!, \ldots, (N_a - 1)!, \ldots$, because we cannot distinguish between $N_j!$ possible rearrangements of identical particles occupying the same, jth, state. (An equivalent way to state this is that there are $(N-1)!/(N_1! N_2! \cdots (N_a - 1)! \cdots)$ ways to choose one-particle wave functions to be acted upon by $f_1(\xi_a)$.) Therefore, we obtain the expression

$$\langle \Phi_B|\mathcal{F}_1|\Phi_B \rangle$$

$$= \sum_a \sum_{p_a} \frac{N_1! N_2! \cdots N_a! \cdots}{N!} \left(\frac{(N-1)!}{N_1! \cdots (N_a - 1)! \cdots}\right) \langle p_a|f_1(\xi_a)|p_a \rangle$$

$$= \sum_a \sum_{p_a} \frac{N_a}{N} \langle p_a|f_1(\xi_a)|p_a \rangle$$

$$= \sum_q N_q \langle q|f_1|q \rangle, \tag{1.112}$$

and we no longer need to show explicitly on the coordinates of what particle the operator f_1 acts.

Now let us calculate the off-diagonal matrix elements of the operator. This time on the left side there will be one extra function $\phi_f^*(\xi)$, and on the right one extra $\phi_i(\xi)$, so that

$$
\langle \Phi_{B'} | \mathcal{F}_1 | \Phi_B \rangle = \sum_a \left(\frac{N_1! \cdots (N_i - 1)! \cdots N_f! \cdots}{N!} \right)^{1/2}
$$

$$
\times \left(\frac{N_1! \cdots N_i! \cdots (N_f - 1)! \cdots}{N!} \right)^{1/2}
$$

$$
\times \int \int \cdots \int d\xi_1 d\xi_1 \cdots d\xi_N \sum_{p,p'} \mathcal{P}_s [\phi_{p_1'}^*(\xi_1) \phi_{p_2'}^*(\xi_2) \cdots \phi_{p_N'}]^*(\xi_N)
$$

$$
\times f_1(\xi_a) \mathcal{P}_s [\phi_{p_1}(\xi_1) \phi_{p_2}(\xi_2) \cdots \phi_{p_N}](\xi_N).
$$

These unmatched functions must then be integrated with the operator to yield $(f | f_1(\xi_a) | i) = \int d\xi_a \phi_f^*(\xi_a) f_1(\xi_a) \phi_i(\xi_a)$, while the rest can be rearranged in $(N - 1)!/(N_1! N_2! \cdots (N_i - 1)! \cdots (N_f - 1)! \ldots)$ ways. The result is

$$
\langle \Phi_{B'} | \mathcal{F}_1 | \Phi_B \rangle = \sum_a \sqrt{N_i N_f} \frac{N_1! N_2! \cdots (N_i - 1)! \ldots (N_f - 1)! \cdots}{N!}
$$

$$
\times \left(\frac{(N - 1)!}{N_1! \cdots (N_i - 1)! \cdots (N_f - 1)! \cdots} \right) \langle f | f_1(\xi_a) | i \rangle
$$

$$
= \sqrt{N_i N_f} \langle f | f_1 | i \rangle. \tag{1.113}
$$

Now we are in a position to employ the creation/annihilation operators described earlier. For the bosons they are often denoted by $b^\dagger$, b. We define them with the following factors:

Bose annihilation operator:

$$
b_j | \ldots, N_j, \ldots \rangle_B = \sqrt{N_j} | \ldots, N_j - 1, \ldots \rangle_B. \tag{1.114}
$$

Thus it has a single nonzero matrix element, $\langle N_j - 1 | b_j | N_j \rangle = \sqrt{N_j}$. It equals its complex conjugate, which according to the rules of quantum mechanics is given by $\langle N_j - 1 | b_j | N_j \rangle^* = \langle N_j | b_j^\dagger | N_j - 1 \rangle$. This means that the creation and annihilation operators as we defined them are indeed Hermitian conjugate (in the Bose case so far), and

Bose creation operator:

$$
b_j^\dagger | \ldots, N_j, \ldots \rangle_B = \sqrt{N_j + 1} | \ldots, N_j + 1, \ldots \rangle_B. \tag{1.115}
$$

The combinations $b_j b_j^\dagger$, $b_j^\dagger b_j$ are evidently diagonal: $\langle N_j | b_j b_j^\dagger | N_j \rangle = N_j + 1$; $\langle N_j | b_j^\dagger b_j | N_j \rangle = N_j$. The latter combination, for evident reasons, is called the *occupation number operator* (or particle number operator):

$$\mathcal{N}_j \equiv b_j^\dagger b_j; \tag{1.116}$$

$$\mathcal{N}_j |N_j\rangle = N_j |N_j\rangle.$$

The commutator of the two operators is then $[b_j, b_j^\dagger] = 1$. It is straightforward to check that generally,

$$\left[b_j, b_k^\dagger\right] = \delta_{jk}; \tag{1.117}$$

$$\left[b_j^\dagger, b_k^\dagger\right] = [b_j, b_k]$$

$$= 0.$$

These are the *Bose commutation relations*; we could start from them, and the demand that they are satisfied would completely determine the structure of the many-particle Bose wave function.

Returning to our one-particle operator, we see that it can be expressed via creation/annihilation operators as follows:

$$\mathcal{F}_1 = \sum_{i,f} \langle f|f_1|i\rangle b_f^\dagger b_j. \tag{1.118}$$

Indeed, the matrix elements of this operator between any two states from Fock's space are the same as we have calculated above. Intuitively, the expression looks evident: a particle is being "teleported" (annihilated in state $|i>$ and then created in state $|f\rangle$), that is, scattered. A less evident and very useful property of the above expression is that the coefficients $\langle f|f_1|i\rangle$ are just matrix elements of a one-particle operator f between corresponding one-particle states. Not only have we expressed the operator via $b^\dagger$, b, but we have also reduced $\mathcal{F} = \sum_a f_1(\xi_a)$ to a constituent operator f_1 in the process. This suggests that we rewrite Eq. (1.118) as

$$\mathcal{F}_1 = \int d\xi \hat{\phi}^\dagger(\xi) f_1 \hat{\phi}(\xi). \tag{1.119}$$

1.4.2.1 Bosonic Field Operators

In the above equation we have introduced the so-called *field operators*, $\hat{\phi}^\dagger(\xi)$, $\hat{\phi}(\xi)$, which are by definition

$$\hat{\phi}(\xi) = \sum_p \phi_p(\xi) b_p;$$

$$\hat{\phi}^{\dagger}(\xi) = \sum_p \phi_p^*(\xi) b_p^{\dagger}. \tag{1.120}$$

Evidently, these operators also act in Fock space.[3] What do they create or annihilate? The operator $b_p^{\dagger}$, e.g., creates a particle with a wave function $\phi_p(\xi')$. The operator $\hat{\phi}^{\dagger}(\xi)$ thus creates a particle with a wave function

$$\sum_p \phi_p^*(\xi) \phi_p(\xi') = \delta(\xi - \xi')$$

(we have used the completeness of the basis of one-particle states). That is, the field operator creates (or annihilates) a particle at a given point. An important *density* operator,

$$\varrho(\xi) = \hat{\phi}^{\dagger}(\xi)\hat{\phi}(\xi) = \sum_p |\phi_p(\xi)|^2 b_p^{\dagger} b_p \equiv \sum_p |\phi_p(\xi)|^2 \mathcal{N}_p, \tag{1.121}$$

evidently gives the density of particles at a given point.

The *commutation relations for bosonic field operators* immediately follow from the definition and Eq. (1.117):

$$[\phi(\xi, t), \phi^{\dagger}(\xi', t)] = \delta(\xi - \xi'),$$
$$\left[\phi(\xi, t), \phi(\xi', t)\right] = [\phi^{\dagger}(\xi, t), \phi^{\dagger}(\xi', t)] \tag{1.122}$$
$$= 0.$$

Time dependence, of course, can enter the above equations either through the operators $b^{\dagger}$, b (Heisenberg representation) or through the basic functions ϕ_p (Schrödinger representation), or both (interaction representation). What is important is that definite commutation relations exist *only* between field operators taken at the same moment of time.

Looking at the definition, Eq. (1.120), one sees that field operators are built of the annihilation/creation operators in the same way as an expansion of a *single-particle* wave function in a generalized Fourier series over some complete set of functions $\{\phi_i\}_{i=0}^{\infty}$:

$$\Psi(\xi) = \sum_i \phi_i(\xi) C_i;$$
$$\hat{\psi}(\xi) = \sum_i \phi_i(\xi) b_i.$$

---

[3] We will omit the hats over the field operators when it does not create confusion.

That is why the method is called *second* quantization: it looks as if we quantize the quantum wave function one more time, transforming it into an operator! (See also Problem 3 to this chapter.)

1.4.2.2 Operators of Observables in the Second Quantization Formalism

Using field operators, we can write any operator in the second-quantization form almost without thinking. The recipe is that any n-particle operator

$$\mathcal{F}_n = \frac{1}{n!} \sum_{j_1 \neq j_2 \neq \cdots \neq j_n} f_n(\xi_{j1}, \xi_{j2}, \ldots)$$

(where $f_n(\xi_{j1}, \xi_{j2}, \ldots)$ acts on the coordinates of particles number $j_1, j_2, \ldots, j_n$) in the formalism of second quantization is given by

$$\mathcal{F}_n = \frac{1}{n!} \int d\xi_1 d\xi_2 \cdots d\xi_n \hat{\phi}^\dagger(\xi_1) \hat{\phi}^\dagger(\xi_2) \cdots \hat{\phi}^\dagger(\xi_n)$$
$$\times f_n(\xi_1, \xi_2, \ldots, \xi_n) \hat{\phi}(\xi_n) \hat{\phi}(\xi_{n-1}) \cdots \hat{\phi}(\xi_1). \tag{1.123}$$

This is exactly the sort of expression that we would obtain *if we calculated the average value of an operator f_n using one-particle wave functions*; here they are substituted by field operators. The factorial simply takes into account that there are $n!$ versions of the above expression differing only by the arrangement of indices $1, 2, \ldots, n$. (Being operators, the $\hat{\phi}$'s are sensitive to their order; for this particular recipe to be correct, the ordering of the field operators should be as shown; the outermost couple of operators should have the same argument, etc.)

This rule can be derived for arbitrary n, though three-and more-particle interactions (collisions) are found rarely. For one-particle operators it is correct; it is enough to look at our previous results. We therefore sketch here a proof for $n = 2$.

In this case

$$\mathcal{F}_2 = \frac{1}{2} \sum_{a \neq b} f_2(\xi_a, \xi_b).$$

(An example: scalar pair interaction, where $f_2(\xi_a, \xi_b)$ is simply a scalar potential energy $U(|\xi_a - \xi_b|)$.) We will calculate its matrix elements in the same way as we did it for a one-particle operator. Besides diagonal ones, there are only following nonzero matrix elements:

$$\langle N_f, N_i - 1, N_j - 1 | \mathcal{F}_2 | N_f - 2, N_i, N_j \rangle;$$

$$\langle N_f, N_g, N_i - 1, N_j - 1 | \mathcal{F}_2 | N_f - 1, N_g - 1, N_i, N_j \rangle;$$

$$\langle N_f, \ N_g, \ N_i - 2|\mathcal{F}_2|N_f - 1, \ N_g - 1, \ N_i\rangle;$$

$$\langle N_f, \ N_i - 2|\mathcal{F}_2|N_f - 2, \ N_i, \ N\rangle$$

(corresponding to all possible transitions of two particles); and, of course,

$$\langle N_f, \ N_i - 1|\mathcal{F}_2|N_f - 1, \ N_i\rangle$$

(affecting only one particle). Evidently, the operator $\mathcal{F}_2$ must be of the form

$$\sum_{m,n,p,q} C_{m,n,p,q} b_m^\dagger b_n^\dagger b_p b_q,$$

and we have only to find the coefficients.

Let us calculate the matrix element $\langle N_f, N_g, \ N_i - 1, \ N_j - 1|\mathcal{F}_2|N_f - 1, N_g - 1,$ $N_i, \ N_j\rangle$, corresponding to scattering of two particles from states $|i\rangle$, $|j\rangle$ in two different states $|f\rangle$, $|g\rangle$.

$$\langle N_f, \ N_g, \ N_i - 1, \ N_j - 1|\mathcal{F}_2|N_f - 1, \ N_g - 1, \ N_i, \ N_j\rangle$$

$$= \frac{1}{2} \sum_{a \neq b} \sqrt{\frac{\dots N_f! \cdots N_g! \cdots (N_i - 1)! \cdots (N_j - 1)! \cdots}{N!}}$$

$$\times \sqrt{\frac{\cdots (N_f - 1)! \cdots (N_g - 1)! \cdots N_i! \cdots N_j! \cdots}{N!}}$$

$$\times \int d\xi_1 \cdots d\xi_N \sum_{p,p'\mathcal{P}[} \phi_{p_1'}^*(\xi_1) \cdots \phi_{p_N'}^*(\xi_N) f_2(\xi_a, \ \xi_b) \mathcal{P}[\phi_{p_1}(\xi_1) \cdots \phi_{p_N}](\xi_N).$$

In this expression there are unmatched states, ϕ_f^*, ϕ_g^* on the left and ϕ_i, ϕ_j on the right, which will be integrated with the operator f_2 to yield:

$$\int d\xi_a \int d\xi_b (\phi_f^*(\xi_a)\phi_g^*(\xi_b) f_2(\xi_a, \ \xi_b)\phi_i(\xi_a)\phi_j(\xi_b)$$

$$+ \phi_g^*(\xi_a)\phi_f^*(\xi_b) f_2(\xi_a, \ \xi_b)\phi_i(\xi_a)\phi_j(\xi_b)$$

$$+ \phi_f^*(\xi_a)\phi_g^*(\xi_b) f_2(\xi_a, \ \xi_b)\phi_j(\xi_a)\phi_i(\xi_b)$$

$$+ \phi_g^*(\xi_a)\phi_f^*(\xi_b) f_2(\xi_a, \ \xi_b)\phi_j(\xi_a)\phi_i(\xi_b))$$

(we have explicitly written all terms following from symmetrization, $\mathcal{P}$). The symmetrization of the rest $(N - 2)$ one-particle states produces the combinatorial factor $(N-2)!/(\dots (N_f-1)! \dots (N_g-1)! \dots (N_i-1)! \dots (N_j-1)! \dots)$, so that gathering these results we obtain

$$\langle N_f, \ N_g, \ N_i - 1, \ N_j - 1|\mathcal{F}_2|N_f - 1, \ N_g - 1, \ N_i, \ N_j\rangle$$

$$
= \frac{1}{2} \sum_{a \neq b} \sqrt{\frac{\cdots N_f! \cdots N_g! \cdots (N_i - 1)! \cdots (N_j - 1)! \cdots}{N!}}
$$

$$
\times \sqrt{\frac{\cdots (N_f - 1)! \cdots (N_g - 1)! \cdots N_i! \cdots N_j! \cdots}{N!}}
$$

$$
\times \frac{(N - 2)!}{\cdots (N_f - 1)! \cdots (N_g - 1)! \cdots (N_i - 1)! \cdots (N_j - 1)! \cdots}
$$

$$
\times (\langle fg| f_2 |ij \rangle + \langle gf | f_2 |ij \rangle + \langle fg| f_2 |ji \rangle + \langle gf | f_2 |ji \rangle)
$$

$$
= \left(\frac{1}{2} \frac{\sum_{a \neq b} 1}{N(N-1)} \right) \sqrt{N_f N_g N_i N_j} (\langle fg| f_2 |ij \rangle
$$

$$
+ \langle gf | f_2 |ij \rangle + \langle fg| f_2 |ji \rangle + \langle gf | f_2 |ji \rangle)
$$

$$
= \frac{1}{2} \sqrt{N_f N_g N_i N_j} (\langle fg| f_2 |ij \rangle + \langle gf | f_2 |ij \rangle + \langle fg| f_2 |ji \rangle + \langle gf | f_2 |ji \rangle).
$$

If we take the same matrix element of $\sum_{m,n,p,q} C_{m,n,p,q} b_m^\dagger b_n^\dagger b_p b_q$, we obtain for each $b^\dagger b^\dagger b b$ the combination:

$$
\langle N_f, \ N_g, \ N_i - 1, \ N_j - 1 | b_m^\dagger b_n^\dagger b_p b_q | N_f - 1, \ N_g - 1, \ N_i, \ N_j \rangle
$$

$$
= \sqrt{N_f N_g N_i N_j} \langle N_f, \ N_g, \ N_i - 1, \ N_j - 1 | N_f, \ N_g, \ N_i - 1, \ N_j - 1 \rangle
$$

$$
\times (\delta_{mf} \delta_{ng} \delta_{pi} \delta_{qj} + \delta_{mg} \delta_{nf} \delta_{pi} \delta_{qj} + \delta_{mf} \delta_{ng} \delta_{pj} \delta_{qi} + \delta_{mg} \delta_{nf} \delta_{pj} \delta_{qi}),
$$

which means that $C_{m,n,p,q} = (1/2) \langle mn| f_2 |pq \rangle$. Now, this is the very coefficient that we obtain if we write the expression for $\mathcal{F}_2$ following the general recipe, and then express field operators through $b^\dagger$, b. You can check that this is the case for the rest of the matrix elements as well.

1.4.3 Number and Phase Operators and Their Uncertainty Relation

We have introduced the occupation number operator $\mathcal{N}$. In the simplest case of a single harmonic oscillator it would describe the number of quanta—that is, the amplitude of the oscillations. This is a well-defined observable in the classical limit. Therefore, it is reasonable to ask, what its conjugate operator is, if such can be constructed, and what would be the corresponding classical variable. The canonical commutation relation between this hypothetical Hermitian operator, $\hat{\varphi}$, and $\mathcal{N}$ should be

$$
[\mathcal{N}, \hat{\varphi}] = -i. \tag{1.124}
$$

(We do not write $\hbar$ here for reasons that will be clear shortly.)

The Hamiltonian of the harmonic oscillator can be written as

$$\mathcal{H} = \hbar\omega_0(\mathcal{N} + \frac{1}{2}),$$ (1.125)

where $\mathcal{N} = b^\dagger b$, and $b^\dagger$, b are Bose operators: $[b,\ b^\dagger] = 1$. (This is a fairly standard exercise in any textbook on quantum mechanics, e.g. [6].) Using (1.124) in the Heisenberg equation of motion for $\hat{\varphi}$, we find

$$i\hbar\dot{\hat{\varphi}}(t) = [\hat{\varphi},\ \mathcal{H}] = i\hbar\omega_0.$$ (1.126)

Therefore, $\hat{\varphi}(t) = \omega^t$ is the *phase* of the oscillation. We come to the conclusion, that the phase and the number of quanta (i.e., the amplitude) look like conjugate observables—something that sounds quite reasonable. An immediate important consequence of this should be the *number–phase uncertainty relation* (but see later):

$$\Delta N \cdot \Delta\varphi \geq \frac{1}{2}.$$ (1.127)

It means that you cannot measure phase and amplitude of an oscillator simultaneously with an arbitrary degree of accuracy. (You may ask, where is the Planck constant in here? Is it a *classical* uncertainty relation? Of course not, as you can check by direct observation of a pendulum. The $\hbar$ is hidden in $\mathcal{N}$. Indeed, the number of quanta of an oscillator is its classical energy E divided by $\hbar\omega_0$, so that $\hbar$ is in the denominator of the left-hand side of (1.127).)

More generally, we can attempt to introduce a Hermitian operator $\hat{\varphi}$ demanding that the creation/annihilation Bose operators could be presented as

$$b = e^{-i\hat{\varphi}}\sqrt{\mathcal{N}};$$ (1.128)
$$b^\dagger = \sqrt{\mathcal{N}}e^{i\hat{\varphi}}.$$

We see that $b^\dagger b = \sqrt{\mathcal{N}}e^{i\hat{\varphi}} \cdot e^{-i\hat{\varphi}}\sqrt{\mathcal{N}} = \mathcal{N}$. On the other hand, $bb^\dagger = e^{-i\hat{\varphi}}\mathcal{N}e^{i\hat{\varphi}}$, and this must equal $b^\dagger b + 1$. Therefore, we must have

$$[e^{-i\hat{\varphi}},\ \mathcal{N}] = e^{-i\hat{\varphi}},$$ (1.129)

which is identically satisfied if

$$[\mathcal{N}, \hat{\varphi}] = -i,$$ (1.130)

exactly the commutation relations we discussed. (This can be verified by expanding the exponent and commuting it with $\mathcal{N}$ term by term.)

The definition (1.128) makes perfect sense in the classical limit. In this case the number of quanta is huge (we are speaking of bosons here). The difference between $b^\dagger b$ and $bb^\dagger$ is only one. Therefore, $\mathcal{N}$ becomes essentially a classical number, n, and so does $\hat{\varphi}$. The operators $b^{\dagger}$, b themselves become complex conjugate numbers, their phase and amplitude given by $\sqrt{n}$ and φ. (Recalling the definition of the second quantized wave function, $\Psi(x) = \sum_n (b_n \phi_n(x) + b^{\dagger} \phi_n^*(x))$, we see that it becomes a classical field – like an electromagnetic field (many photons), or a sound wave (many phonons).)

Unfortunately, the situation is more complicated than that. For example, it is clear that for the eigenstates of $\mathcal{N}$, $\mathcal{N}|n\rangle = n|n\rangle$, where $\Delta N = 0$, Eq. (1.127) would imply $\Delta\varphi = \infty$. This is strange, since the phase is defined only modulo 2π, and its maximum uncertainty should not exceed 2π.

A subtler analysis (see [1]) shows that the phase does not make a good quantum-mechanical "observable," which means that the operator $\hat{\varphi}$ introduced earlier *cannot* be Hermitian.

The sketch of a proof goes as follows. If $\hat{\varphi}$ is Hermitian, then $\hat{U} \equiv e^{-i\hat{\varphi}}$ must be unitary,

$$\hat{U}\hat{U}^{\dagger} = \hat{U}^{\dagger}\hat{U} = \mathcal{I}.$$

Here $\hat{U}^{\dagger} = e^{i\hat{\varphi}}$. On the other hand, acting by (1.129) on $|n\rangle$, we obtain

$$\mathcal{N}\hat{U}|n\rangle = \hat{U}(\mathcal{N} - \mathcal{I})|n\rangle = (n-1)\hat{U}|n\rangle.$$

This means that $\hat{U}|n\rangle = |n-1\rangle$. Note, that $\hat{U}$ must annihilate the vacuum state, $\hat{U}|0\rangle = 0$. (Otherwise we could repeatedly act on $\hat{U}|0\rangle$ by $\hat{U}$, producing states with negative occupation numbers!)

In a similar way we find that $\hat{U}^{\dagger}|n\rangle = |n+1\rangle$.

Now, for any $n = 0, 1, 2, \ldots$ we see that $\langle m|\hat{U}\hat{U}^{\dagger}|n\rangle = \langle m|\hat{U}|n+1\rangle = \langle m|n\rangle = \delta_{mn}$, and therefore, $\hat{U}\hat{U}^{\dagger} = \mathcal{I}$. This holds for $\hat{U}^{\dagger}\hat{U}$ as well, but only for positive n, because $\langle 0|\hat{U}^{\dagger}\hat{U}|0\rangle = 0$. Therefore, $\hat{U}^{\dagger}\hat{U} \neq \mathcal{I}$, and $\hat{\varphi}$ cannot be a well defined Hermitian operator. This is the result of the fact that the states with negative occupation numbers do not exist.

On the bright side, it can be shown that the above approach provides a good approximation if $N \gg 1$ (which is, fortunately, what we are usually dealing with). (The deviation of the occupation number from its average value N is confined to the interval $[-N, \infty)$; at large N, it is "almost" $(-\infty, \infty)$.) The relation (1.127) holds for the states with small $\Delta\varphi$.[4]

Keeping in mind these caveats, we can infer from (1.124) that in the basis of eigenstates of $\hat{\varphi}$ the action of operators $\hat{\varphi}$ and $\mathcal{N}$ on a wave function $\Psi(\varphi) \equiv (\varphi|\Psi)$ is given by

$$\hat{\varphi}\Psi(\varphi) = \varphi\Psi(\varphi), \tag{1.131}$$

$$\mathcal{N}\Psi(\varphi) = \frac{1}{i}\frac{\partial}{\partial\varphi}\Psi(\varphi). \tag{1.132}$$

---

[4] For the eigenstates of $\mathcal{N}$, as one should expect, the results are close to the uniform distribution of φ in the interval $[0, 2\pi)$.

The eigenstates of $\mathcal{N}$ in this basis are evidently

$$\langle\varphi|n\rangle \propto e^{in\varphi}, \tag{1.133}$$

$$\mathcal{N}\langle\varphi|n\rangle = \frac{1}{i}\frac{\partial}{\partial\varphi}\langle\varphi|n\rangle = n\langle\varphi|n\rangle, \tag{1.134}$$

and in the basis of eigenstates of $\mathcal{N}$,

$$\mathcal{N}\Psi(n) \equiv \mathcal{N}\langle n|\Psi\rangle = n\Psi(n), \tag{1.135}$$

we formally have

$$\hat{\varphi}\Psi(n) = -\frac{1}{i}\frac{\partial}{\partial n}\Psi(n). \tag{1.136}$$

These representations are related by a Fourier transform:

$$\Psi(n) = \int_0^{2\pi}\frac{d\varphi}{2\pi}e^{in\varphi}\Psi(\varphi); \tag{1.137}$$

$$\Psi(\varphi) = \sum_n e^{-in\varphi}\Psi(n). \tag{1.138}$$

We will see how useful this proves when dealing with transport in small super-conductors. The phase there is the superconducting phase; the bosons are Cooper pairs of electrons – fermions – which is certainly nontrivial.

1.4.4 Fermions

In the beginning, we state that the formal results of the above section hold for Fermi statistics as well. We can introduce fermionic field operators in the same way as bosonic ones, and use the same recipe to write down a second-quantized expression for any n-particle operator $\mathcal{F}_n$. The only difference (and a world of difference!) is that now instead of Bose creation/annihilation operators we use the Fermi ones, $a^\dagger$, a. Unlike the previous case, due to the Pauli exclusion principle, for any state of the Fermi system and for any one-particle state p, $a_p^\dagger a_p^\dagger|\Phi\rangle_F = 0$ and $a_p a_p|\Phi\rangle_F = 0$; i.e., $(a_\rho^\dagger)^2 = (a_p)^2 = 0$. (In formal mathematical language they are called *nilpotent* operators.) In order to find their matrix elements, we again return to the case of a one-particle operator $\mathcal{F}_1$ and calculate its matrix elements between fermionic states $|\Phi\rangle_F$, which are now given by Slater determinants (1.105),

$$\Phi_F^{N_1,N_2,\dots}(\xi_1, \xi_2, \dots\xi_N) \equiv |N_1, N_2, \rangle_{(F)}$$

$$= \frac{1}{\sqrt{N!}} \begin{vmatrix} \phi_{p_1}(\xi_1) & \phi_{p_1}(\xi_2) & \cdots & \phi_{p_1}(\xi_N) \\ \phi_{p_2}(\xi_1) & \phi_{p_2}(\xi_2) & \cdots & \phi_{p_2}(\xi_N) \\ \cdots & \cdots & \cdots & \cdots \\ \phi_{p_N}(\xi_1) & \phi_{p_N}(\xi_2) & \cdots & \phi_{p_N}(\xi_N) \end{vmatrix}.$$

To fix the sign of the wave function, set $p_1 < p_2 < \cdots < p_N$.

For a one-particle operator the only nonzero off-diagonal matrix elements are of the form $\langle 1_f, 0_i | \mathcal{F}_1 | 0_f, 1_i \rangle$ (here we explicitly use the fact that in the Fermi system any occupation number N_p is either 0 or 1; indices i, f show what occupied or empty state is affected by the operator). Using the definition of a determinant, we can write

$$\langle 1_f, 0_i | \mathcal{F}_1 | 0_f, 1_i \rangle = \sum_a \frac{1}{N!} \int d\xi_1 \cdots d\xi_N \sum_P \sum_{P'} (-1)^P (-1)^{P'}$$
$$\times \phi_{p'_1}^*(\xi_1) \cdots \phi_{p'_a}^* \cdots f_1(\xi_a) \phi_{p_1}(\xi_1) \cdots \phi_{p_a} \cdots ,$$

where the sums are taken over all particles and all permutations of indices $\mathcal{P}[p_1, p_2, \cdots, p_N]$, $\mathcal{P}'[p'_1, p'_2, \cdots, p'_N]$, $(-1)^{P,P'}$ being parities of corresponding permutations.

There is an extra $\phi_i(\xi)$ on the right-hand side and an extra $\psi_f^*(\xi)$ on the right-hand side, and they must be connected by the operator $f_1(\xi_a)$. Therefore, in all terms contributing to the matrix element, the permutations $\mathcal{P}$ and $\mathcal{P}'$ differ only by a single index, f, instead of i:

$$\mathcal{P}: p_1 p_2 \ldots i \ldots p_N;$$
$$\mathcal{P}': p_1 p_2 \ldots f \ldots p_N.$$

These permutations have relative parity $(-1)^Q$, where Q is the number of occupied states between i and f (that is, with $i < p < f$, if $i < f$). Indeed, the parity of the permutation $\mathcal{P}$ is $(-1)^P$, where P is the number of steps you need to achieve it starting from the ordered set of indices $p_a < p_b < \cdots < i < \cdots$ (we can transpose two adjacent indices at each step). To obtain the permutation $\mathcal{P}'$, we replace i in the initial set with the index of the final state, f. Now we first have to put index f in place; if between i and f there is no occupied state, then it is already in place; if not, we have to make exactly Q steps, to order the set of indices, after which they can be rearranged in the same P steps. Therefore, the relative parity of $\mathcal{P}$ and $\mathcal{P}'$ is $(-1)^Q$.

This allows us to write for the matrix element

$$\langle 1_f, 0_i | \mathcal{F}_1 | 0_f, 1_i \rangle = \langle f | f_1 | i \rangle (-1)^Q,$$

while for the diagonal matrix element, of course,

$$\langle|\mathcal{F}_1|\rangle = \sum_j \langle j|f_1|j\rangle N_j.$$

Here we keep the same notation for the matrix elements of operator f_1 as in Bose case.

Evidently, the creation/annihilation operators in the Fermi case, $a^\dagger$, a, will have only one nonzero matrix element each. We define them to be

$$\langle 0_j|a_j|1_j\rangle = \langle 1_j|a_j^\dagger|0_j\rangle = (-1)^{\sum_{s-1}^{j-1} N_s}. \tag{1.139}$$

First let us check whether the particle number operator $\mathcal{N}_j$ in the Fermi case is still $a_j^\dagger a_j$. The answer is positive, since

$$\langle 0_j|a_j^\dagger a_j|0_j\rangle = 0,$$

$$\langle 1_j|a_j^\dagger a_j|1_j\rangle = 1,$$

and N_j can take no other values.

The only nonzero matrix element of a "transmission" operator $a_f^\dagger a_i$ is

$$\langle 1_f, 0_i|a_f^\dagger a_i|0_f, 1_i\rangle = \langle 1_f, 0_i|a_f^\dagger|0_f, 0_i\rangle\langle 0_f, 0_i|a_i|0_f, 1_i\rangle$$

$$= (-1)^{\sum_{s=1}^{f-1} N_s'}(-1)^{\sum_{z=1}^{i-1} N_z}$$

$$= (-1)^Q$$

(the prime means that in the left sum the occupation numbers are calculated *after* a particle in the initial state was annihilated. E.g., if $i < f$, then the resulting expression is $(-1)^{\sum_{s=i}^{f-1} N_s'} = (-1)^{\sum_{s=i+1}^{f-1} N_s} \equiv (-1)^Q$.) This confirms our guess that a one-particle operator can indeed be written via $a^\dagger$, a operators in the same way as in Bose case,

$$\mathcal{F}_1 = \sum_{f,i=1}^{N} \langle f|f_1|i\rangle a_f^\dagger a_i.$$

On the other hand, for the operator $a_i a_f^\dagger$ we obtain

$$\langle 1_f, 0_i|a_i a_f^\dagger|0_f, 1_i\rangle = \langle 1_f, 0_i|a_i|1_f, 1_i\rangle \langle 1_f, 1_i|a_f^\dagger|0_f, 1_i\rangle$$

$$= (-1)^{\sum_{s=1}^{f-1} N_s'}(-1)^{\sum_{z=1}^{i-1} N_z}$$

$$= (-1)^{(Q+1)}.$$

Table 1.3 Second quantization representation of operators

$\mathcal{F}_n = \frac{1}{n!}\sum_{a1\neq a2\neq\cdots\neq an=1}^{N} f_n(\xi_{a1}, \cdots, \xi_{an})$	n-particle operator
$\hat{\phi}(\xi,\, t) = \sum_{j=1}^{N} \phi_j(\xi,\, t)b_j(t)$	Field operator (annihilation)
$\hat{\phi}^\dagger(\xi,\, t) = \sum_{j=1}^{N} \phi_j^*(\xi,\, t)b_j^\dagger(t)$	Field operator (creation)
$\mathcal{F}_n = \frac{1}{n!}\int d\xi_1 \cdots \int d\xi_n \hat{\phi}^\dagger(\xi_n,\, t)\cdots\hat{\phi}^\dagger(\xi_1,\, t)f_n(\xi_1, \cdots,\, \xi_n)\hat{\phi}(\xi_1,\, t)\cdots\hat{\phi}(\xi_n,\, t)$	

This gives us the *Fermi(anti)commutation relations,*

$$\{a_i,\, a_j^\dagger\} = \delta_{ij}, \quad \{a_i,\, a_j\} = \{a_i^\dagger,\, a_j^\dagger\} = 0. \tag{1.140}$$

Here $\{,\, \}$, as usual, denotes *an* anticommutator, $\{\mathcal{A},\, \mathcal{B}\} \equiv \mathcal{AB} + \mathcal{BA}$. In its turn, this immediately yields *Fermi (anti)commutation relations for the field operators,*

$$\{\psi(\xi,\, t),\, \psi^\dagger(\xi',\, t)\} = \delta(\xi - \xi')$$
$$\{\psi(\xi,\, t),\, \psi(\xi',\, t)\} = \{\psi^\dagger(\xi,\, t),\, \psi^\dagger(\xi',\, t)\} = 0. \tag{1.141}$$

(Fermi field operators are built from $a^\dagger$, a and one-particle wave functions in the same way as in the Bose case, and (1.141) follows from the completeness of those functions' set.)

As an exercise, you can check that the same recipe as in the Bose case holds, e.g., for two-particle operators. But we can now finally conclude our considerations on how to express *operators of observables in the second quantization representation* (Table 1.3)

1.5 Problems

• *Problem 1*

Derive the path integral expression for the single-particle propagator in momentum space, starting from the definition:

$$K(\mathbf{p},\, t;\, \mathbf{p}',\, t') = \langle \mathbf{p},\, t|\mathbf{p}',\, t'\rangle\theta(t - t') \rightsquigarrow \int \frac{\mathcal{D}\mathbf{p}}{(2\pi\hbar)^3} \int \mathcal{D}\mathbf{x}\, e^{\frac{i}{\hbar}S(\mathbf{p},\mathbf{x})},$$

where $|\mathbf{p},\, t\rangle$ is an eigenvector of the Heisenberg momentum operator $\hat{p}_H(t) == \mathcal{U}^\dagger(t)\,\hat{p}\,\mathcal{U}(t)$.

What form has the action $S(\mathbf{p},\, \mathbf{x})$ in this case?

• Note that in classical mechanics the Lagrange function "is always uncertain to a total time derivative of any function of the coordinates and time" ([4], pp. 2–5), say $d/dt(\mathbf{p}(\mathbf{x},\, \dot{\mathbf{x}})\mathbf{x})$.

Show that the result coincides with the Fourier transform of the coordinate-space propagator:

$$K(\mathbf{p},\ t;\mathbf{p}',\ t') = \int d^3\mathbf{x} \int d^3\mathbf{x}' e^{-\frac{i}{\hbar}\mathbf{p}\mathbf{x}} K(\mathbf{x},\ t;\mathbf{x}',\ t') e^{\frac{i}{\hbar}\mathbf{p}'\mathbf{x}'}$$

- *Problem 2*

Derive the Feynman rules in momentum space for the scattering by the potential

$$V(\mathbf{x},\ t) = \int \frac{d^3\mathbf{p}\,dE}{(2\pi\hbar)^4} e^{\frac{i}{\hbar}(\mathbf{p}\mathbf{x}-Et)} v(\mathbf{p},\ E).$$

- *Problem 3*

Suppose the interactions of particles in the system are described by a scalar pair potential u,

$$\mathcal{W} = \frac{1}{2} \sum_{a \neq b} u\langle|\mathbf{x}_a - \mathbf{x}_b|\rangle,$$

and the Hamiltonian is thus

$$\mathcal{H} = \mathcal{K}_{(\text{kin.energy})} + \mathcal{W}.$$

Write down the expression for $\mathcal{K}$ and $\mathcal{W}$ in the second quantized form. Write the equations of motion for field operators (Bose and Fermi case) in Heisenberg and interaction representations. Compare them to one-particle Schrödinger equation. What is the difference?

References

1. Carruthers, P., Nieto, M.M.: Phase and angle variables in quantum mechanics. Rev. Mod. Phys. **40**, 411 (1967)
2. Feynman, R.P., Hibbs, A.R.: Quantum Mechanics and Path Integrals. McGraw-Hill, New York (1965)
3. Gardiner, C.W.: Handbook of Stochastic Methods for Physics, Chemistry, and the Natural Sciences. Springer, Berlin (1985)
4. Goldstein, H.: Classical Mechanics. Addison-Wesley, Reading (1980)
5. Imry, Y.: Physics of mesoscopic systems. In: Grinstein, G., Mazenko, G. (eds.) Directions in Condensed Matter Physics: Memorial Volume in Honor of Shang-Keng Ma. World Scientific, Singapore (1986)
6. Landau, L.D., Lifshitz, E.M.: Quantum mechanics, non-relativistic theory. A Course of theoretical physics, vol. III, pp. 64–65. Pergamon Press, New York (1989) (A concise and clear explanation of the second quantization formalism)
7. Ryder, L.: Quantum field theory. Chapter 5, Cambridge University Press, New York (1996) (A discussion of path integrals in quantum mechanics and field theory)

8. Washburn, S., Webb, R.A.: Quantum transport in small disordered samples from the diffusive to the ballistic regime. Rep. Progr. Phys. **55**, 1311 (1992) (A review of theoretical and experimental results on mesoscopic transport)
9. Ziman, J.M.: Elements of advanced quantum theory, Cambridge University Press, Cambridge (1969) (An excellent introduction into the very heart of the method of Green's functions)

Chapter 2
Green's Functions at Zero Temperature

Men say that the Bodhisat Himself drew it with grains of rice upon dust, to teach His disciples the cause of things. Many ages have crystallised it into a most wonderful convention crowded with hundreds of little figures whose every line carries a meaning.

Rudyard Kipling. "Kim."

Abstract Green's functions as a tool for probing the response of a many-body system to an external perturbation. Similarity and difference from a one-particle propagator. Statistical ensembles. Definition of Green's functions at zero temperature. Analytical properties of Green's functions and their relation to quasiparticles. Perturbation theory and diagram techniques for Green's functions at zero temperature. "Dressing" of particles and interactions: Polarization operator and self energy. Many-particle Green's functions.

2.1 Green's Function of The Many-Body System: Definition and Properties

2.1.1 Definition of Green's Functions of the Many-Body System

When we discussed one-particle states in the second quantization representation, from the formally mathematical point of view *any* complete set of functions dependent on the coordinates (spin, etc.) of one particle would work. But from the point of view of physics it is not so: the set must be chosen in a "physically reasonable way." That is, since we virtually never can solve our equations exactly and not always can provide an "epsilon-delta" -style estimate of the approximations involved, the initial setup must be as close as possible to the solution that we are striving to reach.

A. Zagoskin, *Quantum Theory of Many-Body Systems*,
Graduate Texts in Physics, DOI: 10.1007/978-3-319-07049-0_2,
© Springer International Publishing Switzerland 2014

For the one-particle state this means that it should be relatively stable, thus possessing some measurable characteristics and giving us a palatable zero-order approximation: the one of *independent quasiparticles*. (We have discussed this at some length in Chap. 1. Here we are going to elaborate that qualitative discussion and see how it fits the general formalism of perturbation theory developed so far.)

We have seen that the field operator satisfies the "Schrödinger equation"

$$
i\hbar\frac{\partial}{\partial t}\psi(\xi,\ t) = (\mathcal{E}(\xi,\ t) + \mathcal{V}(\xi,\ t))\psi(\xi,\ t)
$$
$$
+ \int d\xi'\psi^\dagger(\xi',\ t)\mathcal{U}(\xi',\ \xi)\psi(\xi',\ t)\psi(\xi,\ t) + \cdots \qquad (2.1)
$$

Here $\mathcal{E}$ and $\mathcal{V}$ are operators of kinetic energy and external potential; $\mathcal{U}$ describes instantaneous particle–particle interactions, and so on. Evidently, if the basic set of one-particle functions is chosen correctly, the leading term in this equation will be the one-particle one; i.e.,

$$
i\hbar\frac{\partial}{\partial t}\psi(\xi,\ t) \approx (\mathcal{E}(\xi,\ t) + \mathcal{V}(\xi,\ t))\psi(\xi,\ t). \qquad (2.2)
$$

This is a mathematical demonstration of the fact that we have a system of weakly interacting objects—"quasiparticles"—which can be approximately described by (2.2). Since the deviations from this description are small, the lifetimes of these objects are large enough to measure their characteristics in some way, and so they are reasonably well defined. And as often as not they are drastically different from the properties of free particles, already due to the presence of the $\mathcal{V}$-term in the above equation. Let us consider this point in more detail.

For example, when we investigate the properties of electrons in a metal, the reasonable first approximation is to take into account the periodic potential of the crystal lattice, neglecting for a while both electron—electron interactions and "freezing" the ions at their equilibrium positions. Even in this crude approximation the properties of these "quasielectrons" are very different from those of a QED electron, with its mass of 9.109×10^{-28} g and electric charge of -4.803×10^{-10} esu. Its mass is now, generally, anisotropic and may be significantly less or more; its charge may become positive; its momentum is no longer conserved due to the celebrated Umklapp processes; and there can exist several different species of electrons in our system! (See Fig. 2.1.)

On the other hand, when we are interested in the properties of the crystal lattice of the very metal, we quantize the motion of the ions, and come to the concept of *phonons*. These are quasiparticles, if there are any, because outside the lattice phonons simply don't exist, while inside it they thrive. They even interact with electrons and each other—through terms like the third one in (2.1).

Of course, our ultimate goal is to take into account these other terms as well. *Then* we will have some other objects, governed by an equation like (2.2) without any extra interaction—and *they* will be the actual quasiparticles in our system.

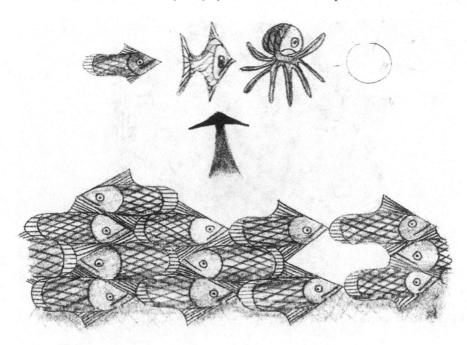

Fig. 2.1 Quasiparticles in the Fermi system

There is no contradiction here. In Chap. 1 we have seen how interactions "dress" a bare particle, making of it a quasiparticle; here we see it being made in two steps. The first step (a right choice of the basic set of one-particle states) is made *without* addressing any perturbation theory, usually based on symmetry considerations (or common physical sense), as when translation symmetry of the crystal lattice forces us to describe otherwise free electrons in terms of Bloch functions instead of simple plane waves, and we use the concept of phonons as a more adequate description of low-energy dynamics of the ions. Once chosen, this set of states ("basic quasiparticles," if you wish) plays the very same role as the states of free particles in the absence of an external potential, and these two sets of states have a lot in common. For example, they live infinitely long (because by definition they have definite energy, $\mathcal{E}\phi_j = E_j\phi_j$. If—as is the case in most books—we are dealing with a "liquid" of interacting fermions on a uniform background, the most natural choice of "basic quasiparticles" is *real* particles.

Therefore, later on we will call them simply "particles", while reserving the term "quasiparticles" (or "elementary excitations") par excellence for the ones "dressed" due to interactions with other particles. (The assumption that the description of the Fermi liquid can be based on the picture of a weakly interacting gas of such quasifermions was in the foundation of Landau's phenomenological theory.)

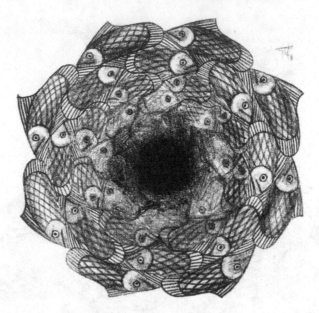

Fig. 2.2 Fermi surface and Fermi sphere

Now we can turn to the building of such a theory in the many-body case. We will begin with the case of single-component normal, uniform, and homogeneous Fermi and Bose system (the above-mentioned textbook case) at zero temperature.

This is the simplest possible *and* practically important case, since it does not involve superfluid (superconducting) condensate. (The discussion of the latter we postpone until Chap. 4.) In the Fermi case it applies to nonsuperconducting metals and semiconductors—if we forget for a while about the subtleties of band structures. Alkali metals are especially good examples (Fig. 2.2).

In the Bose case the example seems purely academic (since bosons must undergo Bose condensation at zero) until we recall that there is at least *one* practically important system of bosons that don't condense: phonons! (This is because the number of phonons is not conserved, but this is not important when we use the grand potential formalism.)

In Chap. 1 we introduced the one-particle propagator

$$K(x,\ t; x',\ t') = \langle x|\mathcal{S}(t,\ t')|x'\rangle = \langle xt|x't'\rangle$$

as a transmission amplitude of a particle between points $(x',\ t')$ and $(x,\ t)$. A straightforward generalization of the former expression is a matrix element of the N-particle S-operator $\langle\Phi|\mathcal{S}(t,\ t')|\Phi'\rangle$. Unfortunately, it is useless, since it a involves transmission amplitude involving $N \approx 10^{23}$ particles. On the other hand, the one-particle propagator in the latter form suggests that we could look at two states with a single particle excited,

$$|x, t\rangle_{\text{state}} \equiv \psi^{\dagger}(x, t)|\text{state}\rangle;$$
$$|x', t'\rangle_{\text{state}} \equiv \psi^{\dagger}(x', t')|\text{state}\rangle$$

(here the Heisenberg field operator $\psi^{\dagger}(x, t)$ creates a particle at a given point), and introduce Green's function as their overlap:

$$\text{Green's function} \rightsquigarrow \langle\text{state}|\psi(x, t)\psi^{\dagger}(x', t')|\text{state}\rangle.$$

Of course, we must average over the states of the many-body system on which our field operators act, in order to get rid of all other nonmacroscopic variables except the two coordinates and moments of time between which the quasiparticle travels. Such an averaging of an operator $\mathcal{A}$ (both quantum and statistical) is achieved by taking its trace with the statistical operator (density matrix) of the system, $\hat{\varrho}$,

$$\langle\mathcal{A}\rangle = \text{tr}(\hat{\varrho}\mathcal{A}). \tag{2.3}$$

Now, the above formula indeed looks like a propagator, describing a process when we add to our system of N identical fermions one extra particle, let it propagate from (x', t') to (x, t), and then take it away. It is a good probe of particle-particle interactions in the system. The other option would be *first* to take away the particle, look at how the resulting *hole* propagates, and then fill it, restoring the particle (Fig. 2.3).

Then the *one-particle causal Green's function*, describing both processes, can be defined by the expression

$$G_{\alpha\alpha'}(\mathbf{x}, t; \mathbf{x}', t')$$
$$= -i\left\langle \psi_\alpha(\mathbf{x}, t)\psi^{\dagger}_{\alpha'}(\mathbf{x}', t') \right\rangle \Theta(t - t') \mp i\left\langle \psi^{\dagger}_{\alpha'}(\mathbf{x}', t')\psi_\alpha(\mathbf{x}, t) \right\rangle \Theta(t' - t)$$
$$\equiv -i\left\langle T\psi_\alpha(\mathbf{x}, t)\psi^{\dagger}_{\alpha'}(\mathbf{x}'; t') \right\rangle. \tag{2.4}$$

Here we write spin indices explicitly. They can take two values (e.g., up and down) for fermions (and only one for phonons).[1]

Averaging defined by (2.3) is a linear operation. Using this property, we can apply the time differentiation operator, $\partial/\partial t$, to $T\psi_\alpha(\mathbf{x}, t)\psi^{\dagger}_{\alpha'}(\mathbf{x}', t')$, and average it to obtain

---

[1] It can be shown that no matter what the spin of real fermions, the (basic) quasiparticles will have spin 1/2 (though there will be *several* types of quasiparticles); see [4], §1.

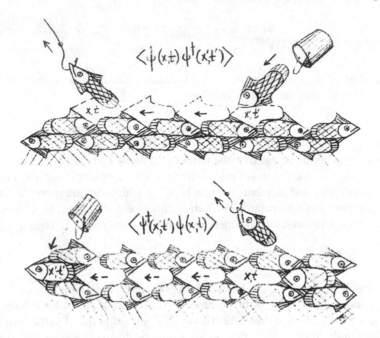

$$\langle \psi(x,t)\, \psi^\dagger(x',t')\rangle$$

$$\langle \psi^\dagger(x',t')\, \psi(x,t)\rangle$$

Fig. 2.3 Causal Green's function: the physical sense

$$i\hbar \frac{\partial}{\partial t_1} G_{\alpha\beta}(\mathbf{x}_1,\ t_1;\mathbf{x}_2,\ t_2) \tag{2.5}$$

$$= \mathcal{E}(\mathbf{x}_1) G_{\alpha\beta}(\mathbf{x}_1,\ t_1;\mathbf{x}_2,\ t_2) + \mathcal{V}(\mathbf{x}_1,\ t_1) G_{\alpha\beta}(\mathbf{x}_1,\ t_1;\mathbf{x}_2,\ t_2)$$

$$-\, i \int d\mathbf{x}_3 U(\mathbf{x}_3,\ \mathbf{x}_1) \left\langle T\psi^\dagger_\gamma(\mathbf{x}_3,\ t_1)\psi_\gamma(\mathbf{x}_3,\ t_1)\psi_\alpha(\mathbf{x}_1,\ t_1)\psi^\dagger_\beta(\mathbf{x}_2,\ t_2)\right\rangle + \cdots$$

$$+\, \hbar\delta(\mathbf{x}_1 - \mathbf{x}_2)\delta(t_1 - t_2). \tag{2.6}$$

We see that the *many-body* Green's function defined above is not a Green's function in the mathematical sense. It is a solution to differential Eq. (2.6). This is not a closed equation for Green's function, since it contains averages of four and more field operators (two-particle, three-particle, etc. Green's functions). Thus (2.6) is only the first equation in the quantum analogue to the well known infinite *BBGKY chain* (Bogoliubov–Born–Green–Kirkwood–Yvon) in classical statistical mechanics. The latter consists of interlinked equations of motion for classical *n*-particle distribution functions (see e.g. in Chap. 3, [2]).

As in the classical case, breaking this chain leads to a *nonlinear* differential equation for Green's function, as distinct from the *linear* Eq. (1.25), which governs the one-particle propagator $K(x,\ t;x',\ t')$.

The averaging procedure in equilibrium can be performed most conveniently using either the *canonical* or *grand canonical* ensemble. Mathematically, the two reflect different choices of independent variables: ($T,\ V,\ N$: temperature, volume,

(a)   **(b)**

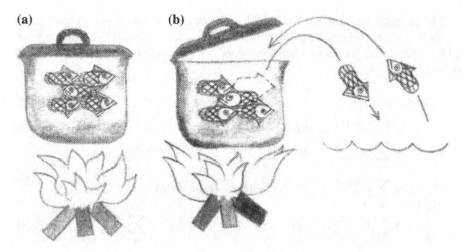

Fig. 2.4 Ensemble averaging. **a** Canonical ensemble; **b** grand canonical ensemble

number of particles) versus (T, V, μ; temperature, volume, chemical potential). The physical difference is that in the former case the system can exchange only energy with its surroundings (thermostat), while in the latter the particles can leave and enter the system as well (Fig. 2.4), the average number of particles being fixed by the chemical potential, μ, as the temperature fixes the average kinetic energy of the particles in both cases.

The statistical operators are of Gibbs form,

$$\hat{\varrho}_{CE} = e^{\beta(F-\mathcal{H})}, \tag{2.7}$$

or

$$\hat{\varrho}_{GCE} = e^{\beta(\Omega-\mathcal{H}')}. \tag{2.8}$$

Here $\beta = 1/T$ (we put $k_B = 1$ to simplify the notation); $F = -(1/\beta) \times \ln \operatorname{tr} e^{-\beta\mathcal{H}}$ is the free energy; $\mathcal{H}' = \mathcal{H} - \mu\mathcal{N}$, $\mathcal{N}$ is the particle number operator; and the *grand potential* $\Omega = -(1/\beta)\ln \operatorname{tr} e^{-\beta\mathcal{H}'} = -PV$. (The Lagrange term $-\mu\mathcal{N}$ evidently commutes with the rest of the Hamiltonian.) We can therefore introduce Heisenberg and interaction representations using $\mathcal{H}'$ instead of $\mathcal{H}$ and come to the same results, with the only difference being in corresponding eigenvalues of the Hamiltonian.

In the thermodynamic limit both approaches are equivalent. The situation may change when the size of the system under consideration becomes small enough, so that the fluctuations of the particle number cannot be ignored, and the two ensembles describe *physically* different systems (such as an isolated conducting grain vs. one connected to massive conductors by leads). At present, we will not deal with such a situation. Since the grand canonical ensemble is easier to work with, we will use it, and—to simplify notation—will henceforth omit primes in $\mathcal{H}'$ and put $\hbar = 1$.

The Hamiltonian now has the complete set of eigenstates $|n\rangle$ with eigenvalues $E'_n = E_n - \mu N_n$, where N_n is the number of particles in state $|n\rangle$. The average of a time-ordered product of two field operators in Heisenberg representation can now be written as

$$
\begin{aligned}
\Big\langle T\psi(t_1)\psi^\dagger(t_2) \Big\rangle & \\
&= \mathrm{tr}(e^{\beta(\Omega-\mathcal{H})}T\psi(t_1)\psi^\dagger(t_2)) \\
&= \sum_{n,m} \langle n|e^{\beta(\Omega-\mathcal{H})}|m\rangle\langle m|T\psi(t_1)\psi^\dagger(t_2)|n\rangle\langle m|m\rangle^{-1}\langle n|n\rangle^{-1} \\
&= \sum_n e^{\beta\Omega-\beta(E_n-\mu N_n)}\langle n|T\psi(t_1)\psi^\dagger(t_2)|n\rangle\langle n|n\rangle^{-1} \\
&= \sum_n e^{-\beta(E_n-\mu N_n)}\langle n|T\psi(t_1)\psi^\dagger(t_2)|n\rangle\langle n|n\rangle^{-1} \Big/ \sum_n e^{-\beta(E_n-\mu N_n)}. \quad (2.9)
\end{aligned}
$$

(We have used the standard trick of inserting the complete set of states, and allow for the possibility that they are not normalized to unity.)

At zero temperature ($\beta \to \infty$) we are left with

$$
G_{\alpha\beta}(\mathbf{x}_1, t_1; \mathbf{x}_2, t_2) = -i\frac{\langle 0|T\psi_\alpha(\mathbf{x}_1, t_1)\psi_\beta^\dagger(\mathbf{x}_2, t_2)|0\rangle}{\langle 0|0\rangle}. \quad (2.10)
$$

Here $|0\rangle$ is the exact *ground state* of the system in *Heisenberg representation*: it is time independent and includes all interaction effects.

In a homogeneous and isotropic system in a stationary state, Green's function can depend only on differences of coordinates and times:

$$
G_{\alpha\beta}(\mathbf{x}_1, t_1; \mathbf{x}_2, t_2) = G_{\alpha\beta}(\mathbf{x}_1 - \mathbf{x}_2, t_1 - t_2). \quad (2.11)
$$

If, moreover, the system is not magnetically ordered and is not placed in and external magnetic field, then spin dependence in (2.11) reduces to a unit matrix:

$$
G_{\alpha\beta} = \delta_{\alpha\beta}G, \quad (2.12)
$$

where $G = \frac{1}{2}\,\mathrm{tr}\,G_{\alpha\beta}$ (otherwise there would be a special direction in space, the axis of spin quantization).

2.1.1.1 Unperturbed Green's Functions

It is straightforward now to calculate the unperturbed Green's functions, starting from the definition. For the fermions,

$$
G^0(\mathbf{x}, t) = \frac{1}{i}\Big\langle T\psi(\mathbf{x}, t)\psi^\dagger(0, 0) \Big\rangle_0. \quad (2.13)
$$

Expanding the field operators in the basis of plane waves, $\psi(\mathbf{x}, t) = \frac{1}{\sqrt{V}}$ $\sum_{\mathbf{k}} a_{\mathbf{k}} e^{i\mathbf{k}\mathbf{x} - i(\epsilon_{\mathbf{k}} - \mu)t}$, and taking into account that at zero temperature in equilibrium the average $\langle a_{\mathbf{k}}^{\dagger} a_{\mathbf{k}'} \rangle_0 = \delta_{\mathbf{k},\mathbf{k}'} n_F(\epsilon_{\mathbf{k}}) = \delta_{\mathbf{k},\mathbf{k}'} \theta(\mu - \epsilon_{\mathbf{k}})$, we find that

$$G^0(\mathbf{x}, t) = \frac{1}{iV} \sum_{\mathbf{k}} [\theta(t)(1 - \theta(\mu - \epsilon_{\mathbf{k}})) - \theta(-t)\theta(\mu - \epsilon_{\mathbf{k}})] e^{i\mathbf{k}\mathbf{x} - i(\epsilon_{\mathbf{k}} - \mu)t}. \quad (2.14)$$

The Fourier transform of this expression yields, finally,

$$G^0(\mathbf{p}, \omega) = \frac{1}{\omega - (\epsilon_{\mathbf{p}} - \mu) + i0\,\mathrm{sgn}(\epsilon_{\mathbf{p}} - \mu)}$$

$$= \frac{1}{\omega - (\epsilon_{\mathbf{p}} - \mu) + i0\,\mathrm{sgn}\omega}. \quad (2.15)$$

The infinitesimal term in the denominator indicates in what half-plane of complex frequency the corresponding integrals will converge, exactly like what we had earlier for the retarded propagator. (For example, the integral $\int_0^\infty dt\, e^{i\omega t}$ due to the first term in (2.14) converges if $\Im\omega \to 0+$.) The difference is that here we have the causal Green's function, which contains both $\theta(t)$ and $\theta(-t)$.

You are welcome to calculate the expression for the unperturbed phonon Green's function, defined as

$$D(\mathbf{x}, t; \mathbf{x}', t') = -i\langle 0|T\varphi(\mathbf{x}, t)\varphi(\mathbf{x}', t')|0\rangle, \quad (2.16)$$

where the phonon field operator is

$$\varphi(\mathbf{x}, t) = \frac{1}{\sqrt{V}} \sum_{\mathbf{k}} \left(\frac{\omega_{\mathbf{k}}}{2}\right)^{1/2} \left\{ b_{\mathbf{k}} e^{i(\mathbf{k}\mathbf{x} - \omega_{\mathbf{k}}t)} + b_{\mathbf{k}}^{\dagger} e^{-i(\mathbf{k}\mathbf{x} - \omega_{\mathbf{k}}t)} \right\}, \quad (2.17)$$

and b, $b^{\dagger}$ are the usual Bose operators. (Since the phonon field is ultimately a quantized sound wave, it should be Hermitian to yield in the classical limit a classical observable, the medium displacement.) You will see that

$$D^0(\mathbf{k}, \omega) = \frac{\omega_{\mathbf{k}}^2}{\omega^2 - \omega_{\mathbf{k}}^2 + i0}. \quad (2.18)$$

The unperturbed Green's function *is* the Green's function in the mathematical sense. For the fermions, e.g., it satisfies a linear equation,

$$\left(i\hbar \frac{\partial}{\partial t_1} - \mathcal{E}(\mathbf{x}_1) \right) G^0_{\alpha\beta}(\mathbf{x}_1, t_1; \mathbf{x}_2, t_2) = \delta(\mathbf{x}_1 - \mathbf{x}_2)\delta(t_1 - t_2). \quad (2.19)$$

Symbolically this can be written as

$$(G^0)^{-1}(1)G^0(1, 2) = \mathcal{I}(1, 2),$$

where the operator $(G^0)^{-1}(1)$ in coordinate space is $(i\partial/\partial t_1 - \mathcal{E}(\nabla_{\mathbf{x}_1}))$, in momentum space $(\omega - \mathcal{E}(\mathbf{k}))$.

The derivation of the corresponding equation for $D^0(1, 2)$ is suggested as one of the problems to this chapter.

2.1.2 Analytic Properties of Green's Functions

When dealing with one-particle problems, we observed that some important properties of propagators could be obtained from general physical considerations, independently of the details of the system. This can be done in the many-body case as well. Specifically, we will derive the *Källen-Lehmann* representation for Green's functions in momentum space (as a function of $(\mathbf{p}, \omega)$), which determines the analytic properties of Green's function in complex ω plane and leads to physically significant consequences.

Our only assumption here will be that our system is in a stationary and homogeneous state, or is space and time invariant. This means that (1) the full Hamiltonian $\mathcal{H}$ is time independent and (2) the momentum operator $\mathcal{P}$ commutes with $\mathcal{H}$ (then the total momentum is by definition conserved); here

$$\mathcal{P} = \sum_\alpha \int d^3\mathbf{x}\psi_\alpha^\dagger(\mathbf{x})(-i\nabla)\psi_\alpha(\mathbf{x}). \tag{2.20}$$

It can be easily seen that for the simultaneous commutator,

$$[\psi_\alpha(\mathbf{x}, t), \mathcal{P}] = -i\nabla\psi_\alpha(\mathbf{x}, t) \tag{2.21}$$

for both Fermi and Bose field operators. (It is enough to substitute the definition of $\mathcal{P}$ and use the canonical (anti)commutation relations.) Equation (2.21) is reminiscent of the Heisenberg equations of motion for a field operator (1.84), and they together imply

$$\psi_\alpha(\mathbf{x}, t) = e^{-i(\mathcal{P}\mathbf{x}-\mathcal{H}t)}\psi_\alpha e^{i(\mathcal{P}\mathbf{x}-\mathcal{H}t)}; \tag{2.22}$$
$$\psi_\alpha \equiv \psi_\alpha(\mathbf{0}, 0).$$

In a more formal language, the Hamiltonian and the operator of total momentum are *generators of temporal and spatial shifts* respectively.

We can substitute the above expression into Green's function, (2.4), and then insert the unity operator constructed of the full set of common eigenstates of the two commuting operators $\mathcal{H}$, $\mathcal{P}$,

$$\mathcal{I} = \sum_s |s\rangle\langle s|,$$

wherever it seems reasonable. The following calculations are tedious but straight-forward.

$$\langle 0|T\psi(x, t)\psi^\dagger(x', t')|0\rangle = \theta(t - t')\sum_s \langle 0|\psi(x, t)|s\rangle\langle s|\psi^\dagger(x', t')|0\rangle$$

$$\mp \theta(t' - t)\sum_s \langle 0|\psi^\dagger(x', t')|s\rangle\langle s|\psi(x, t)|0\rangle$$

$$= \theta(t - t')\sum_s \langle 0|e^{i(\mathcal{H}t-\mathcal{P}x)}\psi e^{-i(\mathcal{H}t-\mathcal{P}x)}|s\rangle\langle s|e^{i(\mathcal{H}t'-\mathcal{P}x')}\psi^\dagger e^{-i(\mathcal{H}t'-\mathcal{P}x')}|0\rangle$$

$$\mp \theta(t' - t)\sum_s \langle 0|e^{i(\mathcal{H}t'-\mathcal{P}x')}\psi^\dagger e^{-i(\mathcal{H}t'-\mathcal{P}x')}|s\rangle\langle s|e^{i(\mathcal{H}t-\mathcal{P}x)}\psi e^{-i(\mathcal{H}t-\mathcal{P}x)}|0\rangle$$

$$= \theta(t - t')\sum_s e^{i(E_0-\mu N_0)t}\langle 0|\psi|s\rangle e^{-i((E_s-\mu N_s)t-P_s x)}$$

$$\times e^{i((E_s-\mu N_s)t'-Px')}\langle s|\psi^\dagger|0\rangle e^{-i(E_0-\mu N_0)t'}$$

$$\mp \theta(t' - t)\sum_s e^{i(E_0-\mu N_0)t'}\langle 0|\psi^\dagger|s\rangle e^{-i((E_s-\mu N_s)t'-P_s x')}$$

$$\times e^{i((E_s-\mu N_s)t - P_s x)}\langle s|\psi|0\rangle e^{-i(E_0-\mu N_0)t}.$$

The momentum of the state $|0\rangle$ is zero. The energy exponents here contain some subtlety. Since the field operators create or annihilate particles one at a time, in the first part of the expression, which contains $\langle 0|\psi|s\rangle\langle s|\psi\dagger|0\rangle \equiv |\langle 0|\psi|s\rangle|^2$ states $|s\rangle$ must contain one particle *more* than the state $|0\rangle$, say $N_s = N_0 + 1 \equiv N + 1$ (otherwise the annihilation operator has nothing to annihilate). On the other hand, in the second half of the expression, with $\langle 0|\psi^\dagger|s\rangle\langle s|\psi|0\rangle \equiv |\langle s|\psi|0\rangle|^2$ the states $|s\rangle$ contain $N - 1$ particles. Since, generally, the eigenvalues of the Hamiltonian depend on the number of particles both via the $-\mu N$ term and directly, we have in the exponents (showing explicitly the dependence of eigenenergies on the particle number)

$$\exp[i(E_0(N) - \mu N)t - i((E_s(N + 1) - \mu(N + 1))t - P_s x)]$$
$$\times \exp[i((E_s(N + 1) - \mu(N + 1))t' - Px') - i(E_0(N) - \mu N)t']$$
$$= \exp[i[(E_0(N) - E_s(N + 1) + \mu)(t - t') + P_s(x - x')]];$$
$$\exp[i(E_0(N) - \mu N)t' - i((E_s(N - 1) - \mu(N - 1))t' - P_s x')]$$
$$\times \exp[i((E_s(N - 1) - \mu(N - 1))t - Px) - i(E_0(N) - \mu N)t]$$
$$= \exp[i[(E_0(N) - E_s(N - 1) - \mu)(t' - t) + P_s(x' - x)]].$$

As expected, everything depends on the differences of coordinates and times.
Denote the *excitation energies* by

$$\epsilon_s^{(+)} = E_s(N+1) - E_0(N) > \mu; \tag{2.23}$$

$$\epsilon_s^{(-)} = E_0(N) - E_s(N-1) < \mu. \tag{2.24}$$

Evidently, the former gives the energy change when a particle is added to the state $|s\rangle$; the latter, when the particle is removed from the state $|s\rangle$. (You can check the above inequalities, if you recall that at zero temperature for the ground-state energy (a thermodynamic observable!) $(\partial E_0/\partial N) = (\partial \Phi/\partial N)$, ($\Phi$ is thermodynamical potential) and that by definition $(\partial \Phi/\partial N) = \mu$.) For the phonons, $\mu = 0$.

Now we can sweep all the details of the system under the carpet by introducing

$$A_s = \left[\frac{1}{2}\right] \langle 0|0\rangle^{-1} \left[\sum_\alpha\right] |\langle 0|\psi_\alpha|s\rangle|^2; \tag{2.25}$$

$$B_s = \left[\frac{1}{2}\right] \langle 0|0\rangle^{-1} \left[\sum_\alpha\right] |\langle s|\psi_\alpha|0\rangle|^2. \tag{2.26}$$

(The operations in the brackets are reserved for the spin degrees of freedom, α.) Those are some functions of index s only. Therefore, we can easily take the Fourier transform of Green's function, using the above results, and finally get the *Källén–Lehmann's representation*:

$$G(\mathbf{p}, \omega) = (2\pi)^3 \sum_s \left(\frac{A_s \delta(\mathbf{p} - \mathbf{P}_s)}{\omega - \epsilon_s^{(+)} + \mu + i0} \pm \frac{B_s \delta(\mathbf{p} + \mathbf{P}_s)}{\omega - \epsilon_s^{(-)} + \mu - i0}\right). \tag{2.27}$$

In this expression, delta functions of momenta arise from the exponential factors $\exp[i\mathbf{P}_s x]$; they indicate the values of momenta, corresponding to single-particle excitations (note that the second term in (2.27) clearly indicates the holes, with momenta $-\mathbf{P}_s$ and energies $\epsilon_s^{(-)}$). The frequency denominators contain the infinitesimal $\pm i0$, due to the presence of theta functions of time in the initial expression, exactly as when we calculated the unperturbed Green's functions. Of course, our present result is consistent with expressions (2.15, 2.18). Mathematically, the Källén–Lehmann representation tells us that Green's function of a finite system is a *meromorphic function* of the complex variable ω; all its singularities are simple poles. Each pole corresponds to a definite excitation energy, $\epsilon_s^{(\pm)}$, and definite momentum of the system, $\pm\mathbf{P}_s$. The poles are infinitesimally shifted into the upper half-plane of ω when $\omega > 0$, and into the lower one when $\omega < 0$. Thus the causal Green's function is not analytic in either half-plane.

In the thermodynamic limit ($N, V \to \infty$, $N/V = $ const) it is more convenient to use a different form of (2.27):

$$G(\mathbf{p}, \omega) = \int_{-\infty}^{\infty} \frac{d\omega'}{\pi} \left(\frac{\rho_A(\omega')}{\omega' - \omega - i0} + \frac{\rho_B(\omega')}{\omega' - \omega + i0}\right), \tag{2.28}$$

where

$$\rho_A(\mathbf{p}, \omega') = -\pi(2\pi)^3 \sum_s A_s \delta(\mathbf{p} - \mathbf{P}_s)\delta(\omega' - \epsilon_s^{(+)} + \mu); \qquad (2.29)$$

$$\rho_B(\mathbf{p}, \omega') = \mp\pi(2\pi)^3 \sum_s B_s \delta(\mathbf{p} + \mathbf{P}_s)\delta(\omega' - \epsilon_s^{(-)} + \mu). \qquad (2.30)$$

Indeed, in this limit we can no longer resolve the individual levels $\epsilon_s^{(\pm)}$. The densities $\rho_{A,B}(\mathbf{p}, \omega')$ become continuous functions, zero at negative (resp. positive) frequencies on the real axis; the latter becomes a *branch cut* in the complex ω plane.

The real and imaginary parts of Green's function (for *real* frequencies) can be easily obtained from (2.27), using the Weierstrass (or Sokhotsky—Weierstrass) formula

$$\frac{1}{x \pm i0} = \mathcal{P}\frac{1}{x} \mp i\pi\delta(x). \qquad (2.31)$$

Here $\mathcal{P}$ means the principal value. This is to be understood as a *generalized function*; that is, the above formula strictly speaking makes sense only under the integral, with an integrable function $F(x)$:

$$\int dx\, F(x)\frac{1}{x \pm i0} = \mathcal{P}\int dx\frac{F(x)}{x} \mp i\pi F(0).$$

The principal value integral $\mathcal{P}\int dx\, F(x)/x$ is defined as $\lim_{\epsilon \to 0}\left[\int^{-\epsilon} dx\, F(x)/x + \int_\epsilon dx\, F(x)/x\right]$. The other term arises from the integration over an infinitesimal semicircle around the pole at $x = 0$. This formula is easy to prove using the technique of residues in the complex analysis.

Using this recipe, we find

$$\Re G(\mathbf{p}, \omega) = (2\pi)^3 \sum_s \mathcal{P}\left(\frac{A_s \delta(\mathbf{p} - \mathbf{P}_s)}{\omega - \epsilon_s^{(+)} + \mu} \pm \frac{B_s \delta(\mathbf{p} + \mathbf{P}_s)}{\omega - \epsilon_s^{(-)} + \mu}\right); \qquad (2.32)$$

$$\Im G(\mathbf{p}, \omega) = (2\pi)^3\pi \begin{cases} -\sum_s A_s \delta(\mathbf{p} - \mathbf{P}_s)\delta(\omega - \epsilon_s^{(+)} + \mu), & \omega > 0; \\ \pm\sum_s B_s \delta(\mathbf{p} + \mathbf{P}_s)\delta(\omega - \epsilon_s^{(-)} + \mu), & \omega < 0. \end{cases} \qquad (2.33)$$

We thus obtain an important relation:

$$\text{sgn}\Im G(\mathbf{p}, \omega) = -\text{sgn}\,\omega \quad \text{for } Fermi \text{ systems};$$
$$\text{sgn}\Im G(\mathbf{p}, \omega) = -1 \quad \text{for } Bose \text{ systems}. \qquad (2.34)$$

The difference reflects the fact that there is no Fermi surface in Bose systems (and thus "particles" and "holes" are the same).

The asymptotic behavior of $G(\omega)$ as $\omega \to \infty$ is very simple:

$$G(\omega) \sim 1/\omega. \tag{2.35}$$

To prove it, note that in this limit we can neglect all other terms in the denominators of (2.27), so that

$$G(\mathbf{p}, \ \omega) \sim \frac{1}{\omega}(2\pi)^3 \sum_s (A_s \delta(\mathbf{p} - \mathbf{P}_s) \pm B_s \delta(\mathbf{p} + \mathbf{P}_s)).$$

The sum can be evaluated by performing an inverse Fourier transform and making use of the canonical (anti)commutation relation between field operators in coordinate space:

$$(2\pi)^3 \sum_s (A_s \delta(\mathbf{p} - \mathbf{P}_s) \pm B_s \delta(\mathbf{p} + \mathbf{P}_s))$$

$$= \sum_s \int d^3(\mathbf{x} - \mathbf{x}') \left(A_s e^{i(\mathbf{p} - \mathbf{P}_s)(\mathbf{x} - \mathbf{x}')} \pm B_s e^{i(\mathbf{p} + \mathbf{P}_s)(\mathbf{x} - \mathbf{x}')} \right)$$

$$= \sum_s \int d^3(\mathbf{x} - \mathbf{x}') \left[\frac{1}{2} \sum_\alpha \right] \left\{ |\langle 0|\psi_\alpha(0)|s\rangle|^2 e^{i(\mathbf{p} - \mathbf{P}_s)(\mathbf{x} - \mathbf{x}')} \right.$$

$$\left. \pm |\langle s|\psi_\alpha(0)|0\rangle|^2 e^{i(\mathbf{p} + \mathbf{P}_s)(\mathbf{x} - \mathbf{x}')} \right\}$$

$$= \int d^3(\mathbf{x} - \mathbf{x}') \left[\frac{1}{2} \sum_\alpha \right] e^{i\mathbf{p}(\mathbf{x} - \mathbf{x}')}$$

$$\times \langle 0|\psi_\alpha(\mathbf{r}, \ t)\psi_\alpha^\dagger(\mathbf{r}', \ t) \pm \psi_\alpha^\dagger(\mathbf{r}', \ t)\psi_\alpha(\mathbf{r}, \ t)|0\rangle$$

$$= 1.$$

Poles of Green's Function and Quasiparticle Excitations

It is easy to see from the Källén–Lehmann representation that in the thermodynamic limit only those poles of $G(\omega)$ survive as distinct poles (and do not merge into the branch cut along the real axis) that correspond to the situation when all energy and momentum of the excited system can be ascribed to one object, a quasiparticle, with a definite dispersion law (i.e., energy-momentum correspondence). Otherwise, for any given energy there is a whole set of corresponding momenta, and the pole will be eliminated by the integration over them. Thus the quasiparticle dispersion law $\omega(\mathbf{p})$ is defined by the equation

$$\frac{1}{G(\mathbf{p}, \omega - \mu)} = 0. \tag{2.36}$$

2.1.3 *Retarded and Advanced Green's Functions*

We will define two more Green's functions: *retarded* and *advanced* ones, G^R and G^A.

$$G^R_{\alpha\beta}(\mathbf{x}_1, t_1; \mathbf{x}_2, t_2)$$
$$= -i \left\langle \psi_\alpha(\mathbf{x}_1, t_1)\psi^\dagger_\beta(\mathbf{x}_2, t_2) \pm \psi^\dagger_\beta(\mathbf{x}_2, t_2)\psi_\alpha(\mathbf{x}_1, t_1) \right\rangle \theta(t_1 - t_2); \quad (2.37)$$

$$G^A_{\alpha\beta}(\mathbf{x}_1, t_1; \mathbf{x}_2, t_2)$$
$$= +i \left\langle \psi_\alpha(\mathbf{x}_1, t_1)\psi^\dagger_\beta(\mathbf{x}_2, t_2) \pm \psi^\dagger_\beta(\mathbf{x}_2, t_2)\psi_\alpha(\mathbf{x}_1, t_1) \right\rangle \theta(t_2 - t_1). \quad (2.38)$$

The definition is chosen in such a way as to guarantee that (1) the retarded (advanced) Green's function is zero for all negative (positive) time differences $t - t'$, and (2) at $t = t'$ both have a $(-i\delta(x - x'))$ discontinuity, exactly as does the causal Green's function. The latter statement is easy to check by taking its time derivative and using the canonical relation

$$\lim_{t \to t'} \frac{\partial}{\partial t} G^{R(A)}(t - t') = \mp i \langle 0| \left[\psi_\alpha(\mathbf{x}, t), \psi^\dagger_\beta(\mathbf{x}', t') \right]_\pm |0\rangle \cdot (\pm\delta(t - t'))$$
$$= -i\delta(\mathbf{x} - \mathbf{x}').$$

Again, in the uniform and stationary case, $G^{R(A)}_{\alpha\beta}(\mathbf{x}_1, t_1; \mathbf{x}_2, t_2) = G^{R(A)}(\mathbf{x}_1 - \mathbf{x}_2, t_1 - t_2)\delta_{\alpha\beta}$. Unperturbed retarded and advanced Green's functions, for example, can be easily found by direct calculation:

$$G^{0R,A}(\mathbf{p}, \omega) = \frac{1}{\omega - (\epsilon_\mathbf{p} - \mu) \pm i0};$$

$$D^{0R,A}(\mathbf{k}, \omega) = \frac{\omega_\mathbf{k}}{2} \left(\frac{1}{\omega - \omega_\mathbf{k} \pm i0} - \frac{1}{\omega + \omega_\mathbf{k} \pm i0} \right) = \frac{\omega_\mathbf{k}^2}{\omega^2 - \omega_\mathbf{k}^2 + i0\,\mathrm{sgn}\omega}.$$
$$(2.39)$$

The Källén–Lehmann representation for $G^{R,A}$ can be obtained in the same way as for the causal Green's function. The result is as follows:

$$G^R(\mathbf{p}, \omega) = (2\pi)^3 \sum_s \left(\frac{A_s\delta(\mathbf{p} - \mathbf{P}_s)}{\omega - \epsilon_s^{(+)} + \mu + i0} \pm \frac{B_s\delta(\mathbf{p} + \mathbf{P}_s)}{\omega - \epsilon_s^{(-)} + \mu + i0} \right); \quad (2.40)$$

$$G^A(\mathbf{p}, \omega) = (2\pi)^3 \sum_s \left(\frac{A_s\delta(\mathbf{p} - \mathbf{P}_s)}{\omega - \epsilon_s^{(+)} + \mu - i0} \pm \frac{B_s\delta(\mathbf{p} + \mathbf{P}_s)}{\omega - \epsilon_s^{(-)} + \mu - i0} \right). \quad (2.41)$$

Taking real and imaginary parts of these expressions, we see that on the real axis

$$\begin{cases} \Re G^A(\mathbf{p}, \omega) = \Re G^A(\mathbf{p}, \omega) = \Re G(\mathbf{p}, \omega); \\ \Im G^R(\mathbf{p}, \omega) = \Im G(\mathbf{p}, \omega); & \omega > 0; \\ \Im G^A(\mathbf{p}, \omega) = \Im G(\mathbf{p}, \omega); & \omega < 0; \\ G^R_{\alpha\beta}(\mathbf{p}, \omega) = \left[G^A_{\beta\alpha}(\mathbf{p}, \omega) \right]^*. \end{cases} \tag{2.42}$$

On the other hand, the retarded (advanced) Green's function is clearly analytic in the upper (lower) ω-half-plane. This means that they are *analytic continuations* of the causal Green's function from the rays $\omega > 0 (\omega < 0)$ respectively. Their asymptotic behavior is, of course, the same as that of the causal Green's function:

$$G^{R,A}(\omega) \sim 1/\omega, \ |\omega| \to \infty.$$

In the thermodynamic limit, as in (2.28), we can write

$$G^{R,A}(\mathbf{p}, \ \omega) = \int_{-\infty}^{\infty} \frac{d\omega'}{\pi} \frac{\rho^{R,A}(\mathbf{p}, \omega')}{\omega' - \omega \pm i0}, \tag{2.43}$$

with *spectral density*

$$\rho^{R,A}(\mathbf{p}, \ \omega') = -\pi(2\pi)^3 \sum_s (A_s \delta(\mathbf{p} - \mathbf{P}_s)\delta(\omega' - \epsilon_s^{(+)} + \mu) \tag{2.44}$$

$$\pm B_s \delta(\mathbf{p} + \mathbf{P}_s)\delta(\omega' - \epsilon_s^{(-)} + \mu)).$$

Evidently,

$$\rho^R(\mathbf{p}, \ \omega') = -\Im G^R(\mathbf{p}, \ \omega'). \tag{2.45}$$

From formula (2.44) it is clear that $\rho^R(\mathbf{p}, \ \omega')$ is proportional to the probability density of an elementary excitation with momentum $\mathbf{p}$ having energy ω. In the non-interacting case, e.g., $-\Im G^{0,R} = \pi\delta(\omega - (\epsilon_\mathbf{p} - \mu))$, because in the absence of interactions, quasiparticles would indeed coincide with "basic" particles, with dispersion law $\epsilon_\mathbf{p}$.

2.1.3.1 Quasiparticle Excitations and Retarded and Advanced Green's Functions

We can now visualize the concept of a quasiparticle excitation and its relation to the existence of isolated poles of $G(\mathbf{p}, \ \omega)$. Suppose that there is such a pole at $\omega = \Omega - i\Gamma$, $\Gamma > 0$. (This corresponds to a particle added in the state $\mathbf{p}$.) To see the evolution of the excitation created by the operator $a_\mathbf{p}^\dagger$, calculate Green's function in the $(\mathbf{p}, \ t)$-representation:

Fig. 2.5 Contour integrations in the complex frequency plane. The poles of $G^R(\omega)$ are marked by crosses, those of $G^A(\omega)$ by circles

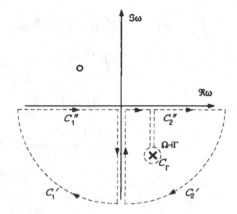

$$G(\mathbf{p},\ t) = \int\limits_{-\infty}^{\infty} \frac{d\omega}{2\pi} e^{-i\omega t} G(\mathbf{p},\ \omega).$$

For negative t the integration contour can be closed in the upper half-plane, and since there are no singularities there, the integral is zero. For positive t the contour closes in the lower half-plane and will contain the pole. We cannot, though, simply calculate the residue, since the causal Green's function is not analytic in the lower half-plane. We must then replace it by its analytic continuations, $G^{R,A}$. In order to do this, we close the integration contour as shown in Fig. 2.5. For $\Re\omega < 0$ we can replace $G(\mathbf{p},\ \omega)$ by $G^A(\mathbf{p},\ \omega)$, and for $\Re\omega > 0$ by $G^R(\mathbf{p},\ \omega)$. Now, the Cauchy theorem of complex analysis tells us that the integral we *are* interested in can be written as follows:

$$G(\mathbf{p},\ t) = -\int\limits_{C_1'+C_1''} \frac{d\omega}{2\pi} e^{-i\omega t} G^A(\mathbf{p},\ \omega) - \int\limits_{C_2'+C_2''+C_\Gamma} \frac{d\omega}{2\pi} e^{-i\omega t} G^R(\mathbf{p},\omega).$$

Watson's lemma, together with the $1/\omega$ asymptotics of Green's functions, ensures that the integrals over infinitely remote quarter circles C_1' and C_2' are zero. Therefore, we are left with two terms: the contribution from the pole, and the integral along the negative imaginary axis:

$$G(\mathbf{p},\ t) = -iZe^{-i\Omega t}e^{-\Gamma t} + \int\limits_{-i\infty}^{0} \frac{d\omega}{2\pi} e^{-i\omega t} \left[G^A(\mathbf{p},\ \omega) - G^R(\mathbf{p},\ \omega)\right], \quad (2.46)$$

where Z is the residue of $G^R(\omega)$ in the pole. The first term describes a free quasi-particle with finite lifetime $\sim 1/\Gamma$. The contribution from the integral is small, if only $\Omega t \gg 1$, $\Gamma t \ll 1$. (This means that the decay rate must be small enough, $\Gamma \ll \Omega$.) Indeed, invoking the Källén-Lehmann representation, we see that

$$\int_{-i\infty}^{0} \frac{d\omega}{2\pi} e^{-i\omega t} \left[G^A(\mathbf{p},\, \omega) - G^R(\mathbf{p},\, \omega) \right]$$

$$\approx \int_{-i\infty}^{0} \frac{d\omega}{2\pi} e^{-i\omega t} \left[\frac{A}{\omega - \epsilon_p + \mu - i\Gamma} - \frac{A}{\omega - \epsilon_\rho + \mu + i\Gamma} \right]$$

$$= -2i\Gamma A \int_{-i\infty}^{0} \frac{d\omega}{2\pi} \frac{e^{-i\omega t}}{(\omega - \epsilon_p + \mu)^2 + \Gamma^2}$$

$$\approx \frac{-\Gamma A}{\pi t} \frac{e^{-i\mu t}}{(\mu - \epsilon_\rho)^2} \ll Z e^{-i\Omega t} e^{-\Gamma t}$$

if $\Gamma \ll (\epsilon_\rho - \mu) = \Omega$.

2.1.3.2 Kramers-Kronig Relations

From the Källén–Lehmann representation for the retarded and advanced Green's functions (2.40, 2.41) follows a beautiful (and important) relation between the real and imaginary parts of those functions of *real* frequencies: the *Kramers–Kronig relation*

$$\Re G^{R,A}(\mathbf{p},\, \omega) = \pm \mathbf{P} \int_{-\infty}^{\infty} \frac{d\omega'}{\pi} \frac{\Im G^{R,A}(\mathbf{p},\, \omega')}{\omega' - \omega}. \tag{2.47}$$

(This can be established directly by taking the imaginary and real parts of (2.40, 2.41). It can be shown that the reason why this relation holds is the *causality*, that is, the property of advanced and retarded Green's functions of *time* to be zero for $t > (<)t'$. The proof (which is quite straightforward) is like our calculation of the Fourier transform of the propagator in Chap. 1, where we established for the first time that the poles of $K(\omega)$ must be infinitesimally displaced from the real axis in order to provide for the $\theta(t)$-like behavior of $K(t)$. Then, knowing that $G^{(R,A)}(\omega)$ is an analytic function in the corresponding half-plane, we can calculate the integral $\int_{-\infty}^{\infty} \frac{d\omega'}{\pi} \frac{\Im G^{R,A}(\mathbf{p},\omega')}{\omega'-\omega}$ along the real axis, using the Cauchy theorem, and come to the above relations. In mathematics this relation is known as the Plemelj theorem [6].

2.1.4 Green's Function and Observables

The way of expressing observables (average values of quantum-mechanical operators) in terms of Green's function (once the latter is known) directly follows from its definition. (Since it contains only an average of two field operators, it is clear that only one-particle operators can be treated this way.)

For example, the particle density in the system (meaning real or "basic" particles) is by definition

$$n(\mathbf{r}) = \sum_\alpha \langle \psi_\alpha^\dagger(\mathbf{r}) \psi_\alpha(\mathbf{r}) \rangle \equiv \mp i \sum_\alpha G_{\alpha\alpha}(\mathbf{r}, t - 0; \mathbf{r}, t). \tag{2.48}$$

So, for a uniform system of spinful fermions we have

$$n = \frac{N}{V} = -2iG(\mathbf{r} = 0, t = -0) = 2\Im G(\mathbf{r} = 0, t = -0). \tag{2.49}$$

The relation between density and Green's function allows us, in turn, to express the thermodynamic properties of the system at $T = 0$ through its Green's function. Indeed, the grand potential of the system satisfies (see, e.g., [4])

$$d\Omega = -SdT - Nd\mu = -Nd\mu$$

at $T = 0$ (since the entropy $S(0) = 0$). This equation can thus be integrated (remembering that $\Omega(\mu = 0) = 0$):

$$\Omega = - \int_0^\mu d\mu N(\mu),$$

where we can substitute the expression for $N(\mu)$ from (2.48) or (2.49) (where Green's function is explicitly μ-dependent; see Problem 2.1).

A slightly more sophisticated example is presented by current, which in agreement with Sect. 1.4 is expressed through the field operators as

$$\mathbf{j}(\mathbf{r}) = \frac{ie}{2m} \sum_\alpha \langle (\nabla \psi_\alpha^\dagger(\mathbf{r})) \psi_\alpha(\mathbf{r}) - \psi_\alpha^\dagger(\mathbf{r})(\nabla \psi_\alpha(\mathbf{r})) \rangle.$$

We can express the current through Green's function, using a "hair splitting" trick that allows us to separate the differentiation over two coinciding coordinates:

$$\mathbf{j}(\mathbf{r}) = \frac{ie}{2m} \sum_\alpha \lim_{\mathbf{r}' \to \mathbf{r}} (\nabla_{\mathbf{r}'} - \nabla_{\mathbf{r}}) \langle \psi_\alpha^\dagger(\mathbf{r}') \psi_\alpha(\mathbf{r}) \rangle$$

$$= \frac{ie}{2m} \sum_\alpha \lim_{t \to -0} \lim_{\mathbf{r}' \to \mathbf{r}} (\nabla_{\mathbf{r}'} - \nabla_{\mathbf{r}})(\mp i G_{\alpha\alpha}(\mathbf{r}', t; \mathbf{r}, 0)). \tag{2.50}$$

2.2 Perturbation Theory: Feynman Diagrams

The beautiful formulae of complex analysis and the physical insight we have achieved so far unfortunately do not provide more specific information about Green's functions of a realistic system—the one with interactions. On this level of generalization all Fermi systems look the same, and all Bose systems look the same (which is, of course,

Fig. 2.6 Different ways of drawing Feynman diagrams

true, but not sufficient). On the other hand, we cannot look into the differences due to interactions, because, e.g., we have no way of determining the matrix elements of field operators in the Källén-Lehmann representation.

As always, we have to apply perturbation theory. The great achievement by Feynman was to build *the* perturbation theory formalism, in which the whole perturbation expansion, including its most cumbersome expressions, is reduced to a set of sometimes spectacular and always physically understandable graphs—*Feynman diagrams*.

As in medieval paintings, there is a strict set of rules for both drawing and reading those images (determined by the Hamiltonian of the system). This makes diagram techniques a highly symbolic form of art. (Of course, there are differences between various schools and books, as in Fig. 2.6.)

When calculating the Green's function of a system with interactions, we meet the usual obstacle of not knowing the wave function (state) of the system over which the average is to be taken. We don't know the ground state, $|0\rangle$. We don't know excited states. Moreover, *any* approximation we are going to make will be virtually *orthogonal* to the proper many-particle state. Luckily, the approximate *matrix elements* (like Grren's function) can be quite accurate. The seeming paradox is just a reflection of the fact that while the wave function involves all N particle states, Green's function deals with only two one-particle states (initial and final). As Thouless has noted [7], if the one-particle state is approximated with a small mistake α, the projection of the corresponding many-particle state on the exact state will be of the order of $(1 - \alpha)^N \sim e^{-N\alpha} \to 0$ as $N \to \infty$ for my finite α. On the other hand, the average of a one-particle operator (like Green's function) will contain only a small mistake α !

Two conclusions can be drawn. (1) Green's functions provide a physically sensible method of approaching the many-body problem. (Something we could guess after a short glance cast at library shelves filled with numerous folios on the subject, to which we dare add a book of our own.) (2) The results are to be expressed in terms of averages over the unperturbed state of the system, rather than corrections to many-body wave functions. (In usual quantum mechanics both would be equivalent—*because* it is a one-body theory.)

2.2.1 Derivation of Feynman Rules. Wick's and Cancellation Theorems

In order to fulfill our program, let us turn to the interaction representation. We have seen before that this is a natural way to deal with perturbation theory. To avoid unnecessary subscripts, here we denote the field operators in the interaction representation by uppercase Greek letters. The connection between them and Heisenberg operators is given by

$$\psi(\mathbf{x},\, t) = \mathcal{U}^\dagger(t)\psi_S(\mathbf{x})\mathcal{U}(t) = \mathcal{U}^\dagger(t)e^{-i\mathcal{H}_0 t}\Psi(\mathbf{x},\, t)e^{i\mathcal{H}_0 t}\mathcal{U}(t).$$

Inserting this into the definition of Green's function, we obtain

$$G_{\alpha\alpha'}(\mathbf{x},\, t;\mathbf{x}',\, t') = -i\left[\langle 0|\mathcal{U}^\dagger(t)e^{-i\mathcal{H}_0 t}e^{i\mathcal{H}_0 t}\mathcal{U}(t)|0\rangle\right]^{-1}$$
$$\times\left\{\langle 0|\mathcal{U}^\dagger(t)e^{-i\mathcal{H}_0 t}\Psi_\alpha(\mathbf{x},\, t)S_{(I)}(t,\, t')\Psi_{\alpha'}^\dagger(\mathbf{x}',\, t')e^{i\mathcal{H}_0 t'}\mathcal{U}(t')|0\rangle\theta(t-t')\right.$$
$$\left.\pm\langle 0|\mathcal{U}^\dagger(t')e^{-i\mathcal{H}_0 t'}\Psi_{\alpha'}^\dagger(\mathbf{x}',\, t')S_{(I)}(t',\, t)\Psi_\alpha(\mathbf{x},\, t)e^{i\mathcal{H}_0 t}\mathcal{U}(t)|0\rangle\theta(t'-t)\right\}.$$
$$(2.51)$$

Later on we suppress the (I) subscript in the S-matrix as well; we use it only in the interaction representation anyway; that is,

$$S(t,\, t') = T e^{-i\int_{t'}^{t}dt\mathcal{W}(t)}\qquad,\, t > t'.$$

Here $|0\rangle$ is a Heisenberg ground-state vector. Then $e^{i\mathcal{H}_0 t'}\mathcal{U}(t')|0\rangle = e^{i\mathcal{H}_0 t'}|0(t')\rangle_S = |0(t')\rangle_I$, i.e., the ground-state vector in interaction representation, and

$$|0(t')\rangle_I = S(t',\, t'')|0(t'')\rangle_I(t' > t'') = S(t',\, -\infty)|0(-\infty)\rangle_I, \qquad (2.52)$$

because we can take the moment t'' to minus infinity.

We can repeat this argumentation for $\langle 0|\mathcal{U}^\dagger(t)e^{-i\mathcal{H}_0 t}$ to get eventually

$$e^{i\mathcal{H}_0 t'}\mathcal{U}(t')|0\rangle = \mathcal{S}(t', -\infty)|0(-\infty)\rangle_I; \tag{2.53}$$

$$\langle 0|\mathcal{U}^\dagger(t)e^{-i\mathcal{H}_0 t} = \langle 0(\infty)|_I \mathcal{S}(\infty, t). \tag{2.54}$$

Now we introduce a very important *adiabatic hypothesis*. First, let us assume that the perturbation was absent very long ago and was turned on in an infinitely slow way, say $\mathcal{W}(t \leq t_1) = \mathcal{W}\exp(\alpha(t - t_1))$, $\alpha \to 0+$. It will be turned off in some very distant future, say $\mathcal{W}(t \geq t_2) = \mathcal{W}\exp(-\alpha(t - t_2))$, $\alpha \to 0+$. Here $[t_1, t_2]$ is the time interval within which we investigate our system (and we don't care what happens later or previously: *aprés mois le déluge*).

Of course, physically it is rather easy to turn on and off the external potential, while we don't have such a free hand when the perturbation is due to particle—particle interaction. But nothing a priori prohibits such a property of the perturbation term in the Hamiltonian, and finally we take $\alpha = 0$ anyway. (A mathematically inclined reader may recognize that we are actually going to use so-called Abel regularization of conditionally convergent integrals, which appear a little later.)

Now, since at minus infinity there is no perturbation, we can write instead of $|0(-\infty)\rangle_I$ the *unperturbed ground state* vector $|\Phi_0\rangle$ (which is time independent, because $i\partial|\Phi_0(t \to -\infty)\rangle/\partial t = \mathcal{W}\exp(\alpha(t - t_1))|\Phi_0(t \to -\infty)\rangle \to 0$). It is convenient to choose a normalized state, $\langle\Phi_0|\Phi_0\rangle = 0$.

Now, it seems natural to think that since we had an unperturbed ground state at minus infinity, when there was no interaction, we should have the same state at plus infinity, when there will be no interaction. This is not true, though: it is known from usual quantum mechanics that the adiabatically slow perturbation can actually switch the system to a different state *with the same energy*. Our good luck is that the ground state of a quantum-mechanical system is always non-degenerate, and it is the ground state averages that we deal with! Therefore, the only difference between the states at minus and plus infinity may be a phase factor: $|0(+\infty)\rangle_I = (\exp(iL))|\Phi_0\rangle$, and this factor anyway cancels from the numerator and denominator of (2.51)).

Thus we have derived the key formula

$$iG_{\alpha\alpha'}(x, t; x', t') = \frac{\langle\Phi_0|T\mathcal{S}(\infty, -\infty)\Psi_\alpha(x, t)\Psi^\dagger_{\alpha'}(x', t')|\Phi_0\rangle}{\langle\Phi_0|\mathcal{S}(\infty, -\infty)|\Phi_0\rangle} \tag{2.55}$$

(we rearranged the terms under the sign of time ordering to gather all the parts of the S-operator into $\mathcal{S}(\infty, -\infty) = T\exp\{-i\int_{-\infty}^{\infty}dt\mathcal{W}(t)\}$. This formula is the basis for the perturbation theory: we have only to expand the exponent and obtain the series like the one we obtained for the one-particle propagator, containing terms

$$\langle\Phi_0|T\mathcal{W}(t_1)\mathcal{W}(t_2)\cdots\mathcal{W}(t_m)\Psi_\alpha(x, t)\Psi^\dagger_{\alpha'}(x', t')|\Phi_0\rangle.$$

The difference is that (1) we need the matrix elements not between the coordinate (momentum) eigenstates, but between unperturbed ground state vectors of a many-body system, and (2) now we have the denominator $\langle\Phi_0|\mathcal{S}(\infty, -\infty)|\Phi_0\rangle$.

2.2.1.1 Wick's Theorem

This is *the* theorem of quantum field theory, because it makes all the formalism tick. After the exponent in the S-operator is expanded, we must calculate the matrix elements of the sort $\langle \Phi_0 | T \phi_1 \phi_2 \cdots \phi_m | \Phi_0 \rangle$. Here ϕ_1, ϕ_2, $\ldots$, ϕ_m are (Fermi or Bose) field operators in interaction representation, and it is Wick's theorem that allows us to do this.

For the sake of clarity, from now on we denote the set of variables $(\mathbf{x}, t, \alpha)$ by a single number or capital letter. For example:

$$\Psi_\alpha(x, t) \rightsquigarrow \Psi_X;$$
$$\Psi_{\gamma_1}(\mathbf{x}_1, t_1) \rightsquigarrow \Psi_1;$$
$$\sum_\alpha \int d^3\mathbf{x} \int dt \rightsquigarrow \int dX;$$
$$\sum_{\gamma_1} \int d^3\mathbf{x}_1 \int dt_1 \rightsquigarrow \int d1.$$

Wick's theorem states that

The time-ordered product of field operators in interaction representation equals to the sum of their normal products with all possible contractions:

$$T\phi_1\phi_2 \cdots \phi_m\phi_{m+1}\phi_{m+2} \ldots \phi_n =: \phi_1\phi_2 \cdots \phi_m\phi_{m+1}\phi_{m+2} \cdots \phi_n :$$
$$+ : \overbrace{\phi_1\phi_2} \cdots \phi_m\phi_{m+1}\phi_{m+2} \cdots \phi_n :$$
$$+ : \overbrace{\phi_1\phi_2 \cdots \phi_m\phi_{m+1}}\phi_{m+2} \cdots \phi_n :$$
$$+ \cdots + : \phi_1 \underbrace{\phi_2 \cdots \phi_m} \phi_{m+1} \phi_{m+2} \cdots \phi_n : .$$

$$(2.56)$$

Now some definitions.

The *normal ordering* of field operators, $: \phi_1\phi_2 \ldots \phi_m :$, means that all "destruction" operators stay to the right of the "construction" ones. We label "destruction" those operators that give zero when acting on the unperturbed ground state (vacuum state); and the "construction" operators are their conjugates. In the Fermi case, e.g., the vacuum is the filled Fermi sphere, and therefore "destruction" operators are *annihilation* operators $a_\mathbf{p}$ with $p > p_F$, and *creation* operators $a_\mathbf{p}^\dagger$ with $p < p_F$. We can, if we wish, explicitly split the Fermionic field operator in two corresponding parts,

$$\psi_X = \psi_X^{(-)} + \psi_X^{(+)}$$
$$= \frac{1}{\sqrt{V}} \sum_{p > p_F} e^{i(\mathbf{px} - \epsilon_p t)} a_\mathbf{p} + \frac{1}{\sqrt{V}} \sum_{\rho < \rho_F} e^{i(\mathbf{px} - \epsilon_p t)} a_\mathbf{p}. \qquad (2.57)$$

The bosonic (e.g., phonon) field is already presented in this form,

$$\varphi_X = \varphi_X^{(-)} + \varphi_X^{(+)}$$
$$= \frac{1}{\sqrt{V}} \sum_{\mathbf{k}} \left(\frac{\omega_{\mathbf{k}}}{2}\right)^{1/2} b_{\mathbf{k}} e^{i(\mathbf{kx} - \omega_{\mathbf{k}}t)} + \frac{1}{\sqrt{V}} \sum_{\mathbf{k}} \left(\frac{\omega_{\mathbf{k}}}{2}\right)^{1/2} b_{\mathbf{k}}^{\dagger} e^{-i(\mathbf{kx} - \omega_{\mathbf{k}}t)}. \quad (2.58)$$

Since both time ordering and normal ordering are distributive, we can deal with the $(+)$ and $(-)$ parts separately, and therefore all the intermediate manipulations can be on the field operators ψ, φ themselves.

By definition, the normal product of any set of the field operators A, B, C, $\ldots$ has zero ground-state average,

$$\langle \Phi_0 | : ABC \cdots : | \Phi_0 \rangle = 0. \quad (2.59)$$

The *contraction*, or *pairing*, of two operators, $\overbrace{\phi_m \phi_n}$, is the difference between their time- and normally ordered products

$$\overbrace{\phi_m \phi_n} = T \phi_m \phi_n - : \phi_m \phi_n : . \quad (2.60)$$

If both operators here are of the same sort (both creation or both annihilation), the contraction is identically zero. Indeed, then the normal ordering does not affect their product, and

$$\overbrace{\phi_1 \phi_2} = \theta(t_1 - t_2)\phi_1 \phi_2 \mp \theta(t_2 - t_1)\phi_2 \phi_1 - \phi_1 \phi_2$$
$$= (\theta(t_1 - t_2) + \theta(t_2 - t_1))\phi_1 \phi_2 - \phi_1 \phi_2$$
$$= 0.$$

On the other hand, a contraction of conjugate field operators is a *number*: taking into account that the operators are in interaction representation and their time dependence is trivial, we see that, for example,

$$\overbrace{\phi_1^{\dagger} \phi_2} = \sum_{k} \sum_{q} \left(e^{iE_k t_1} \phi_k^{\dagger}\right)\left(e^{-iE_q t_2} \phi_q\right)$$
$$= \sum_{k} \sum_{q} e^{iE_k t_1} e^{-iE_q t_2} \left(\theta(t_1 - t_2)\phi_k^{\dagger}\phi_q \mp \theta(t_2 - t_1)\phi_q \phi_k^{\dagger} - \phi_k^{\dagger}\phi_q\right)$$
$$= \sum_{k} \sum_{q} e^{iE_k t_1} e^{-iE_q t_2} \left[(\theta(t_1 - t_2) + \theta(t_2 - t_1))\phi_k^{\dagger}\phi_q - \phi_k^{\dagger}\phi_q + \theta(t_2 - t_1)\delta_{kq}\right],$$

and all the operator terms cancel. This is an important fact, that the contraction of Fermi/Bose field operators is a usual number, because then we can write

$$\overbrace{\phi_1\phi_2} = \langle\Phi_0|\overbrace{\phi_1\phi_2}|\Phi_0\rangle$$
$$= \langle\Phi_0|T\phi_1\phi_2|\Phi_0\rangle - \langle\Phi_0|:\phi_1\phi_2:|\Phi_0\rangle$$
$$= \langle\Phi_0|T\phi_1\phi_2|\Phi_0\rangle = iG^0(12). \tag{2.61}$$

Contraction of Fermi/Bose operators is actually an *unperturbed* Green's function, which we know how to find! Look how lucky we are: *if* the commutation relations for our field operators contained an operator instead of a delta function (as they do for spin operators), we would have no use for Wick's theorem when calculating the averages of many-operator products. This is why there is no really handy diagrammatic approach to the corresponding problems.

But with bosons and fermions we can use the theorem to deal with the average $\langle\Phi_0|\phi(1)\cdots\phi(N)|\Phi_0\rangle$. Note that we can extract all the contractions from under the symbol of normal ordering. (We have only to commute the paired operators with their neighbors in order to bring them together, and then simply calculate the contraction (a number) and take it outside. This operation will give us at most a factor of $(-1)^P$, the parity of the permutation of Fermi operators that we did on our way.) Therefore only the contribution due to the sum of all possible *fully contracted* terms survives (the other terms contain some normally ordered operators and thus have zero vacuum average): evidently, only terms with an even number of operators can be fully contracted. This sum is actually a sum of products of unperturbed Green's functions, corresponding to all possible ways of picking pairs of conjugate field operators from the general crowd, taken with corresponding parity factors. Each of these terms cm be presented by a distinctive Feynman diagram, of which the rules for drawing, reading, and calculating we are going to establish.

But before doing so, let us check that Wick's theorem is plausible. (The proof involves some tiresome algebra and can be found, e.g., in [2].) Following [4], we will make instead a simpler argument valid not for the operators themselves, but for their matrix elements (a so-called *weak* statement), and only in the thermodynamic limit. On the other hand, it is valid for averages over an arbitrary state of the system, not only its ground state.

The argument goes as follows. Write an Another example: average in Fourier components $\langle X|\phi_1\phi_2\cdots|X\rangle = \Omega^{-1/2}\sum_{k_1}\Omega^{-1/2}\sum_{k_2}\cdots e^{ik_1x_1-iE_1t_1}e^{ik_2x_2-iE_2t_2}\cdots$ $\langle X|c_{k_2}c_{k_2}\cdots|X\rangle$. Ω is the volume of the system. In this expression there must be an even number of operators: $N/2$ creation operators and $N/2$ annihilation operators, with the same values of k (otherwise the matrix element is zero). There can be no more than $N/2$ *different* values of k. If indeed they are all different, we have, e.g.,

$$\Omega^{-N/2}\sum_{k_1}\sum_{k_2}\cdots e^{ik_1(x_1-x_1')-iE_1(t_1-t_1')}\cdots\langle X|c_{k_1}^+c_{k_1}|X\rangle$$

$$\times \langle X|c_{k_2}^+c_{k_2}|X\rangle\cdots\langle X|c_{k_{N/2}}^+c_{k_{N/2}}|X\rangle(-1)^P$$

(we could here insert $|X\rangle\langle X|$ instead of the complete expression for the unit operator, $\sum_s |s\rangle\langle s|$, because the rest of it does not contribute anything.) The above expression

Fig. 2.7 The art of physics

is actually the fully contracted term of Wick's theorem. In the thermodynamic limit, $\Omega \to \infty$, it stays finite, because every power of Ω in the normalization term is compensated by the summation over k (both are proportional to the number of particles in the system).

On the other hand, there are other terms in the expression for $\langle X | \phi_1 \phi_2 \cdots | X \rangle$, but they contain $N/2 - 1$, $N/2 - 2$, etc. different values of k, and thus independent summations. Because of that, some powers of volume in the denominator will not be canceled by summations, and all these terms vanish, which concludes our reasoning (Fig. 2.7).

Now the path is straightforward. (1) Expand the time-ordered exponent in the expression for Green's function; (2) Take all averages over the ground state, using Wick's theorem, thus factoring all terms in products of unperturbed Green's functions (with appropriate integrations); (3) Represent these terms by graphs—Feynman diagrams. After the correspondence between those graphs and the analytic terms in the expansion series is established, it is much simpler to work with the diagrams, which give much clearer understanding of the structure of the expressions involved.

The rules of drawing and reading Feynman diagrams in some detail, of course, depend on the interaction. What is even worse, they depend on tastes and preferences of the author whose book or chapter you are reading (Fig. 2.6); there are at least three popular schools. Here we will take one of those approaches, where time flows from right to left and so it is also from right to left that the lines symbolizing Green's functions are drawn. This has at least the advantage, that the order of letters labeling the diagram is the same as that in the analytic formulas, where later moments stand to the left.

For illustration, we derive the rules for the simple case of scalar electron—electron interaction, which by definition involves only one sort of particles (electrons) interacting via instantaneous spin-independent potential:

$$W(t) = \frac{1}{2} \sum_{\alpha_1} \sum_{\alpha_2} \int d^3\mathbf{x}_1 \int d^3\mathbf{x}_2 \Psi_{\alpha_1}^{\dagger}(\mathbf{x}_1, t)\Psi_{\alpha_2}^{\dagger}(\mathbf{x}_2, t)U(\mathbf{x}_1 - \mathbf{x}_2)$$
$$\times \Psi_{\alpha_2}(\mathbf{x}_2, t)\Psi_{\alpha_1}(\mathbf{x}_1, t).$$

For convenience we introduce

$$U(1-2) \equiv U(\mathbf{x}_1 - \mathbf{x}_2)\delta(t_1 - t_2);$$

and now we will integrate over space and time coordinates indiscriminately, and the whole set $(\mathbf{x}, t, \alpha)$ (α is spin index) is written down as X.

Expanding the exponent in the general expression for $iG(X, X')$ up to the first order in interactions and substituting out scalar electron—electron interaction, we find

$$iG(XX') \tag{2.62}$$

$$\approx \frac{\langle\Phi_0|\Psi_X\Psi_{X'}^{\dagger}|\Phi_0\rangle + (-\frac{i}{\hbar})\frac{1}{2}\int d1\int d2 U(1-2)\langle\Phi_0|T\,\Psi_1^{\dagger}\Psi_2^{\dagger}\Psi_2\Psi_1\Psi_X\Psi_{X'}^{\dagger}|\Phi_0\rangle}{1 + (-\frac{i}{\hbar})\frac{1}{2}\int d1\int d2 U(1-2)\langle\Phi_0|T\Psi_1^{\dagger}\Psi_2^{\dagger}\Psi_2\Psi_1|\Phi_0\rangle}.$$

This was step one. In step two, the average of the six operators in the numerator and that of the four operators in the denominator must be evaluated using Wick's theorem.

The expression $\langle\Phi_0|T\Psi_1^{\dagger}\Psi_2^{\dagger}\Psi_2\Psi_1\Psi_X\Psi_{X'}^{\dagger}|\Phi_0\rangle$ can be fully contracted in six different ways. For example, $\langle\Phi_0|T\,\Psi_1^{\dagger}\,\Psi_2^{\dagger}\Psi_2\,\Psi_1\,\Psi_X\Psi_{X'}^{\dagger}|\Phi_0\rangle = n^0(1)n^0(2)iG^0$ (X, X'). Here $n^0(1) = \langle\Psi_1^{\dagger}\Psi_1\rangle_0$ is simply unperturbed electronic density in the system. As you see, in this term the "probe" particle (traveling from X' to X) is decoupled—disconnected—from the rest of the system, that is, does not interact with it.

Another example:

$$\langle\Phi_0|T\Psi_1^{\dagger}\,\Psi_2^{\dagger}\Psi_2\Psi_1\,\Psi_X\Psi_{X'}^{\dagger}|\Phi_0\rangle = -iG^0(1, 2)iG^0(2, 1)iG^0(X, X')$$

also produces a disconnected term, with minus sign, because in order to put together the field operators to form corresponding pairings we had to change the places of an odd number of Fermi operators.

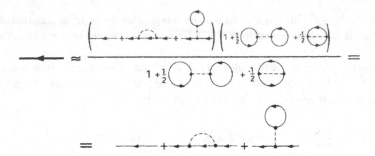

Fig. 2.8 Green's function to first order in scalar e–e interaction

Still another choice gives us at last a connected term, where the probe particle interacts with the system:

$$\langle\Phi_0|T\,\overbrace{\Psi_1^\dagger\,\underbrace{\Psi_2^\dagger\Psi_2}\,\underbrace{\Psi_1\Psi_\chi}\,\Psi_{\chi'}^\dagger}|\Psi_0\rangle = iG^0(X,\,2)iG^0(2,\,1)iG^0(1,\,X').$$

All this is rather boring, but if we mark all the elements of the above expression—points where interaction occurs, unperturbed Green's functions, etc.–as suggested in Table 2.1, we can draw Green's function to first order in interaction as shown in Fig. 2.8. The resulting diagrams are like the ones we obtained earlier for the one-particle propagator. Once again, the probe particle interacts with the particles in the system, and propagates freely between those acts of scattering. But now the "background" particles interact with each other, and this is expressed in the structure of the graphs, which is now far richer.

You see that the first two terms in the numerator are indeed disconnected; they literally fall apart. On the other hand, the remaining four terms are connected, and they show that the probe particle is scattered (that is, interacts) with the other panicles in the system.

By definition, the connected diagrams are the ones that do not contain parts disconnected from the external ends, that is, the coordinates of the "external" particle (in our case, external ends are $X,\,X'$). Only the external ends of the diagram carry significant coordinates (spins, etc.), the ones that actually appear as the arguments of the exact Green's function that we wish to calculate. All the rest are dummy labels, because there will be integrations (summations) over them. Of course, it does not matter how we denote a dummy variable, and all the diagrams that differ only by the mute labels are the same.

Here we see that connected terms contain integrations over the dummy variables 1 and 2. Therefore, of four connected terms there are only two that are different, and we can get rid of the factor one half before them.

Expanding the denominator, we find that

$$\langle\Phi_0|\,\mathcal{S}(-\infty,\infty)\,|\Phi_0\rangle = 1 + \frac{1}{2}\bigcirc\!\!\!-\!-\!\bigcirc + \frac{-1}{2}\bigcirc \;.$$

Within the same accuracy, we can factor the numerator (neglecting the higher-order terms in the interaction), and see that the denominator actually cancels the disconnected terms from the numerator!

This observation is actually a strict mathematical statement, and since the proof is very simple and general, let us prove.

2.2.1.2 Cancellation Theorem

All disconnected diagrams appearing in the perturbation series for the Green's function exactly cancel from its numerator and denominator. Therefore Green's function is expressed as a sum over all connected diagrams.

No need to specify the interaction term, $\mathcal{W}$. Let us consider the νth order term in the numerator of Green's function:

$$\sum_{n=0}^{\infty}\sum_{m=0}^{\infty}\delta_{m+n,\nu}(-i)^{m\,+\,n}\frac{1}{\nu!}\left(\frac{\nu!}{m!n!}\right)\int_{-\infty}^{\infty}dt_1\cdots dt_m\langle\Phi_0|\mathcal{T}\mathcal{W}(t_1)\cdots\mathcal{W}(t_m)$$

$$\times\;\Psi(X)\Psi^{\dagger}(X')|\Phi_0\rangle_{\mathbf{connected}}$$

$$\times\int_{-\infty}^{\infty}dt_{m+1}\cdots dt_{\nu}\langle\Phi_0|\mathcal{T}\mathcal{W}(t_{m+1})\cdots\mathcal{W}(t_{m+n})|\Phi_0\rangle.$$

In this expression each term is explicitly presented as a product of a connected part (of mth order) and a disconnected part (of nth order), m and n, adding up to ν. We included here a combinatorial factor ($\frac{\nu!}{m!n!}$), which is the number of ways to distribute ν interaction operators $\mathcal{W}(t_i)$ in these two groups (connected and disconnected), consisting of m and n operators respectively. (Since interaction terms contain an even number of Fermi operators, no sign change occurs from such a redistribution.) This factor combines with the $\frac{1}{\nu!}$ (from the exponential series) and leaves us with $\frac{1}{m!}\frac{1}{n!}$.

Summation ν from 0 to ∞ simply eliminates the δ-symbol, and we have for a product of two series, of which the second, due to disconnected diagrams, is, after trivial relabeling of the integration variables,

Fig. 2.9 Example of topologically equivalent diagrams

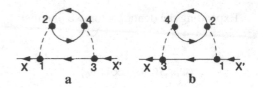

Table 2.1 Feynman rules for scalar electron-electron interaction

$X \overleftarrow{\quad\quad} X'$	$iG(XX') \equiv iG_{\alpha\alpha'}(\mathbf{x}, t; \mathbf{x}', t')$	Causal Green's function
$x \overleftarrow{\quad\quad} x'$	$iG^0(XX') \equiv iG^0(\mathbf{x} - \mathbf{x}', t - t')\delta_{\alpha\alpha'}$	Unperturbed causal Green's function
$\overset{1}{\bullet} \text{-}\text{-}\text{-} \overset{2}{\bullet}$	$-iU(1-2) \equiv -iU(\mathbf{x}_1 - \mathbf{x}_2)\delta(t_1 - t_2) \times \delta(t_1 - t_2)$	Interaction potential
$\overset{1}{\bigcirc}$	$n^0(1) \equiv \langle \Phi_0 | \Psi_1^\dagger \Psi_1 | \Phi | 0 \rangle$	Unperturbed electron density

The integration over all intermediate coordinates and times and summation over mute spin indices is implied

$$\frac{1}{n!} \sum_{n=0}^{\infty} \int_{-\infty}^{\infty} dt_1 \ldots dt_n (-i)^n \langle \Phi_0 | \mathcal{T} \mathcal{W}(t_1) \ldots \mathcal{W}(t_n) | \Phi_0 \rangle$$

$$= \langle \Phi_0 | \mathcal{T} e^{-i \int_{-\infty}^{\infty} dt \mathcal{W}(t)} | \Phi_0 \rangle,$$

that is, the denominator of the expression for Green's function! This contribution cancels, which proves the theorem.

We see as well that in any connected term of mth order there will be exactly $m!$ identical contributions due to rearrangements of $t_1, \ldots, t_m$ in $\langle \Phi_0 | \mathcal{T} \mathcal{W}(t_1) \cdots \mathcal{W}(t_m) \Psi(X) \Psi^\dagger(X') | \Phi_0 \rangle_{\textbf{connected}}$. This cancels the $\frac{1}{m!}$ factors and allows us to deal only with topologically different graphs. An example (Fig. 2.9): these two second order diagrams are the same diagram, because they differ only by labels of the interaction lines, $12 \leftrightarrow 34$. Returning to our specific form of the interaction, we will see that in our case there is also a 2^{-n} factor associated with diagram, due to the one half in the two-particle interaction term. This factor also cancels, this time because we don't distinguish between the ends of the interaction line, $1 \bullet\text{-}\text{-}\text{-}\text{-}\bullet 2$ being the same as $2 \bullet\text{-}\text{-}\text{-}\text{-}\bullet 1$. (As we said, only the labels of external ends matter. The rest are just dummy integration variables!) Then we finally come to

2.2.1.3 General Rules

1. Draw all topologically distinct connected Feynman diagrams,
2. Decode them according to Table 2.1,
3. Multiply every diagram by $(-1)^F$, where F is the number of closed loops with more than one vertex, consisting of fermionic lines. ("Bubbles" (n^0) do not count here.)

The origin of rule 3 is self-evident. When a fermionic loop is formed, we have to contract the Fermi operators like this: $\psi(1)\psi^\dagger(2)\psi(2)\psi(3)\cdots\psi(N)\psi^\dagger(1)$. Since in any Hamiltonian we have an arrangement $\psi^\dagger(1)\psi(1)$ etc., this means that the operator $\psi^\dagger(1)$ must have been dragged to the rightmost place through all the rest of the operators, that is, through an odd number of Fermi operators: its own conjugate, and no matter how many $\psi^\dagger\psi$ pairs. This yields the overall minus sign and explains the rule. The bubbles don't count, because such a bubble corresponds to a $\psi^\dagger\psi$ pair, and no rearrangement is necessary.

If we now perform a Fourier transformation to the momentum representation, we will see that the same rules apply, but the decoding table is somewhat different (here we denote by P the set $(\mathbf{p}, \omega)$), see Table 2.2. The energy and momentum conservation law in each vertex (which reduces the number of integrations in each vertex by one) has a simple origin. In coordinate representation, an intermediate integration in a vertex $Y = (\mathbf{y}, t_y)$ involves the expression $\int d^4Y G^0(.. - Y)G^0(Y - ..)U(Y - ..)$. (We take into account that the unperturbed Green's function and interaction potential are spatially uniform.) Rewriting this in Fourier components, we obtain

$$
\int \frac{d^4K}{(2\pi)^4} \int \frac{d^4K_1}{(2\pi)^4} \int \frac{d^4K_2}{(2\pi)^4} \int d^4Y e^{iK(..-Y)+iK_1(Y-..)+iK_2(Y-..)}
$$
$$
\times\ G^0(K)G^0(K_1)U(K_2).
$$

The integral over Y can be taken immediately; it is a simple exponential integral yielding delta functions:

$$
\int d^4Y e^{iY(-K+K_1+iK_2)} = \int dt_y e^{-it_y(-\omega+\omega_1-\omega_2)} \int d^3y e^{iy(-\mathbf{k}+\mathbf{k}_1+\mathbf{k}_2)}
$$
$$
= (2\pi)^4\delta(-\omega + \omega_1 - \omega_2)\delta(-\mathbf{k} + \mathbf{k}_1 + \mathbf{k}_2).
$$

Thus the energy (frequency) and momentum (wave vector) are conserved in each vertex. The physical reason for this is clear: each vertex of the diagram describes a scattering process. The Hamiltonian of our problem (which describes such scattering) is spatially uniform and time independent, which in agreement with general principles yields momentum and energy conservation.

Besides scalar electron–electron interaction, another important interaction in solid-state systems is electron–phonon interaction. We will not derive here the corresponding Hamiltonian in terms of electron and phonon field operators: this is rather

Table 2.2 Feynman rules for scalar electron–electron interaction (momentum representation)

	$iG(P) \equiv iG_{\alpha\alpha'}(\mathbf{p}, \omega)$	Causal Green's function
	$iG^0(P) \equiv iG^0(\mathbf{p}, \omega)\delta_{\alpha\alpha'}$	Unperturbed causal Green's function
	$-iU(Q) \equiv -iU(\mathbf{q})$	Fourier transform of the interaction potential
	$n^0(\mu)$	Unperturbed electron density

The integration over all intermediate momenta and frequencies $(dP/(2\pi)^4)$ and summation over dummy spin indices is implied, taking into account energy (frequency)/momentum conservation in every vertex

a subject for a course in solid state physics. It is enough for us to note that the electron–phonon interaction is described by terms in the Hamiltonian proportional to $\psi^\dagger(X)\psi(X)\phi(X)$ (this expression is Hermitian, since the phonon operator ϕ (as we defined it) is real). It is clear then that only even-order terms in electron–phonon interaction enter the perturbation expansion, because otherwise there will be unpaired phonon operators, giving zero vacuum average. In the even-order terms, phonon operators pair to form unperturbed phonon Green's functions (propagators) $D^0(\mathbf{k}, \omega)$. The definition of the vertex and of the phonon propagator depends on convention; we give here for your convenience the rules used in two basic monographs on the subject. The following discussion will not actually depend on such details, but each time you perform or follow specific calculations, it pays to check all the conventions beforehand.

2.2.2 Operations with Diagrams. Self Energy. Dyson's Equation

One of reasons why Green's function are so widely used is that the corresponding diagrams have a very convenient property: *The value of any Feynman diagram for Green's function can be found as the composition of expressions corresponding to its parts, independently of the structure of the diagram as a whole.*

This means that any part of the diagram (subdiagram) can be calculated separately once and for all and then inserted into an arbitrary diagram containing such a part. (This is not so, e.g., in the case of diagram expansion for the grand potential.)

What does this mean? Let us look at two different diagrams shown in Fig. 2.10, of second and sixth order respectively. The expressions for them are easily written, and we underline the terms that correspond to the marked parts of the diagrams:

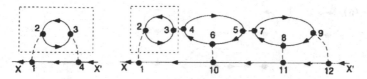

Fig. 2.10 Two diagrams

Table 2.3 Feynman rules for electron-phonon interaction (momentum representation)

$\underset{[1]}{\overset{K}{\wedge\wedge\wedge}}$	$iD(K) \equiv iD(\mathbf{k}, \omega)$	Exact phonon propagator
$\overset{K}{\wedge\wedge\wedge}$	$iD^0(K) \equiv iD^0(\mathbf{k}, \omega) = i\frac{\omega_k^2}{\omega^2 - \omega_k^2 + i0}$	Unperturbed phonon propagator
$\underset{p}{\overset{p+k}{\searrow}}\wedge\wedge k$ [3]	$-ig$	Electron-phonon coupling constant
$\overset{K}{\wedge\wedge\wedge}$	$iD^0(K) \equiv iD^0(\mathbf{k}, \omega) = i\frac{2\omega_k}{\omega^2 - \omega_k^2 + i0}$	Unperturbed phonon propagator
$\underset{p}{\overset{p+k}{\searrow}}\wedge\wedge k$	$-i|M_{\mathbf{k}}|$	Electron-phonon matrix element

$$\int d1d2d3d4[iG^0(X1)(-1)iG^0(23)iG^0(32)$$
$$\times iG^0(14)iG(4X')(-iU(12))(-iU(34))];$$

$$\int d1d2d3d4d5d6d7d8d9d10d11d12[iG^0(X1)(-1)iG^0(23)iG^0(32)$$
$$\times iG^0(54)iG^0(46)iG^0(65)(-1)iG^0(97)iG^0(78)iG^0(89)$$
$$\times (-1)iG^0(1, 10)iG^0(10, 11)iG^0(11, 12)iG^0(12, X')$$
$$\times (-iU(12))(-iU(34))(-iU(57))(-iU(6, 10))(-iU(8, 11))$$
$$(-iU(9, 12))].$$

The final expression is simply constructed of elementary blocks like

$$\overset{2\bullet\quad\bullet 3}{\bigcirc} = -iG^0(23)iG^0(32).$$

This is very different from the diagrammatic series for the grand potential, Ω, where a factor $1/n$ in each nth-order diagram prohibits such *partial summation* of diagram series (Table. 2.3).

The idea of this summation is simple and mathematically shaky. Suppose we have a diagram, for example, —⟶⟶⟶⟶—. In the infinite series for Green's function there is an infinite subset of diagrams like ⟶⟶⟶⟶⟶ , ⟶⟶⟶⟶⟶ , ⟶⟶⟶⟶⟶ , ..., which include all possible corrections to the inner line. Due to

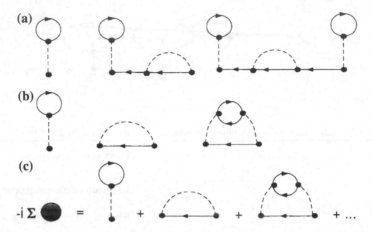

Fig. 2.11 Self energy diagrams: **a** self energy parts, **b** irreducible self energy parts, **c** proper self energy

the fact that there is no explicit dependence of the expression on the order of the diagram, we can forget about everything that lies beyond these interaction points and concentrate on the inside of the graph. The corrections here should transform the thin line (unperturbed Green's function, G^0) into a solid line (exact Green's function, G) in

the same way, as the whole series gives the exact Green's function G: ⎯⎯⎯⎯⎯⎯⎯! We have partially summed the diagram series for Green's function!

This is not yet a victory, though. First, the summation of this sort still gives us an equation: a self-consistent equation for the exact Green's function, usually a nonlinear integral or integro-differential one. To solve it would be really tough! Second, there is absolutely no guarantee that this equation is correct. Indeed, we know from mathematics that only for a very restricted class of convergent series (*absolutely* convergent) the sum is independent of the order of the terms. What we have done here is to redistribute the terms of the perturbation series, about which we even do not know (and usually cannot know) whether it converges at all! The justification here comes from the results: if they are wrong, then something is wrong in our way of partial summation (evidently, there are many, and each is approximate, since some classes of diagrams are neglected). Or maybe something funny occurs to the system, and this is already useful information. We will meet such a case later, when discussing application of the theory to superconductivity. In most cases the results are right if the partial summation is made taking into account the physics of the problem. Usually we can show, with physical if not mathematical rigor, that a certain class of diagrams is more important than the others, and therefore the result of its summation reflects essential properties of the system.

To approach such partial summations systematically, let us make some definitions.

The *self energy part* is called any part of the diagram connected to the rest of it only by two particle lines (Fig. 2.11).

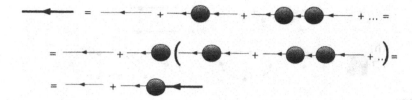

Fig. 2.12 Dyson's equation

The *irreducible*, or *proper, self energy part* is the one that cannot be separated by breaking one particle line, like the one in Fig. 2.11b.

Finally, the *proper self energy*, or *self energy* par excellence, or *mass operator*, is called the sum of all possible irreducible self energy parts and is denoted by $\Sigma_{\alpha\alpha'}(X, X')$. The name is given for historical field-theoretical reasons, and its meaning will become clear a little later.

It is convenient to include a $(-i)$ factor into the definition (Fig. 2.11). Then the series for Green's function can be read and drawn as follows (Fig. 2.12):

$$iG = iG^0 + iG^0 \Sigma G^0 + tG^0 \Sigma G^0 \Sigma G^0 + \cdots \qquad (2.63)$$

Here the terms in the infinite series are redistributed in such a way as to make it a simple series (a geometric progression!) over the powers of self energy and unperturbed Green's function only. (Of course, all necessary integrations and matrix multiplications with respect to spin indices are implied, so that this is an *operator* series.)

Separating the iG^0 factor, we obtain the celebrated *Dyson's equation* (see Fig. 2.12), which is exactly of the self-consistent form we anticipated:

$$G(X, X') = G^0(X, X') \qquad (2.64)$$
$$+ \int dX'' \int dX''' G^0(X, X'') \Sigma(X'', X''') G(X''', X').$$

(Of course, we could take iG^0 from the other side, and get $G(P) = G^0 + G(P)\Sigma(P)G^0(P)$.) In a homogeneous stationary and nonmagnetic system (the last condition means that G and Σ are diagonal with respect to spin indices) we can make a Fourier transformation, reducing the above equation to $G(P) = G^0 + G^0(P)\Sigma(P)G(P)$. Then we see that

$$G(\mathbf{p}, \omega) = \left[(G^0(\mathbf{p}, \omega))^{-1} - \Sigma(\mathbf{p}, \omega) \right]^{-1} = \frac{1}{\omega - \epsilon(\mathbf{p}) + \mu - \Sigma(\mathbf{p}, \omega)}. \qquad (2.65)$$

Symbolically this can be written as

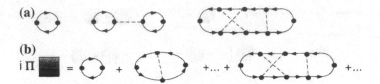

Fig. 2.13 Interaction renormalization: **a** polarization insertions, **b** polarization operator

$$G = \left[i\frac{\partial}{\partial t} - \mathcal{E} - \hat{\Sigma} \right]^{-1}. \tag{2.66}$$

The latter equation holds even if G and Σ are nondiagonal (e.g., in the nonhomogeneous case), understanding $[\ldots]^{-1}$ as an inverse operator.

An important feature of (2.65) is that if we substitute there some finite-order approximation for the self energy, the resulting approximation for G will be equivalent to calculating an *infinite* subseries of the perturbation series, and this gives a much better result than the simple-minded calculation of the initial series term by term.

This is a natural consequence of a self-consistent approach. Another, less pleasant, one is that any approximate self energy is to be checked, lest it violates the general analytic properties of Green's function (which follow from the general causality principle and should not be toyed with). Returning to the simple case (2.65) and recalling the Källén–Lehmann representation, we see that necessarily

$$\begin{cases} \Im\Sigma(\mathbf{p}, \, \omega) \geq 0, \; \omega < 0, \\ \Im\Sigma(\mathbf{p}, \, \omega) \leq 0, \; \omega > 0. \end{cases} \tag{2.67}$$

We see as well that $\Im\Sigma$ is the inverse lifetime of the elementary excitation, while $\Re\Sigma$ defines the change of dispersion law due to interaction. (In quantum field theory this leads to the change of the particle mass, which is why Σ is also called the mass operator.)

2.2.3 Renormalization of the Interaction. Polarization Operator

Following the same approach, we can consider the insertions into the interaction line as well, like those shown in Fig. 2.13.

The *polarization insertion* is called the part of the diagram that is connected to the rest of it only by two interaction lines. The *irreducible polarization insertion* is one that cannot be separated by breaking of a single interaction line. Finally, the *polarization operator* is the sum of all irreducible polarization insertions, Π, and is a direct analogue to the self energy.

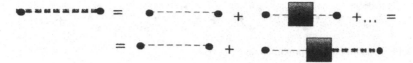

Fig. 2.14 Equation for the polarization operator

Since there is a $(-i)$ factor in the definition of the interaction line, it is convenient to introduce an (i) factor into the polarization operator. The analogue to Dyson's equation is readily obtained and reads (see Fig. 2.14)

$$U_{\text{eff}}(P) = U(P) + U(P)\Pi(P)U_{eff}(P). \tag{2.68}$$

Then we find the *generalized dielectric function*, $\kappa(\mathbf{p}, \omega)$:

$$U_{\text{eff}}(\mathbf{p}, \omega) \equiv \frac{U(\mathbf{p}, \omega)}{\kappa(\mathbf{p}, \omega)} = \frac{U(\mathbf{p}, \omega)}{1 - U(\mathbf{p}, \omega)\Pi(\mathbf{p}, \omega)}, \tag{2.69}$$

which describes the effect of the polarization of the medium on particle–particle interaction. A good example of such an effect is the following.

2.2.3.1 Screening of Coulomb Interaction

The Thomas–Fermi result concerning the screening of the Coulomb potential by the charged Fermi gas can be reproduced if we use the random phase approximation (RPA), which here means taking the lowest-order term in the polarization operator:

$$i\,\Pi_0(\mathbf{p}, \omega) = \quad = 2 \int \frac{d^3q\,d\zeta}{(2\pi)^4} G^0(\mathbf{p} + \mathbf{q}, \omega + \zeta) G^0(\mathbf{q}, \zeta). \tag{2.70}$$

The calculations give the following result for the static screening:

$$\Re\Pi_0(\mathbf{p}, 0) = -\frac{mp_F}{2\pi^2}\left(1 + \frac{p_F^2 - p^2/4}{p_F\,p}\ln\left|\frac{p_F + p/2}{p_F - p/2}\right|\right); \tag{2.71}$$

$$\Im\Pi_0(\mathbf{p}, 0) = 0. \tag{2.72}$$

For the long-range screening ($p \ll p_F$),

$$\Pi_0 \approx -2\mathcal{N}(\mu),$$

Fig. 2.15 Random phase approximation

Fig. 2.16 Ladder approximation

where $\mathcal{N}(\mu) \equiv \frac{m p_F}{2\pi^2}$ is the density of states on the Fermi surface. Thus the Fourier transform of the interaction is

$$U_{\text{efff}}(q) \approx \frac{4\pi e^2/q^2}{1 + 2\mathcal{N}(\mu)4\pi e^2/q^2} = \frac{4\pi e^2}{q^2 + 8\pi e^2 \mathcal{N}(\mu)}. \tag{2.73}$$

The quantity

$$q_{TF}^2 = 8\pi e^2 \mathcal{N}(\mu)$$

is the squared Thomas–Fermi wave vector, and the potential indeed takes the Yukawa form:

$$U_{\text{efff}}(r) = \frac{e^2}{r} \exp(-q_{TF} r). \tag{2.74}$$

Thus, the presence of other charged particles leads to screening of initial long- range Coulomb interactions, and limits it to a finite Thomas–Fermi radius. How this happens is graphically clear from the simplest polarization diagram. The interaction creates a *virtual* electron–hole pair. (Virtual, of course, because the energy-momentum relation for every internal line of a diagram is violated: we integrate over *all* energies and *all* momenta *independently*! For a real particle, $E = \frac{p^2}{2m}$ or something like this.) The approximation we used included only independent events of such virtual electron-hole creation: because the energy and momentum along the interaction line are conserved, the quantum-mechanical phase of the electron–hole pair is immediately lost and does not affect the next virtual pair. This is the reason it is called RPA, random phase approximation (Fig. 2.15). As we discussed in the very beginning of the book, this kind of approach works well if there is a large number of particles within the interaction radius: then indeed it is much more probable to interact with two different particles consecutively than with the same one twice. In the opposite case, when the density of particles is low, RPA naturally fails, while the *ladder approximation* is relevant: Here a virtual pair (quasiparticle–quasihole) interacts repeatedly before disappearing (Fig. 2.16). This is again reasonable, because when density is low, it is improbable to find some other quasiparticle close at hand to interact with.

Unfortunately, on our path to the Thomas-Fermi screening, Eq. (2.74), from the random phase approximation, Eqs. (2.71), (2.72), we made one simplification too many when replaced the static polarization $\Pi_0(\mathbf{p},\ 0)$ with its value at $p = 0$. The logarithmic term in (2.71) is non-analytical at $p = p_F$, and—as it turns out—it produces instead of the exponential screening (2.74) a qualitatively different potential, which far away from the charge behaves as

$$U_{\text{efff}}(r) \propto \frac{e^2}{r^3}\cos(2p_F r).$$

Not only it does not fall off exponentially, but it also demonstrates *Friedel oscillations*. Both effects are due to the sharp step of the Fermi distribution function at $T = 0$, which produced the non-analytical term in (2.71) in the first place (see Appendix A). At finite temperature the step is smeared, and the above expression is multiplied by an exponentially decaying factor, thus reverting to the Yukawa–type screening with Friedel oscillations superimposed.

2.2.4 Many-Particle Green's Functions. Bethe–Salpeter Equations. Vertex Function

We have seen that Green's functions give a convenient apparatus for a description of many-body systems. So far we have used a *one particle* Green's function, dealing with a single quasiparticle excitation, though against the many-body background. They don't apply, e.g., to the case of the bound state of *two* such excitations. Indeed, in a Fermi system such a state would be a boson, while the one-particle Green's function describes a fermion.

This problem can be easily solved. Nobody limits us to consideration of averages $\langle \psi\psi^\dagger \rangle$ only. The "Schrödinger equation" for $G(X,\ X')$ included terms $\langle \psi\psi\psi^\dagger\psi^\dagger \rangle$. Therefore it is natural to introduce n-particle Green's functions. (As usual, there is no common convention here, so when reading a chapter be careful what definition is actually used.)

The n-particle (or $2n$-point) Green's function (Fig. 2.17) is defined as follows:

$$G_{(n)\alpha_1\alpha_2\ldots\alpha_n,\alpha_1'\alpha_2'\ldots\alpha_n'}(\mathbf{x}_1 t_1,\ \mathbf{x}_2 t_2,\ \ldots,\ \mathbf{x}_n t_n;\ \mathbf{x}_1' t_1',\ \mathbf{x}_2' t_2',\ \ldots,\ \mathbf{x}_n' t_n')$$

$$\equiv G_{(n)}(12\cdots n;\ 1'2'\cdots n')$$

$$= \frac{1}{(i)^n}\langle T\psi(1)\psi(2)\cdots\psi(n)\psi^\dagger(n')\cdots\psi^\dagger(2')\psi^\dagger(1')\rangle \qquad (2.75)$$

The rules of drawing and decoding Feynman diagrams stay intact and can be easily derived from the expansion of the S-operator in the average $\langle T\cdots\psi^\dagger\psi^\dagger \rangle$. There is only one additional rule.

The diagram is multiplied by $(-1)^S$, where S is the parity of the permutation of the fermion lines' ends $(1'2'\cdots n') \leftrightarrow (12\cdots n)$ (see Fig. 2.18).

Fig. 2.17 Many-particle Green's function (This convenient "stretched skin" graphics are introduced in [5]; ots mark the outgoing ends)

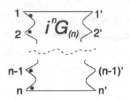

Fig. 2.18 Sign rule for a many-particle Green's function: **a** $(1, 2, 3)$ $\Leftrightarrow$ $(3', 2', 1')$, sign $+1$, **b** $(1, 2, 3)$ $\Leftrightarrow$ $(2', 3', 1')$, sign -1

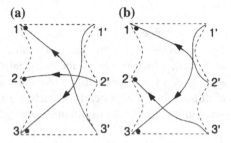

The origin of this rule is easy to see applying Wick's theorem to the lowest order expression for the two-particle Green's function:

$$G_{(2)}(12, 1'2') \equiv (-i)^2 \langle T\psi_1\psi_2\psi_{2'}^\dagger\psi_{1'}^\dagger \rangle \tag{2.76}$$
$$\approx (-i)\langle T\psi_1\psi_{1'}^\dagger \rangle_0 (-i)\langle T\psi_2\psi_{2'}^\dagger \rangle_0 \mp (-i)\langle T\psi_1\psi_{2'}^\dagger \rangle_0 (-i)\langle T\psi_2\psi_{1'}^\dagger \rangle_0$$
$$= G^0(11')G^0(22') \mp G^0(12')G^0(21').$$

The cancellation theorem removes only the diagrams with loose parts disconnected from the *external ends*. This means that not every diagram looking disconnected is actually disconnected! For example, the diagrams corresponding to (2.76) (see the two diagrams in Fig. 2.19) are not disconnected and are not canceled. As a matter of fact they provide a Hartree–Fock approximation for the two-particle Green's function (direct and exchange terms, as is evident from their structure).

The two-particle Green's function is most widely used and therefore has its own letter, K:

$$K(12; 1'2') = -\langle T\psi(1)\psi(2)\psi^\dagger(2')\psi^\dagger(1') \rangle. \tag{2.77}$$

Its diagram expansion to the second order is given in Fig. 2.19.

The importance of the two-particle Green's function is that (1) it determines the scattering amplitude of quasiparticles, that is, their interactions, and thus (2) its poles define the dispersion law of two-particle excitations (e.g., bosonic excitations in a normal Fermi system—say zero sound), as well as appearance of boundstate of two quasiparticles—and therefore the superconducting transition point.

We can define the *irreducible* two-particle Green's function by separating all "seemingly disconnected" diagrams (Fig. 2.20). The first two of this set give the

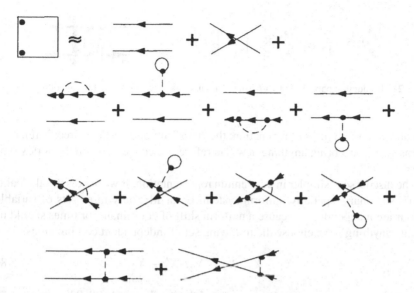

Fig. 2.19 Second-order expansion of the two-particle Green's function

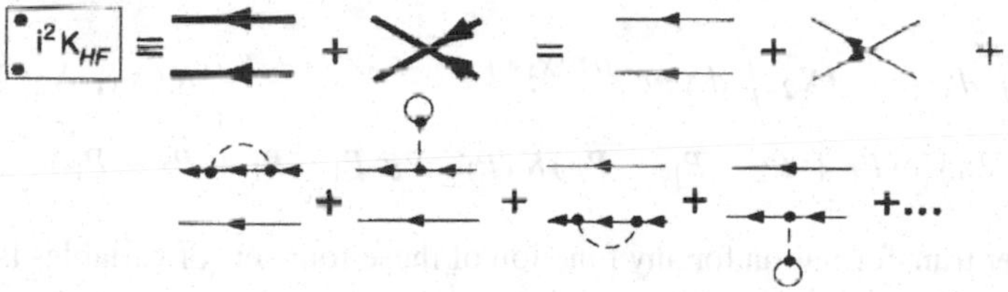

Fig. 2.20 Generalized Hartree–Fock approximation for the two-particle Green's function

self-consistent Hartree–Fock approximation for the two-particle Green's function (self-consistent, because it contains *exact* one-particle Green's functions):

$$G(11')G(22') \mp G(12')G(21') = G^0(11')G^0(22') \mp G^0(12')G^0(21') + \cdots .$$
(2.78)

The rest is the *irreducible two-particle Green's function* and is expressed through the *vertex function* Γ (Fig. 2.21):

$$\tilde{K}(12; 1'2') = K(12; 1'2') - [G(11')G(22') \mp G(12')G(21')]$$
$$= \int d3 \int d3' \int d4 \int d4' G(13)G(24)i\Gamma(33'; 44')G(3'1')G(4'2'). \quad (2.79)$$

The poles of the two-particle Green's function define *two-particle excitations* of the system in the same fashion as the poles of the one-particle Green's function defined quasiparticles. Examples are such excitations in the Fermi systems as zero sound and plasmons. Evidently, the relevant poles of the two-particle Green's func-

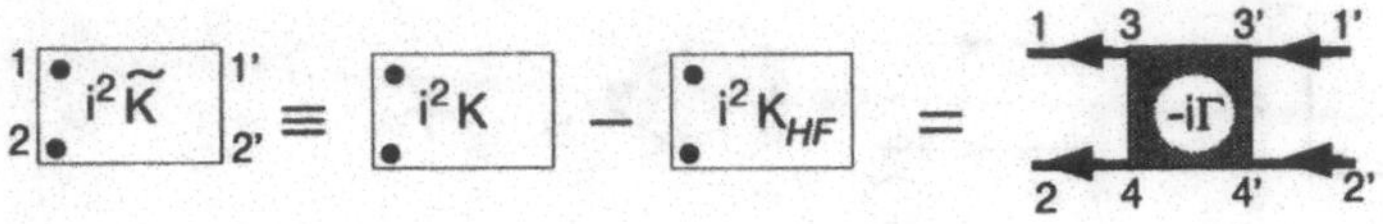

Fig. 2.21 Irreducible part of the two-particle Green's function and the vertex function

tion appear only in the vertex function: the "tails" are one-particle Green's functions and as such don't bring anything new. Therefore, we concentrate on the vertex function.

The discourse is simpler in momentum representation, if we are (as usual) dealing with a stationary, spatially uniform system. Evidently, only three sets of variables of four are independent (because a uniform shift of coordinates or times should not change anything). We choose the following set of independent combinations:

$$X_1 - X_1', \ X_2 - X_2', \ X_1' - X_2. \tag{2.80}$$

Here and later on $X = \mathbf{x}, \ t; \ P = \mathbf{p}, \ \omega$; and the "scalar product" $PX = \mathbf{p} \cdot \mathbf{x} - \omega t$.

Then the two-particle Green's function in momentum space is defined as

$$\int dX_1 \int dX_{1'} \int dX_2 \int dX_{2'} e^{-i(P_1 X_1 + P_2 X_2 - P_{1'} X_{1'} - P_{2'} X_{2'})} K(X_1, \ X_2; X_1', \ X_2')$$

$$= (2\pi)^4 \delta(P_1 + P_2 - P_{1'} - P_{2'}) K(P_1, \ P_2; P_{1'}, \ P_1 + P_2 - P_{1'}). \tag{2.81}$$

The Fourier transformation for any function of these four sets of variables is defined by

$$K(P_1, \ P_2; P_{1'}, \ P_1 + P_2 - P_{1'})$$
$$= \int d(X_1 - X_1') \int d(X_2 - X_2') \int d(X_{1'} - X_2')$$
$$\times e^{-i P_1 (X_1 - X_1') - i P_2 (X_2 - X_2') + i P_1' (X_1' - X_2)} K(X_1, \ X_2; X_1', \ X_2');$$
$$K(X_1, \ X_2; X_1', \ X_2')$$
$$= \int \frac{dP_1}{(2\pi)^4} \int \frac{dP_2}{(2\pi)^4} \int \frac{dP_1'}{(2\pi)^4}$$
$$e^{i P_1 (X_1 - X_{2'}) + i P_2 (X_2 - X_{2'}) - i P_1' (X_1' - X_2')} K(P_1, \ P_2; P_{1'}, \ P_1 + P_2 - P_{1'}).$$

Then Eq. (2.79) can be rewritten as

$$\tilde{K}(P_1, \ P_2; P_{1'}, \ P_1 + P_2 - P_{1'}) = G(P_1)G(P_2)$$
$$\times i\Gamma(P_1, \ P_2; P_{1'}, \ P_1 + P_2 - P_{1'})$$
$$G(P_1')G(P_1 + P_2 - P_{1'}). \tag{2.82}$$

Now we are ready to derive (for scalar electron–electron interaction, our standard guinea pig) an important general relation between the vertex function and self energy. That such a relation should exist is reasonable, since both Σ and Γ have in common, besides being uppercase Greek letters, that they result from summation of all somehow irreducible diagrams. First, we present a very graphic proof, which will be then supported by more rigorous calculation (which, on the other hand, is only a translation of graphs into equations).

We start from writing down the equation of motion for the one-particle Green's function, in position space. As we observed much earlier, such an equation will contain the two-particle Green's function:

$$\left(i\frac{\partial}{\partial t} + \frac{1}{2m}\nabla_{\mathbf{x}}^2 + \mu\right) G_{\alpha\alpha'}(X, X') = \delta_{\alpha\alpha'}\delta(X - X')$$

$$-i\int d^4Y\, U(X - Y) K_{\alpha\gamma,\alpha'\gamma}(X, Y; X', Y) \tag{2.83}$$

(we have made use of the definition: $\langle T\Psi_\gamma^\dagger(Y)\Psi_\gamma(Y)\Psi_\alpha(X)\Psi_{\alpha'}^\dagger(X')\rangle = K_{\alpha\gamma,\alpha'\gamma}(X, Y; X', Y))$. Since $G = G^0 + G^0\Sigma G$, the relation in question is indeed here, and we have only to extract it.

Graphically, it is simple: the equation can be symbolically written as $[iG^0]^{-1}iG = T - (-iU)(i^2K)$, that is,

$$[\!\longleftarrow\!]^{-1}\!\longleftarrow\ =\ I\ -$$

Then we do a series of transformations:

$$[\!\longleftarrow\!]^{-1}\!\longleftarrow\ =\ [\!\longleftarrow\!]^{-1}[\!\longleftarrow + \!\longleftarrow\!\bullet\!\longleftarrow\!]\ =\ I + \!\bullet\!\longleftarrow$$

$$\bullet\!\longleftarrow\ =\ -\quad i^2K\quad =\ -\left[\quad i^2K_{HF} + \quad i^2\tilde{K}\quad\right] =$$

$$=\ -\left[\quad -\quad +\quad -i\Gamma\quad\right]$$

$$(\times(-1))$$

The result is shown in Fig. 2.22. Notice that we used the specific for $(n > 2)$-particle Green's function sign convention in order to determine the signs of the first two terms on the right-hand side of Fig. 2.22: if "decoded" following the one-particle rules, they would lack a (-1)-factor due to the exchange of tails of the two-particle diagram.

Fig. 2.22 Relation between
the self energy and vertex
function

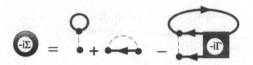

In analytical form, this equation (sometimes called Dyson's equation, but less
often than *the* Dyson equation we encountered earlier) reads

$$\Sigma(P)\delta_{\alpha\beta} = U(0)n(\mu)\delta_{\alpha\beta} + i\delta_{\alpha\beta} \int \frac{dP_1}{(2\pi)^4} U(P - P_1)G(P_1)$$

$$+ \int \frac{dP_1}{(2\pi)^4} \int \frac{dP_2}{(2\pi)^4} G(P_1)G(P_2) \tag{2.84}$$

$$\times \Gamma_{\alpha\gamma,\beta\gamma}(P_1, P_2; P, P_1 + P_2 - P)G(P_1 + P_2 - P)U(P - P_1).$$

Now let us derive it without graphs, or rather write down each step instead of
drawing it. Again, assume a uniform, stationary, and isotropic system. Then, in
momentum space, (2.83) looks like

$$\left[(G^0(P))^{-1}G(P) - 1\right]\delta_{\alpha\alpha'} = -i \int \frac{dP_1 dP_2}{(2\pi)^8} K_{\alpha\gamma,\alpha'\gamma}(P_1, P_2; P, P_1 + P_2 - P)U(P - P_1).$$

(Here $(G^0(P))^{-1} \equiv \left(\omega - \frac{p^2}{2m} + \mu\right)$ is a function, not an operator, and simply equals
$1/G^0(P)$.) Now substitute in this equation the definition (2.79) and divide by $G(P)$.
After this messy operation we obtain

$$[1/G^0(P) - 1/G(P)]\delta_{\alpha\alpha'} = -i\delta_{\alpha\alpha'}U(0) \int \frac{dP_2}{(2\pi)^4} G(P_2)$$

$$\pm i\delta_{\alpha\alpha'} \int \frac{dP_1}{(2\pi)^4} U(P - P_1)G(P_1)$$

$$+ \int \frac{dP_1 dP_2}{(2\pi)^8} \Gamma_{\gamma\alpha,\gamma\alpha'}(P_1, P_2; P, P_1 + P_2 - P)$$

$$\times G(P_1)G(P_2)G(P_1 + P_2 - P)U(P - P_1).$$

Since by virtue of the Dyson equation $1/G^0(P) - 1/G(P) = \Sigma(P)$, then we eventually recover Eq. (2.84). See how much easier it was with the diagrams? By the way,
the graphs immediately show the physical sense of this relation. The first two terms in
(2.84) give the self-consistent Hartree–Fock approximation with initial (bare) potential: they take into account the interaction of the test particle with the medium, and
with itself (exchange term). The rest must contain the effects of renormalization of
the interaction, and indeed, the third graph can be understood as containing the *renormalized interaction vertex* (Fig. 2.23). As you see, it contains, in particular, all the
polarization insertions in the interaction line. This is the reason we had a bare potential line in Fig. 2.22 and Eq. (2.84): otherwise certain diagrams would be included

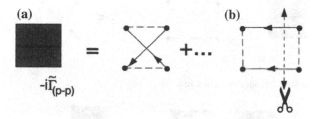

Fig. 2.23 Vertex function and renormalized interaction

Fig. 2.24 Particle–particle irreducible vertex function

Fig. 2.25 Particle-particle irreducible vertex function and two–particle Green's function

twice. In all operations with diagrams we must pay special attention to avoiding the double count.

2.2.4.1 The Bethe–Salpeter Equation

Earlier we introduced the irreducible self energy as a sum of all diagrams that cannot be separated by severing one fermion line. Let us generalize this and introduce the *particle-particle irreducible* vertex function, $\tilde{\Gamma}_{(PP)}$, which includes all diagrams that cannot be separated by severing *two* fermion lines between in- and outcoming ends. (In Fig. 2.24 diagram (a) is particle–particle irreducible, but diagram (b) is not.)

Then the diagram series for the particle–particle irreducible vertex part (or the particle–particle irreducible two–particle Green's function, if drop the external tails) can be drawn as in Fig. 2.25.

For the vertex function we thus obtain the *Bethe–Salpeter equation*, which is a direct analogue of the Dyson equation for the one-particle Green's function[2] (Fig. 2.26):

---

[2] Of course, this equation can as well be written for the two–particle Green's function itself, instead of the vertex function.

Fig. 2.26 Bethe-Salpeter
equation (*particle-particle
channel*)

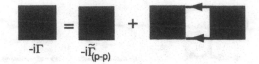

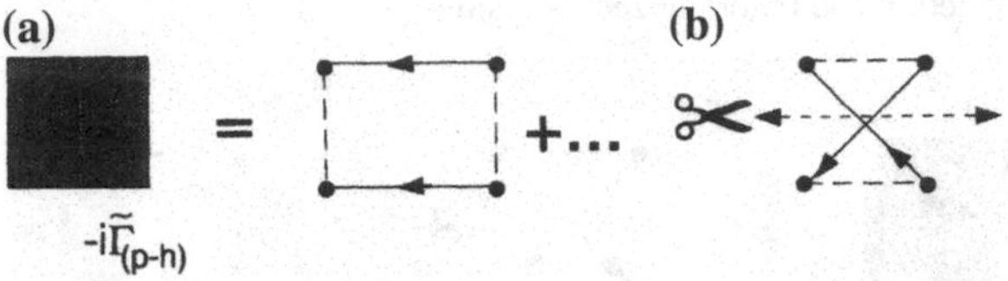

$-i\Gamma$ $-i\tilde{\Gamma}_{(p\text{-}p)}$

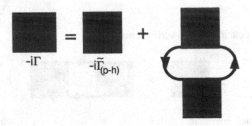

$-i\tilde{\Gamma}_{(p\text{-}h)}$

Fig. 2.27 Particle–hole irreducible vertex function

$-i\Gamma$ $-i\tilde{\Gamma}_{(p\text{-}h)}$

Fig. 2.28 Bethe-Salpeter equation (*particle–hole channel*)

$$\Gamma(12; 1'2') = \tilde{\Gamma}_{(PP)}(12; 1'2')$$

$$+i \int d3 \int d3' \int d4 \int d4' \tilde{\Gamma}_{(PP)}(12; 3'4') G(33') G(44') \Gamma(3'4'; 1'2'). \quad (2.85)$$

Two-particle functions allow for more possibilities: there are more loose ends
in a diagram! Thus, we have a different *particle–hole irreducible* vertex, $\tilde{\Gamma}_{(PH)}$,
(Fig. 2.27, where diagram (b) is now (particle–hole) irreducible, while diagram (a)
is not). This yields another version of the Bethe–Salpeter equation (Fig. 2.28):

$$\Gamma(12; 1'2') = \tilde{\Gamma}_{(PH)}(12; 1'2')$$

$$+i \int d3 \int d3' \int d4 \int d4' \tilde{\Gamma}_{(PH)(42;4'2')} G(43) G(3'4') \Gamma(13;1'3').$$

$$(2.86)$$

Of course, both versions are equivalent mathematically, but not physically. Since
there is little hope that either can be solved exactly, some approximations are in
order, and we should choose, as usual, the version that is better as a starting point.
The latter one, e.g., proves to be useful for investigation of the processes with small
momentum transfer between quasiparticles, but this is beyond the scope of this book.

2.3 Problems

- *Problem 1*

starting from the expression for the grand potential, $\Omega = -PV$,

$$\Omega = \int_0^\mu d\mu (2iV) \lim_{t \to -0} \int \frac{d\mathbf{p} d\omega}{(2\pi)^4} e^{-i\omega t} G(\mathbf{p}, \omega),$$

find the pressure of the ideal Fermi gas at zero temperature

$$\left(G(\mathbf{p}, \omega) = G^0(\mathbf{p}, \omega) = \frac{1}{\omega - (\epsilon_\mathbf{p} - \mu) + i0\mathrm{sgn}\omega} \right).$$

Compare to the classical expression $P = nk_BT$ and find the "effective pressure temperature." How is it related to the "effective energy temperature" $T_F = \mu/k_B$?

- *Problem 2*

Reduce the *ladder approximation* series for the two-particle Green's function to an integral equation:

- *Problem 3*

Calculate the lowest order diagram for the polarization operator:

$p{+}q,\omega{+}\zeta$

q,ζ

and reproduce the results of Eqs. (2.71), (2.72).

- *Problem 4*

Starting from the definition (2.16), derive the equation of motion for the unperturbed phonon propagator.

References

Book and Reviews

1. Abrikosov, A.A., Gorkov, L.P., Dzyaloshinski, I.E.: Methods of quantum field theory in statistical physics. Ch. 2. Dover Publications, New York (1975) (An evergreen classic on the subject.)
2. Fetter, A.L., Walecka, J.D.: Quantum theory of many-particle systems. McGraw-Hill, San Francisco (1971)
3. Mahan, G.D.: Many-particle physics. Plenum Press, New York (1990) ([2] and [3] are high-level, very detailed monographs: the standard references on the subject.)
4. Lifshitz, E.M., Pitaevskii, L.P.: Statistical physics pt. II. (Landau and Lifshitz Course of theoretical physics, v. IX.) Pergamon Press, New York (1980) (Ch. 2. A comprehensive, but very compressed account of the zero-temperature Green's functions techniques.)
5. Mattuck, R.: A guide to Feynman diagrams in the many-body problem. McGraw-Hill, New York (1976) (Green's functions techniques are presented in a very instructive and intuitive way.)
6. Nussenzvejg, H.M.: Causality and dispersion relations. Academic Press, New York (1972). (A very good book for the mathematically inclined reader)
7. Thouless, D.J.: Quantum mechanics of many-body systems. Academic Press, New York (1972)
8. Ziman, J.M.: Elements of advanced quantum theory. Ch. 3, 4. Cambridge University Press, Cambridge (1969)

Chapter 3
More Green's Functions, Equilibrium and Otherwise, and Their Applications

Such, such were the joys
When we all, girls and boys,
In our youth time were seen
On the Ecchoing Green.
William Blake
"Songs of Innocence"

Abstract Equilibrium Green's functions at finite temperature. Formal analogy between the equilibrium statistical operator and the evolution operator in imaginary time. Temperature (Matsubara) Green's functions and their relation to equilibrium Green's functions. Diagram technique for the temperature Green's functions (Matsubara formalism). Linear response theory, Kubo formulas, and fluctuation-dissipation theorem. Nonequilibrium Green's functions. Keldysh formalism: time-ordering along the Keldysh contour and diagram technique for matrix Green's functions. Quantum kinetic equation and its quasiclassical limit. Quantum transport: Landauer formula and conductance quantization. Method of tunneling Hamiltonian.

3.1 Analytic Properties of Equilibrium Green's Functions

The formalism we have developed so far is limited to zero temperature (i.e., to the ground state) properties of many-body systems. As you remember, this is because the ground state is always nondegenerate, so that we could pull off the trick with adiabatic hypothesis: if you slowly turn interactions on, and then off, the worst that can happen is some phase factor, which anyway cancels. This, in turn, allowed us to build up the diagrammatic technique.

Physically, it is rather awkward to be confined to the case of $T = 0$. In principle, the average in the definition of Green's function could be taken over any quantum state, or set of states, and we would be able at least to determine its analytic properties, following the same steps as at $T = 0$. For example, we can define equilibrium Green's

A. Zagoskin, *Quantum Theory of Many-Body Systems*,
Graduate Texts in Physics, DOI: 10.1007/978-3-319-07049-0_3,
© Springer International Publishing Switzerland 2014

functions at finite temperature. Moreover, it turns out that diagram techniques exist that can be used to actually calculate such Green's functions. In this chapter, we will discuss how and why this can be done.

3.1.1 Statistical Operator (Density Matrix): The Liouville Equation

If a quantum system is in some definite quantum state, $|\Phi\rangle$, one says it is in a *pure state*; otherwise (i.e., when the quantum state of the system is known only statistically) it is in a *mixed state* and is described not by the single state vector, but by the *statistical operator*, $\hat{\rho}$;

$$\hat{\rho} = \sum_m |\Phi_m\rangle W_m \langle \Phi_m|. \tag{3.1}$$

Here W_m is the probability of finding the system in the quantum state $|\Phi_m\rangle$; evidently,

$$\sum_m W_m = 1 \tag{3.2}$$

(the states $\{\Phi_m\}_{m=1}^{\infty}$ are supposed to be normalized, but not necessarily orthogonal). The idea is that the statistical operator allows one to find the average value of any operator $\mathcal{O}$ in this mixed state via

$$\langle \mathcal{O} \rangle \equiv \sum_m W_m = \langle \Phi_m | \mathcal{O} | \Phi_m \rangle = \mathrm{tr}(\hat{\rho}\mathcal{O}). \tag{3.3}$$

(In a mixed state we have to do averaging twice: first over each constituent quantum state, and then over the set of these states with weights W_m; the trace with the statistical operator in the above formula takes care of both.) Equation (3.2) ensures that the probabilities add up to one—that is, the unitarity.

If we choose some *orthonormal basis*, $\{|n\rangle\}$, then the statistical operator can be rewritten as follows:

$$\hat{\rho} = \sum_n \sum_{n'} |n\rangle \rho_{nn'} \langle n'|; \tag{3.4}$$

$$\rho_{nn'} = \sum_m W_m \langle n | \Phi_m \rangle \langle \Phi_m | n' \rangle. \tag{3.5}$$

In this form the statistical operator (as a set of matrix elements $\{\rho_{nn'}\}$) is often called the *density matrix*.[1]

The diagonal elements of the density matrix, $\rho_{nn} \geq 0$, give the probabilities of finding the system in the state $|n\rangle$, while the off-diagonal terms describe the *quantum correlations* between different states.

Useful properties of the trace of the statistical operator directly follow from its definition:

$$\mathrm{tr}(\hat{\rho}) = 1;$$

(3.6)

$$\mathrm{tr}\left(\hat{\rho}^2\right) \leq \left(\mathrm{tr}(\hat{\rho})\right)^2$$

(3.7)

(an equality is achieved if and only if the system is in a pure state). The former equality ensures probability conservation and directly follows from (3.2): the trace of a matrix (or an operator) is an invariant under unitary transformations of coordinates, and since in one special basis it equals one (3.2), so will it under any choice of a basic set of states.

The time evolution of the statistical operator can be determined if write it in the form Eq. (3.1) and recall that $|\Phi(t)\rangle = \mathcal{U}(t)|\Phi(0)\rangle$:

$$\hat{\rho}(t) = \sum_m |\Phi_m(t)\rangle W_m \langle \Phi_m(t)| = \mathcal{U}(t)\hat{\rho}(0)\mathcal{U}^\dagger(t).$$

(3.8)

Therefore, the statistical operator satisfies the *Liouville equation* (called so because it is a direct analogue to the classical Liouville equation for the distribution function):

$$i\dot{\hat{\rho}}(t) = [\mathcal{H}(t), \hat{\rho}(t)].$$

(3.9)

Note that this is an equation in Schrödinger, not Heisenberg, representation. The Hamiltonian is a *Schrödinger* operator, and its dependence on time (if any) can *only* be explicit, e.g., due to an alternate external field.

3.1.2 Definition and Analytic Properties of Equilibrium Green's Functions

The general definitions of the causal, retarded, and advanced one-particle Green's functions,

---

[1] In the case of a single particle we can take the basic set of coordinate eigenfunctions $\{|\mathbf{x}\rangle\}$, so that $\langle n|\Phi_m\rangle \Rightarrow \langle \mathbf{x}|\Phi_m\rangle = \Phi_m(\mathbf{x})$, and the result takes the familiar form

$$\rho(\mathbf{x}, \mathbf{x}') = \sum_m W_m \Phi_m(\mathbf{x})\Phi_m^*(\mathbf{x}').$$

$$G_{\alpha\beta}(\mathbf{x}_1, \, t_1; \mathbf{x}_2, \, t_2) = -i\mathrm{tr}\left(\hat{\rho}\mathcal{T}\psi_\alpha(\mathbf{x}_1, \, t_1)\psi_\beta^\dagger(\mathbf{x}_2, \, t_2)\right), \qquad (3.10)$$

$$G_{\alpha\beta}^R(\mathbf{x}_1, \, t_1; \mathbf{x}_2, \, t_2)$$
$$= -i\mathrm{tr}\left(\hat{\rho}\left\{\psi_\alpha(\mathbf{x}_1, \, t_1)\psi_\beta^\dagger(\mathbf{x}_2, \, t_2) \pm \psi_\beta^\dagger(\mathbf{x}_2, \, t_2)\psi_\alpha(\mathbf{x}_1, \, t_1)\right\}\right)\theta(t_1 - t_2), \quad (3.11)$$

$$G_{\alpha\beta}^A(\mathbf{x}_1, \, t_1; \mathbf{x}_2, \, t_2)$$
$$= +i\,\mathrm{tr}\left(\hat{\rho}\left\{\psi_\alpha(\mathbf{x}_1, \, t_1)\psi_\beta^\dagger(\mathbf{x}_2, \, t_2) \pm \psi_\beta^\dagger(\mathbf{x}_2, \, t_2)\psi_\alpha(\mathbf{x}_1, \, t_1)\right\}\right)\theta(t_2 - t_1),$$
$$\qquad (3.12)$$

are, of course, valid for the equilibrium state at finite temperature, when the statistical operator has standard Gibbs form,

$$\hat{\rho} = e^{-\beta(\mathcal{H}-\Omega)}$$
$$= \sum_s e^{-\beta(E_s - \mu N_s - \Omega)}|s\rangle\langle s|$$
$$= \sum_s \rho_s|s\rangle\langle s|, \qquad (3.13)$$
$$\beta \equiv \frac{1}{k_B T}.$$

We choose the basic set of common energy and particle number eigenstates, $|s\rangle$, and work, as usual, with the grand canonical ensemble:

$$\mathcal{H}_{GCE}|s\rangle \equiv (\mathcal{H}_{CE} - \mu\mathcal{N})|s\rangle = (E_s - \mu N_s)|s\rangle.$$

Therefore, the statistical operator is diagonal, $\hat{\rho}_{ss'} \equiv \delta_{ss'}\rho_s = e^{\beta(\Omega - E_s - \mu N_s)}$. The normalization factor, $e^{\beta\Omega} = \mathrm{tr}e^{\beta'H}$, contains the grand potential, Ω.

In the isotropic uniform case, of course,

$$G_{\alpha\beta}(\mathbf{x}_1, t_1; \mathbf{x}_2, t_2) = \delta_{\alpha\beta}G(\mathbf{x}_1 - \mathbf{x}_2, t_1 - t_2), \text{ etc.}$$

Now we will quickly repeat, mutatis mutandis, the calculations we made when discussing analytic properties of zero-temperature Green's functions.

3.1.2.1 The Generalized Källén–Lehmann Representation

The generalized Källén–Lehmann representation is derived in the same way as at zero temperature (Sect. 2.1.2), with the only difference that we must include matrix elements of field operators between *all* states of the system, $\langle s|\psi(X)|s'\rangle$, because now all excited states enter with nonzero weight. As a result, we obtain for the causal Green's function

$$G(\mathbf{p}, \omega) = \left[\frac{1}{2}\right](2\pi)^3 \sum_m \sum_n \rho_n A_{mn} \delta(\mathbf{p} - \mathbf{P}_{mn})$$

$$\times \left\{\frac{1}{\omega - \omega_{mn} + i0} \pm \frac{e^{-\beta\omega_{mn}}}{\omega - \omega_{mn} - i0}\right\}; \qquad (3.14)$$

$$A_{mn} = \left[\sum_\alpha\right]|\langle n|\psi_\alpha|m\rangle|^2; \qquad (3.15)$$

$$\omega_{mn} = E_m - \mu N_m - (E_n - \mu N_n). \qquad (3.16)$$

Separating real and imaginary parts of (3.14) at real frequencies (using the ubiquitous Weierstrass formula (2.31)), we find

$$\Re G(\mathbf{p}, \omega) = (2\pi)^3 \left[\frac{1}{2}\right] \mathcal{P} \sum_m \sum_n \rho_n A_{mn} \delta(\mathbf{p} - \mathbf{P}_{mn})$$

$$\times (1 \pm e^{-\beta\omega_{mn}})\frac{1}{\omega - \omega_{mn}}, \qquad (3.17)$$

$$\Im G(\mathbf{p}, \omega) = -\pi(2\pi)^3 \left[\frac{1}{2}\right] \sum_m \sum_n \rho_n A_{mn} \delta(\mathbf{p} - \mathbf{P}_{mn})$$

$$\times (1 \mp e^{-\beta\omega_{mn}})\delta(\omega - \omega_{mn}). \qquad (3.18)$$

On the other hand, for the retarded and advanced Green's functions we obtain by the same method

$$G^R(\mathbf{p}, \omega) = \left[\frac{1}{2}\right](2\pi)^3 \sum_m \sum_n \rho_n A_{mn} \delta(\mathbf{p} - \mathbf{P}_{mn}) \left\{\frac{1 \pm e^{-\beta\omega_{mn}}}{\omega - \omega_{mn} + i0}\right\}; \qquad (3.19)$$

$$G^A(\mathbf{p}, \omega) = \left[\frac{1}{2}\right](2\pi)^3 \sum_m \sum_n \rho_n A_{mn} \delta(\mathbf{p} - \mathbf{P}_{mn}) \left\{\frac{1 \pm e^{-\beta\omega_{mn}}}{\omega - \omega_{mn} - i0}\right\}. \qquad (3.20)$$

In the thermodynamic limit (N, $V \to \infty$, $N/V = \text{const}$) it is more convenient to use the generalized Källén–Lehmann representation in the continuum form:

$$G^{R,A}(\mathbf{p}, \omega) = \int_{-\infty}^{\infty} \frac{d\omega'}{\pi} \frac{\rho^{R,A}(\mathbf{p}, \omega')}{\omega' - \omega \pm i0}; \qquad (3.21)$$

where the weight function (*spectral density*) is

$$\rho^{R,A}(\mathbf{p},\ \omega') = -\pi(2\pi)^3 \left[\frac{1}{2}\right] \sum_m \sum_n \rho_n A_{mn} \left(1 \pm e^{-\beta\omega_{mn}}\right)$$

$$\times \ \delta(\mathbf{p} - \mathbf{P}_{mn})(\omega' - \omega_{mn}). \tag{3.22}$$

After applying the Weierstrass formula once again, we see that

$$\Re G^{R,A}(\mathbf{p}, \omega) = \Re G(\mathbf{p}, \omega); \tag{3.23}$$

$$\Im G^{R,A}(\mathbf{p}, \omega) = \pm \Im G(\mathbf{p}, \omega) \times \begin{cases} \coth \frac{\beta\omega}{2} \text{Fermi statistics}, \\ \tanh \frac{\beta\omega}{2} \text{Bose statistics}. \end{cases} \tag{3.24}$$

In the limit $\beta \to \infty$ this, of course, reduces to (2.42).

Thus we have derived an important expression of the retarded/advanced equilibrium Green's function through the causal Green's function at finite temperature (for *real* frequencies):

$$G^{R,A}(\mathbf{p}, \omega) = \Re G(\mathbf{p}, \omega) + \begin{cases} \pm i \coth \frac{\beta\omega}{2} \Im G(\mathbf{p}, \omega) \ \text{Fermi statistics}, \\ \pm i \tanh \frac{\beta\omega}{2} \Im G(\mathbf{p}, \omega) \ \text{Bose statistics}, \end{cases} \tag{3.25}$$

Relation (3.25) allows us to find the $G^{R,A}(\omega)$ if we know $G(\omega)$. Note that the latter is *not* an analytic function, so that now the quasiparticle excitations are rather defined by the poles of $G^{R,A}(\omega)$ in the lower (upper) half-plane of complex frequency, respectively.

This comes as no surprise, since we already know that these two Green's functions have direct physical meaning. They are involved, e.g., in calculations of the kinetic properties of the system in *linear response theory*, which we will consider later. But since there is no regular perturbation theory to calculate $G^{R,A}$ directly, we will use an easy detour. There *is* a regular way to find the causal Green's function (the so called *Matsubara formalism*), after which retarded and advanced Green's functions can be directly obtained with the help of (3.25).

We still have a safeguard against mistakes that can be caused by inadequate approximations, the *Kramers–Kronig relations*, which, of course, hold at any temperature (as causality itself):

$$\Re G^{R,A}(\mathbf{p}, \omega) = \pm \mathcal{P} \int_{-\infty}^{\infty} \frac{\ulcorner \to' \ \Im \mathcal{G}^{R,A}(\mathbf{p}, \to')}{\approx \quad \to' - \to},$$

as well as the asymptotic formula

$$G(\omega),\ G^{R,A}(\omega)|_{|\omega|\to\infty} \sim \frac{1}{\omega};$$

Fig. 3.1 Spectral density of
the retarded Green's function

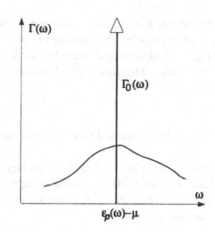

the latter, as we remember, is a result of canonical commutation relations and probability conservation.

It follows from the Kramers–Kronig relations and (3.21) that for the real frequencies the spectral density in the thermodynamic limit, $\rho^R(\mathbf{p}, \omega)$, is

$$\rho^R(\mathbf{p}, \omega) = \Im G^R(\mathbf{p}, \omega) = -\frac{1}{2}\Gamma(\mathbf{p}, \omega). \qquad (3.26)$$

The latter function, $\Gamma(\mathbf{p}, \omega)$, is also frequently called spectral density, which (hopefully) will not lead to any confusion.

3.1.2.2 The Sum Rule for the Retarded Green's Function

Here is another very useful safeguard: the sum rule for the spectral density Γ (here is the opportunity not to get confused!),

$$\int \frac{d\omega}{2\pi}\Gamma(\mathbf{p}, \omega) = 1. \qquad (3.27)$$

Indeed,

$$\Gamma(\mathbf{p}, \omega) = (2\pi)^4 \left[\frac{1}{2}\right] \sum_m \sum_n \rho_n A_{mn}(1 \pm e^{-\beta\omega_{mn}})\delta(\mathbf{p} - \mathbf{P}_{mn})(\omega - \omega_{mn}), \qquad (3.28)$$

and we can integrate over frequency and then roll the calculations back to canonical commutation relations between field operators, in the same manner as we did when we calculated the $1/\omega$-asymptotics of Green's functions.

What is the physical meaning of this formula? $\Gamma(\mathbf{p}, \omega)$ gives the probability that a quasiparticle with energy ω has momentum $\mathbf{p}$ (or vice versa). We have already discussed that due to interactions there is always some momentum and energy exchange between particles, broadening the $(\epsilon_{\mathbf{p}} - \mu - \omega)$ peak of a noninteracting system (Fig. 3.1). Since a quasiparticle *must* have some energy given momentum, the integral (with appropriate normalization) *must* yield unity. Which it does, as we have seen.

3.1.2.3 Unperturbed Green's Functions

Unperturbed Green's functions can be easily calculated directly from the definition. Here it is easier, though, to calculate retarded and advanced Green's functions first, and then obtain the causal Green's function from (3.25). If you perform this useful exercise, you will find

$$G^{R,A(0)}(\mathbf{p}, \omega) = \frac{1}{\omega - \epsilon_{\mathbf{p}} + \mu \pm i0}; \tag{3.29}$$

$$G^{(0)}(\mathbf{p}, \omega) = \mathcal{P}\frac{1}{\omega - \epsilon_{\mathbf{p}} + \mu}$$

$$- i\pi\delta(\omega - \epsilon_{\mathbf{p}} + \mu) \begin{cases} \tanh\left(\frac{\beta\omega}{2}\right) & \text{Fermi statistics,} \\ \coth\left(\frac{\beta\omega}{2}\right) & \text{Bose statistics,} \end{cases} \tag{3.30}$$

in agreement with general analytic properties.

3.1.2.4 Particle Density for Fermi/Bose Particles

The particle density in the momentum space (per spin) is given by

$$n_{\mathbf{p}} = \left[\frac{1}{2}\sum_{\alpha}\right]\left\langle c^{\dagger}_{\mathbf{p}[\alpha]}c_{\mathbf{p}[\alpha]}\right\rangle. \tag{3.31}$$

Here $c^{\dagger}_{\mathbf{p}[\alpha]}$, $c_{\mathbf{p}[\alpha]}$ are Fermi (Bose) creation/annihilation operators. Note also a useful relation

$$\left\langle c^{\dagger}_{\mathbf{p}[\alpha]}c_{\mathbf{p}'[\alpha']}\right\rangle = (2\pi)^3\delta(\mathbf{p} - \mathbf{p}')[\delta_{\alpha\alpha'}]n_{\mathbf{p}}. \tag{3.32}$$

Then we can write

$$(2\pi)^3\delta(\mathbf{p} - \mathbf{p}')[\delta_{\alpha\alpha'}]n_{\mathbf{p}}$$

$$= \left[\frac{1}{2}\sum_{\alpha}\right]\sum_{m}\rho_m\,\langle m|\int d^3\mathbf{x}e^{i\mathbf{p}\mathbf{x}}\psi^{\dagger}_{[\alpha]}(\mathbf{x})\int d^3\mathbf{x}'e^{-i\mathbf{p}'\mathbf{x}'}\psi_{[\alpha']}(\mathbf{x}')|m\rangle,$$

and

$$n_{\mathbf{p}} = \left[\frac{1}{2}\right] \sum_m \sum_n \rho_m \delta(\mathbf{p} - \mathbf{P_{mn}}) A_{mn}. \tag{3.33}$$

Comparison of this expression to (3.28) immediately leads to the following beautiful formula:

$$n_{\mathbf{p}} = \int \frac{d\omega}{2\pi} \Gamma(\mathbf{p}, \ \omega) n_{F,B}(\omega); \tag{3.34}$$

$$n_F(\omega) = \frac{1}{e^{\beta\omega} + 1}; \tag{3.35}$$

$$n_B(\omega) = \frac{1}{e^{\beta\omega} - 1}. \tag{3.36}$$

It has an evident physical meaning: the statistical Fermi (Bose) distribution determines the probability for the particle to have energy ω at given temperature, while the spectral density $\Gamma(\mathbf{p}, \ \omega)$ gives the probability that the particle with this energy has the momentum $\mathbf{p}$.

3.2 Matsubara Formalism

3.2.1 Bloch's Equation

After learning a lot about the analytic properties of equilibrium Green's functions at finite temperatures, we once again meet the nasty question of how to calculate the actual Green's function?

We cannot use directly the results of zero-temperature diagram technique. The reason is that now we have to average over all excited states of the system, not only its ground state. And while the latter is unique, the former are highly degenerate (infinitely degenerate in the thermodynamic limit). Therefore our previous reasoning employing the adiabatic hypothesis no longer works: the adiabatic turning on and off of the interaction can leave the system at $t = +\infty$ in *any* linear combination of excited states, very different from the one present at $t = -\infty$, depending on the interaction, initial state, and exact way of turning this interaction on and off. This, in its turn, means that we cannot separate the $1/\langle S \rangle$-term, and the entire scheme fails. A clear indication of this fact is that the causal Green's function is essentially nonanalytic, and thus cannot be obtained by summation of a series.

There are different ways of dealing with this trouble. First, we could write down an equation of motion for the Green's function, like (2.83), then decouple the higher-order Green's function and find an approximation (checking that Kramers–Kronig relations are satisfied, etc.). The setback here is that you don't have a regular procedure and must rely on a happy guess.

Second, we could calculate Green's function directly from the general formula (3.10) for the average of Heisenberg operators, $\langle n|\psi\psi^\dagger|n\rangle = \langle n|\mathcal{S}^{-1}\mathcal{T}(\Psi\Psi\mathcal{S})|n\rangle$. There is an ingenious way to actually succeed (the *Keldysh formalism*), and it has a bonus of being naturally applicable to any nonequilibrium state of the system as well. We will discuss it later. The setback of this method is that all the Green's functions, self energies etc. become 2×2 matrices, which does not make calculations easier. If we do not exactly need to deal with an essentially nonequilibrium situation, we had better opt for something handier.

Third, we can use the remarkable analogy between the evolution operator in conventional time, $\mathcal{U} = e^{-iHt}$, and the (non-normalized) equilibrium statistical operator $\hat{\rho} = e^{-\beta H}, \beta = 1/T$. The idea of Matsubara was to use this analogy to define some new, Matsubara, or temperature, Green's functions, closely related to conventional causal Green's functions in real time. It turns out that for temperature Green's functions a simple and useful diagrammatics can be developed [1, 6].

If we introduce the variable $\tau, 0 < \tau < \beta$, we see that $\hat{\rho}$ satisfies the *Bloch equation*,

$$\frac{\partial}{\partial\tau}\hat{\rho}(\tau) = -\mathcal{H}\hat{\rho}(\tau), \tag{3.37}$$

with the initial condition $\hat{\rho}(0) = \mathcal{I}$. If we perform the transformation

$$t \leftrightarrow -i\tau, \tag{3.38}$$

this equation transforms into the Schrödinger equation for $\hat{\rho}(it)$ on the imaginary interval $0 > t > -i\beta$:

$$i\frac{\partial}{\partial t}\hat{\rho}(it) = \mathcal{H}\hat{\rho}(it). \tag{3.39}$$

The statistical operator is a generalization of the wave function, and it is not surprising that it satisfies some sort of Schrödinger equation. What is mildly surprising is that imaginary time enters the picture; but this is not totally exotic, because a vaguely similar situation with imaginary frequencies we meet when the evolution to equilibrium is discussed (e.g., in the classical theory of a damped oscillator). Here it is more convenient to rotate the time axis by $\pi/2$ in the complex time plane (see Fig. 3.2).

This so-called *Wick's rotation* transforms the Heisenberg operators into *Matsubara operators*:

$$\psi(\mathbf{x}, t) = e^{i\mathcal{H}t}\psi(\mathbf{x})e^{-i\mathcal{H}t} \rightarrow \psi^M(\mathbf{x}, \tau)$$
$$= e^{\mathcal{H}\tau}\psi(\mathbf{x})e^{-\mathcal{H}\tau}; \tag{3.40}$$
$$\psi^\dagger(\mathbf{x}, t) \rightarrow \bar{\psi}^M(\mathbf{x}, \tau) = e^{\mathcal{H}\tau}\psi^\dagger(\mathbf{x})e^{-\mathcal{H}\tau}. \tag{3.41}$$

Let us stress that the conjugated Matsubara field operator is *not* the Hermitian conjugate of the Matsubara field operator:

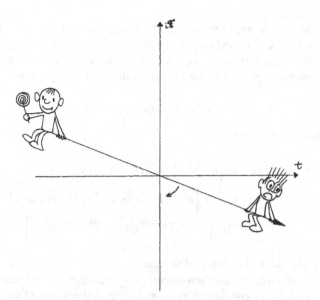

Fig. 3.2 Wick rotation in the complex time plane

$$\bar{\psi}^M(\mathbf{x}, \tau) \not= \left(\psi^M(\mathbf{x}, \tau)\right)^\dagger !$$

These operators satisfy the equations of motion, which are an "analytic continuation" of Heisenberg equations (1.84) at imaginary times:

$$\frac{\partial}{\partial \tau}\psi^M(\mathbf{x}, \tau) = \left[\mathcal{H}, \psi^M(\mathbf{x}, \tau)\right];\qquad(3.42)$$

$$\frac{\partial}{\partial \tau}\bar{\psi}^M(\mathbf{x}, \tau) = \left[\mathcal{H}, \bar{\psi}^M(\mathbf{x}, \tau)\right].\qquad(3.43)$$

3.2.2 Temperature (Matsubara) Green's Function

Now we can define the temperature Green's functions. First, we introduce the *temperature ordering operator*, $\mathcal{T}_\tau$, which, as usual, orders the operators so that the larger is the argument τ the further to the left it stands. The temperature Green's function is then, in direct analogy to (3.10),

$$\mathcal{G}_{\alpha\alpha'}(\mathbf{x}, \tau; \mathbf{x}', \tau') = -\left\langle \mathcal{T}_\tau \psi_\alpha^M(\mathbf{x}, \tau)\bar{\psi}_{\alpha'}^M(\mathbf{x}', \tau')\right\rangle$$

$$= -\mathrm{tr}\left\{e^{-\beta(\mathcal{H}-\Omega)}\mathcal{T}_\tau \psi_\alpha^M(\mathbf{x}, \tau)\bar{\psi}_{\alpha'}^M(\mathbf{x}', \tau')\right\}\qquad(3.44)$$

As usual, we consider the uniform state without magnetic ordering, so that the Green's function depends on $\mathbf{x} - \mathbf{x}'$, and spin dependence (if any) reduces to $\delta_{\alpha\alpha'}$.

Following the usual drill, we now explore the analytic properties of this function. First we show that it depends only on the difference of its τ-arguments:

$$\mathcal{G}(\mathbf{x}, \tau; \mathbf{x}', \tau') = \mathcal{G}(\mathbf{x}, \mathbf{x}'; \tau - \tau'). \tag{3.45}$$

If, for example, $\tau > \tau'$, then

$$\mathcal{G}(\mathbf{x}, \tau; \mathbf{x}', \tau') = -\mathbf{tr}\left\{e^{-\beta(\mathcal{H}-\Omega)}\psi^M(\mathbf{x}, \tau)\bar{\psi}^M(\mathbf{x}', \tau')\right\}$$

$$= -e^{\beta\Omega}\,\mathbf{tr}\left\{e^{-\beta\mathcal{H}}e^{\mathcal{H}\tau}\psi(\mathbf{x})e^{-\mathcal{H}\tau}e^{\mathcal{H}\tau'}\bar{\psi}(\mathbf{x}')e^{-\mathcal{H}\tau'}\right\}$$

$$= e^{\beta\Omega}\mathbf{tr}\left\{e^{-(\beta-\tau+\tau')\mathcal{H}}\psi(\mathbf{x})e^{-\mathcal{H}(\tau-\tau')}\bar{\psi}(\mathbf{x}')\right\}$$

(we have used the cyclic invariance of the trace).

So, temperature Green's functions depend on a variable $\tau - \tau'$, which changes from $-\beta$ to β and can be considered as 2β-periodic on the whole of the real axis τ. Therefore, it can be expanded in a Fourier series:

$$\mathcal{G}(\tau) = \frac{1}{\beta}\sum_{n=-\infty}^{\infty}\mathcal{G}(\omega_n)e^{-i\omega_n\tau}, \tag{3.46}$$

where the Matsubara frequencies are

$$\omega_n = \frac{\pi n}{\beta}. \tag{3.47}$$

Now we will show that depending on the statistics of particles involved, the series contains either odd or even Matsubara frequencies,

$$\omega_v^F = \frac{(2v+1)\pi}{\beta}; \tag{3.48}$$

$$\omega_v^B = \frac{2v\pi}{\beta}. \tag{3.49}$$

To see this, let us take some $\tau < 0$ and calculate $\mathcal{G}(\tau)$ and $\mathcal{G}(\tau + \beta)$:

Fig. 3.3 Periodic and antiperiodic functions of imaginary time τ

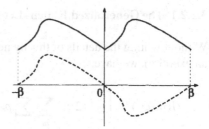

$$\mathcal{G}(\tau) = \pm tr \left\{ e^{\beta(\Omega) - \mathcal{H}} \psi^\dagger e^{\mathcal{H}\tau} \psi e^{-\mathcal{H}\tau} \right\}$$

$$= \pm e^{\beta\Omega} tr \left\{ e^{-\mathcal{H}(\tau+\beta)} \psi^\dagger e^{\mathcal{H}\tau} \psi \right\};$$

$$\mathcal{G}(\tau + \beta) = -tr \left\{ e^{\beta(\Omega - \mathcal{H})} e^{\mathcal{H}(\tau+\beta)} \psi e^{-\mathcal{H}(\tau+\beta)} \psi^\dagger \right\}$$

$$= e^{\beta\Omega} tr \left\{ e^{\mathcal{H}\tau} \psi e^{-\mathcal{H}(\tau+\beta)} \psi^\dagger \right\}$$

$$= \mp \mathcal{G}(\tau).$$

We have used here the cyclic invariance of the trace. The upper sign, as usual, corresponds to the Fermi statistics (Fig. 3.3). Thus temperature Green's functions are periodic (for bosons) or antiperiodic (for fermions) with period β. This is exactly what we obtain in keeping only even or odd Matsubara frequencies in series (3.46), because

$$e^{-i\omega_\nu^F(\tau+\beta)} = e^{-i\omega_\nu^F \tau} e^{-i(2\nu+1)\pi}$$

$$= -e^{-i\omega_\nu^F \tau};$$

$$e^{-i\omega_\nu^B(\tau+\beta)} = e^{-i\omega_\nu^B \tau} e^{-i2\nu\pi}$$

$$= e^{-i\omega_\nu^B \tau}.$$

Finally, expanding $\mathcal{G}(\mathbf{x}, \omega_n)$ in a Fourier integral over momenta, we come to the following form:

$$\mathcal{G}(\mathbf{x}, \tau) = \frac{1}{\beta} \sum_{\nu=-\infty}^{\infty} \int \frac{d^3\mathbf{p}}{(2\pi)^3} e^{i(\mathbf{px} - \omega_\nu \tau)} \mathcal{G}(\mathbf{p}, \omega_\nu), \tag{3.50}$$

with the inverse transformation

$$\mathcal{G}(\mathbf{p}, \omega_\nu) = \int_0^\beta d\tau \int d^3\mathbf{x} e^{-i(\mathbf{px} - \omega_\nu \tau)} \mathcal{G}(\mathbf{x}, \tau). \tag{3.51}$$

3.2.2.1 The Generalized Källen–Lehmann Representation

Without writing the details of this by now routine calculations (which you can do as an exercise), we have

$$\mathcal{G}(\mathbf{p}, \, \omega_v) = \left[\frac{1}{2}\right] (2\pi)^3 \sum_m \sum_n \rho_n A_{mn} \delta(\mathbf{p} - \mathbf{P}_{mn}) \left\{ \frac{1 \pm e^{-\beta \omega_{mn}}}{i\omega_v - \omega_{mn}} \right\}. \qquad (3.52)$$

The coefficients A_{mn} here are the same ones that enter the formulas for real time Green's functions in equilibrium, Eqs. (3.14), (3.19), (3.20). Comparing (3.52) to these equations immediately leads to the relation between temperature and realtime Green's functions:

$$\mathcal{G}(\mathbf{p}, \omega_v) = G^R(\mathbf{p}, i\omega_v); \quad \omega_v > 0; \qquad (3.53)$$

$$\mathcal{G}(\mathbf{p}, \omega_v) = G^A(\mathbf{p}, i\omega_v); \quad \omega_v < 0. \qquad (3.54)$$

If we know temperature Green's function, we can find real-time ones by simple analytic continuation to imaginary frequencies! (A word of warning is in order here: analytic continuation is simple as a concept, but may prove notoriously difficult in actual calculations. On the other hand, static properties of the system can be directly obtained from temperature Green's functions.)

3.2.3 Perturbation Series and Diagram Techniques for the Temperature Green's Function

Now, at last, we can return to drawing pictures. To begin with, we present the system's Hamiltonian in the standard form

$$\mathcal{H} = \mathcal{H}_0 + \mathcal{H}_1$$

(now both terms must be time independent in Schrödinger representation; otherwise, we will not be in equilibrium state: there will be no equilibrium state!). The "Matsubara interaction representation" is then defined by

$$\Psi^M(\mathbf{x}, \tau) = e^{\mathcal{H}_0 \tau} \psi(\mathbf{x}) e^{-\mathcal{H}_0 \tau}, \qquad (3.55)$$

so the "Heisenberg" Matsubara field operator equals

$$\psi^M(\mathbf{x}, \tau) = e^{\mathcal{H}\tau} e^{-\mathcal{H}_0 \tau} \Psi^M(\mathbf{x}) e^{\mathcal{H}_0 \tau} e^{-\mathcal{H}\tau}. \qquad (3.56)$$

Repeating essentially the same steps as before, (Sects. 1.3, 2.2.1), let us introduce the imaginary-time S-matrix in "interaction representation":

$$\sigma(\tau_1, \tau_2) = e^{\mathcal{H}_0\tau_1} e^{-\mathcal{H}(\tau_1 - \tau_2)} e^{-\mathcal{H}_0\tau_2}. \tag{3.57}$$

It satisfies self-evident conditions:

$$\sigma(\tau_2, \tau_1) = \sigma^{-1}(\tau_1, \tau_2); \tag{3.58}$$

$$\sigma(\tau_1, \tau_3)\sigma^{-1}(\tau_2, \tau_3) = \sigma(\tau_1, \tau_2). \tag{3.59}$$

We can also write down a differential equation for $\sigma(\tau, \tau_2)$:

$$\frac{\partial}{\partial \tau}\sigma(\tau, \tau_2) = -\mathcal{H}_1(\tau)\sigma(\tau, \tau_2),$$

where

$$\mathcal{H}_1(\tau) = e^{\mathcal{H}_0\tau} \mathcal{H}_1 e^{-\mathcal{H}_0\tau}. \tag{3.60}$$

Iterating it, we find the analogue to Dyson's expansion for σ, which is (for $\tau_1 > \tau_2$)

$$\sigma(\tau_1, \tau_2) = T_\tau \exp\left\{ -\int_{\tau_2}^{\tau_1} d\tau \mathcal{H}_1(\tau) \right\}. \tag{3.61}$$

You already know what to do next, but we will nevertheless explicitly derive the basic expression for $\mathcal{G}$. Using the "Matsubara interaction" representation for Matsubara field operators, we find that the temperature Green's function can be expressed through the "interaction" field operators as follows (omitting the "M" superscripts for brevity):

$$\mathcal{G}(\mathbf{x}_1, \mathbf{x}_2; \tau_1 - \tau_2)$$

$$= -e^{\beta\Omega}\left[tr\left(e^{-\beta\mathcal{H}} e^{\mathcal{H}\tau_1} e^{-\mathcal{H}_0\tau_1} \Psi(\tau_1) e^{\mathcal{H}_0\tau_1} e^{-\mathcal{H}\tau_1} \right.\right.$$

$$\left. \times e^{\mathcal{H}\tau_2} e^{-\mathcal{H}_0\tau_2} \bar{\Psi}(\tau_2) e^{\mathcal{H}_0\tau_2} e^{-\mathcal{H}\tau_2} \right)\theta(\tau_1 - \tau_2)$$

$$\mp tr\left(e^{-\beta\mathcal{H}} e^{\mathcal{H}\tau_2} e^{-\mathcal{H}_0\tau_2} \bar{\Psi}(\tau_2) e^{\mathcal{H}_0\tau_2} e^{-\mathcal{H}\tau_2} \right.$$

$$\left.\left. \times e^{\mathcal{H}\tau_1} e^{-\mathcal{H}_0\tau_1} \Psi(\tau_1) e^{\mathcal{H}_0\tau_1} e^{-\mathcal{H}\tau_1} \right)\theta(\tau_2 - \tau_1)\right].$$

Take, for example, the first term in this equation. Using the definition of $\sigma(\tau_1, \tau_2)$ (3.57), we can rewrite it as

$$-e^{\beta\Omega} tr\left(e^{-\beta\mathcal{H}_0}\sigma(\beta, \tau_1)\Psi(\tau_1)\sigma(\tau_1, \tau_2)\bar{\Psi}(\tau_2)e^{\mathcal{H}_0\tau_2}\sigma(\tau_2, 0)\right)\theta(\tau_1 - \tau_2).$$

Therefore, we can write

$$\mathcal{G}_{\alpha\gamma}(\mathbf{x}_1, \mathbf{x}_2; \tau_1 - \tau_2) = -e^{\beta\Omega} \, tr \left(e^{-\beta\mathcal{H}_0} T_\tau(\sigma(\beta, 0)\Psi(\tau_1)\bar{\Psi}(\tau_2)) \right)$$

$$= -e^{\beta(\Omega-\Omega_0)}\langle T_\tau(\sigma(\beta, 0)\Psi(\tau_1)\bar{\Psi}(\tau_2))\rangle_0$$

(here $\langle \cdots \rangle_0$ is the average over the unperturbed normalized statistical operator $e^{\beta(\Omega_0-\mathcal{H})}$). This would be an exact counterpart of our zero-temperature formula (2.55) if not for the $e^{\beta(\Omega-\Omega_0)}$-factor instead of the $\langle S(\infty, -\infty)\rangle_0$- denominator. But after noticing that

$$e^{\beta(\Omega-\Omega_0)} = \left[e^{\beta(\Omega_0-\Omega)} \right]^{-1}$$

$$= \left[e^{\beta\Omega_0} tr \, e^{-\beta\mathcal{H}} \right]^{-1}$$

$$= \left[tr \, e^{\beta(\Omega_0-\mathcal{H}_0)}\sigma(\beta, 0) \right]^{-1}$$

$$= \langle \sigma(\beta, 0)\rangle_0^{-1},$$

we see that actually the formula is a direct analogue to the zero-temperature case (where we have explicitly written spin indices):

$$\mathcal{G}_{\alpha_1,\alpha_2}(\mathbf{x}_1, \mathbf{x}_2; \tau_1 - \tau_2) = -\frac{\langle T_\tau \Psi_{\alpha_1}(\mathbf{x}_1, \tau_1)\bar{\Psi}_{\alpha_2}(\mathbf{x}_2, \tau_2)\sigma(\beta, 0)\rangle_0}{\langle \sigma(\beta, 0)\rangle_0}. \qquad (3.62)$$

This formula provides the basis for the Matsubara diagram techniques. We again expand the S-matrix $\sigma(\beta, 0)$ in series over the interaction $\mathcal{H}_1$ and express the terms as averages over the unperturbed ground state. Wick's and cancellation theorems are still valid in this case, but we will not bother to rewrite the proof for imaginary times. You can easily check that e.g., the "thermodynamic" proof of Wick's theorem holds after the substitution $it \to \tau$. Therefore we can present the terms as Feynman diagrams; all disconnected diagrams cancel, and we are left with the usual connected lot. The rules are given in Table 3.1.

The only difference is that in Fourier representation, instead of integrating over dummy frequencies in the vertices from minus to plus infinity we will sum over the discrete set of Matsubara frequencies. This is generally more troublesome than integration (as all discrete mathematics goes), but there are many useful tricks. I will give here the most basic one, which in many cases does the job.

3.2.3.1 Summation over Frequencies

If a function of a complex variable z satisfies $f(z) \sim |z|^{-(1+\epsilon)}$, when $|z| \to \infty$, ϵ being a positive infinitesimal, then the following identities hold:
Fermi frequencies

Table 3.1 Feynman rules for temperature Green's function (scalar electron–electron interaction)

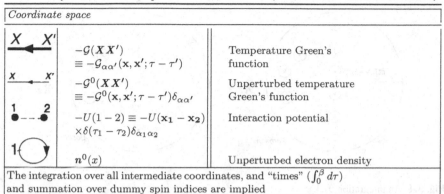

Coordinate space		
	$-\mathcal{G}(XX')$ $\equiv -\mathcal{G}_{\alpha\alpha'}(\mathbf{x},\mathbf{x}';\tau-\tau')$	Temperature Green's function
	$-\mathcal{G}^0(XX')$ $\equiv -\mathcal{G}^0(\mathbf{x},\mathbf{x}';\tau-\tau')\delta_{\alpha\alpha'}$	Unperturbed temperature Green's function
	$-U(1-2)\equiv -U(\mathbf{x_1}-\mathbf{x_2})$ $\times\delta(\tau_1-\tau_2)\delta_{\alpha_1\alpha_2}$	Interaction potential
	$n^0(x)$	Unperturbed electron density

The integration over all intermediate coordinates, and "times" ($\int_0^\beta d\tau$) and summation over dummy spin indices are implied

Momentum space		
	$-\mathcal{G}(\mathbf{p},\omega_v)$	Temperature Green's function
	$-\mathcal{G}^0(\mathbf{p},\omega_v)$	Unperturbed temperature Green's function
	$-U(\mathbf{p})$	Interaction potential
	$n^0(\mu)$	Unperturbed electron density

The integration over all intermediate momenta ($\int d^3\mathbf{p}/(2\pi)^3$) and summation over discrete frequencies $\frac{1}{\beta}\sum_{v=-\infty}^{\infty}$ and over dummy spin indices are implied, taking into account energy (frequency)/momentum conservation in every vertex

$$\frac{1}{\beta}\sum_{v=-\infty}^{\infty} f\left(i\omega_v^F\right) = -\frac{1}{2}\sum_s \tanh\frac{\beta z_s}{2}\operatorname*{Res}_{z=z_s} f(z). \tag{3.63}$$

Bose frequencies

$$\frac{1}{\beta}\sum_{v=-\infty}^{\infty} f\left(i\omega_v^B\right) = -\frac{1}{2}\sum_s \coth\frac{\beta z_s}{2}\operatorname*{Res}_{z=z_s} f(z). \tag{3.64}$$

The origin of these formulae is clear if we recall that the function of complex variable z, $\tanh\frac{\beta z}{2} = \frac{e^{\beta z}-1}{e^{\beta z}+1}$, has poles exactly at the points $z_v = i\pi(2v+1)/\beta = i\omega_v^F$, and its residue at any of these points equals $2/\beta$.

The contour integral $\oint dz f(z)\tanh\frac{\beta z}{2}$ along the infinitely large circle of Fig. 3.4 vanishes (this is ensured by the condition that $f(z)$ vanishes faster than $1/|z|$ at infinity). On the other hand, by Cauchy theorem, the integral is proportional to the sum of the residues of the integrand; the residues of tanh give the left-hand side of (3.63), while the rest gives its right-hand side. The same considerations lead to (3.64) if we use $\coth(\beta z/2)$ instead of $\tanh(\beta z/2)$. Since $\tanh(\beta z/2) = n_B(z)/n_F(z)$, we

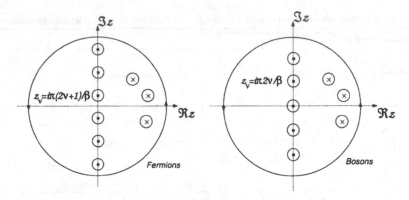

Fig. 3.4 An integration trick

see that equilibrium distribution functions naturally enter the calculations from this
(seemingly) formal side.

3.3 Linear Response Theory

3.3.1 Linear Response Theory: Kubo Formulas

We have developed some approaches that allow us, we hope, to calculate equilibrium real-time Green's functions, both at zero and at finite temperatures. We have
repeatedly referred to $G^R(t)$ as the function that naturally describes the reaction
of the system to external perturbation. This could seem a contradiction, since in
equilibrium—where only we can calculate Green's functions using the methods
developed so far—there can be *no* external perturbation. Nevertheless, we still can
use equilibrium Green's functions in order to find the *linear* reaction of the system
to a *weak* perturbation. This constitutes so-called linear response theory. The main
idea behind it is, well, that of linear response: we can neglect the higher powers of
the perturbation, as long as it is small enough. The two famous examples of this
approach are Hooke's law ($F = kx$) and Ohm's law ($V = IR$): the constants k, R
are taken in equilibrium, at zero strain or current, and we neglect the higher-order
corrections in x or I.

Suppose that the system is affected by a weak external perturbation (which is
generally time dependent, say an external electromagnetic field). Its Hamiltonian is
thus

$$\mathcal{H}(t) = \mathcal{H}_0 + \mathcal{H}_1(t)$$

(here $\mathcal{H}_0$ includes *all* interactions in the system except the external perturbation under
consideration). We are interested in some observable, represented by an operator $\mathcal{A}$

(say, the electric current). We measure its average value, $\langle \mathcal{A} \rangle_t$, as a function of the perturbation strength.

In accordance with our usual approach, we introduce the statistical operator in interaction representation:

$$\tilde{\rho}(t) = e^{i\mathcal{H}_0 t} \hat{\rho}(t) e^{-i\mathcal{H}_0 t}, \tag{3.65}$$

which satisfies the Liouville equation

$$i\dot{\tilde{\rho}}(t) = \left[\tilde{\mathcal{H}}_\infty(t), \tilde{\rho}(t) \right]. \tag{3.66}$$

Here $\tilde{\mathcal{H}}_\infty(t) = e^{i\mathcal{H}_0 t} \mathcal{H}_1(t) e^{-i\mathcal{H}_0 t}$.

We will use the statistical operator (3.66) in order to calculate $\langle \mathcal{A} \rangle_t$. Due to the cyclic invariance of the trace it is given by

$$\langle \mathcal{A} \rangle_t = \mathrm{tr} \left(\hat{\rho}(t) \mathcal{A} \right) = \mathrm{tr} \left(\tilde{\rho}(t) \tilde{\mathcal{A}}(t) \right). \tag{3.67}$$

The Liouville equation for $\tilde{\rho}(t)$ rewritten in integral form gives

$$\tilde{\rho}(t) = -i \int_{-\infty}^{t} dt' \left[\tilde{H}_1(t'), \tilde{\rho}(t') \right] + \tilde{\rho}(-\infty).$$

The idea of *Kubo's approach* to linear response theory is as follows: Assume that at $t = -\infty$ the perturbation was absent (and later adiabatically switched on), so that the system is in equilibrium:

$$\tilde{\rho}(-\infty) = \hat{\rho}_0 \equiv \exp\left[\beta(\Omega - \mathcal{H}_0)\right].$$

Then the *linear response* of the system to the perturbation is given by the following expression:

$$\Delta\tilde{\rho}(t) \equiv \tilde{\rho}(t) - \hat{\rho}_0 = -i \int_{-\infty}^{t} dt' \left[\tilde{H}_1(t'), \hat{\rho}_0 \right] + \mathcal{O}\left((H_1)^2\right). \tag{3.68}$$

In other words, we use the first-order perturbation theory for the nonequilibrium statistical operator. Of course, we could go further and find the second, third, etc. orders in perturbation. The unpleasant feature of such a series is that it is a series in nth-order commutators, $\left[\tilde{H}_1(t'), \left[\tilde{H}_1(t''), \left[\tilde{H}_1(t'''), \ldots \left[\tilde{H}_1(t^{(n)}), \hat{\rho}_0 \right] \ldots \right] \right] \right]$, which cannot be expressed in such convenient way as the higher-order time-ordered products of field operators. But as the basis for linear theory it is all right.

The shift of the average value of an operator $\mathcal{A}$ in the first order in perturbation (*linear response*) is given by

$$\Delta A(t) = \text{tr}\left(\Delta\tilde{\rho}(t)\tilde{\mathcal{A}}(t)\right)$$

$$= -i\int_{-\infty}^{t} dt'\,tr\left(\left[\tilde{H}_1(t'),\hat{\rho}_0\right]\tilde{\mathcal{A}}(t)\right)$$

$$= -i\int_{-\infty}^{t} dt'\left\langle\left[\tilde{\mathcal{A}}(t),\tilde{H}_1(t')\right]\right\rangle \tag{3.69}$$

(here $\langle\cdots\rangle \equiv \text{tr}(\hat{\rho}_0\cdots)$).

It is usually convenient to write the perturbation in the form

$$\mathcal{H}_1(t) = -f(t)\mathcal{B}, \tag{3.70}$$

where the c-number function $f(t)$ is the so-called *generalized force*, and $\mathcal{B}$ is some operator defined for the system under consideration. This is usually described as the perturbation being coupled to $\mathcal{B}$ through $f(t)$. Examples are $-\mathbf{S}\cdot\mathbf{H}(t)$ (the spin coupled to the external magnetic field), or $-\frac{1}{c}\hat{\mathbf{j}}\cdot\mathbf{A}(t)$ (the current coupled to the vector potential). It is not always this easy to tell what we should consider as the generalized force. A useful recipe is as follows. Since the only time dependent term in (3.70) (as well as in the whole Hamiltonian $\mathcal{H}$) is $f(t)$, we can figure out the proper expression for the generalized force if we write down the *energy dissipation* of the external field in the system per unit time:

$$Q = \dot{\mathcal{E}}$$
$$= \langle\dot{\mathcal{H}}\rangle$$
$$= q(t)\langle\mathcal{B}\rangle \tag{3.71}$$
$$\Rightarrow q(t) = -\dot{f}(t). \tag{3.72}$$

Now let us introduce the *retarded Green's function of two operators*:

$$\ll \mathcal{A}(t)\mathcal{B}(t') \gg^R = \frac{1}{i}\langle[\mathcal{A}(t),\mathcal{B}(t')]\rangle\theta\left(t-t'\right). \tag{3.73}$$

This construction may seem a little strange, but it agrees with our earlier retarded Green's function: evidently, $G^R(t,\,t') = \ll \psi(t)\psi^\dagger(t') \gg^R$. After all, we can always rewrite the operators $\mathcal{A}(t)$, $\mathcal{B}(t')$ in terms of the field operators, thus reducing this retarded Green's function to the ones we are accustomed to. This expression, though, has more direct physical meaning: it defines the system's response (already in terms of the observable $\mathcal{A}(t)$ we are interested in) at time t to an external perturbation at earlier moments of time, $t' < t$ (coupled to the operator $\mathcal{B}(t')$). We can as well introduce the advanced Green's function, $\ll A(t)B(t') \gg^A = -\frac{1}{i}\left([A(t), B(t')]\right)\theta(t'-t)$, though it does not have a straight-forward physical sense.

Please notice that we no longer write tildes ($\sim$) over the operators: they all are assumed to be in interaction representation in relation to external perturbation, $A(t) = \exp(i\mathcal{H}_0 t)A(0)\exp(-i\mathcal{H}_0 t)$. But the *only* thing not included in H_0 is an external perturbation term (3.70). Therefore, the averages are to be calculated using the perturbation series *over interaction terms*, and then the operator can be regarded as taken in Heisenberg representation.

Now we can rewrite Eq. (3.69) in the form of the *Kubo formula*:

$$\Delta A(t) = -\int\limits_{-\infty}^{\infty} dt'\, f(t') \ll A(t)B(t') \gg^R . \tag{3.74}$$

This is a very transparent formula: the change in the value of the observable is determined by the (first order of the) external force $f(t')$ applied at all earlier moments of time, and the kernel of this integral operator is exactly the "AB"-Green's function. Generally it is a tensor, since the operators A, B don't have to be scalars.

The equilibrium state of the system is, of course, time independent. Therefore,

$$\ll A(t)B(t') \gg^R = \ll A(0)B(t'-t) \gg^R = \ll A(t-t')B(0) \gg^R, \tag{3.75}$$

and in Fourier components we find

$$\Delta A(\omega) = -f(\omega) \ll AB \gg^R_\omega . \tag{3.76}$$

The *generalized susceptibility* is defined by

$$\chi(\omega) = \frac{\Delta A(\omega)}{f(\omega)}. \tag{3.77}$$

Then

$$\chi(\omega) = -\ll AB \gg^R_\omega .$$

Examples are many: electric conductivity, $j_a(\omega) = \sigma_{ab}(\omega)E_b(\omega)$; magnetic susceptibility, $m_g(\omega) = \chi_{ab}(\omega)H_b(\omega)$; and so on. Here tensor indices $a, b = x, y, z$, and Einstein's summation rule is implied.

There are various ways of writing Kubo formulas. The above one seems quite general. *If* you can write the perturbation as coupled to the same operator, the average value of which you investigate, you will come to a more often met expression of the Kubo formula. For example, if we are calculating the electrical conductivity, the operator A is the current operator, $\hat{j}$. (I don't want to bother with tensor indices here; suppose we have a thin wire, with only one current component allowed.) On the other hand, the external field is coupled to the system through $-\frac{1}{c}\hat{j}A$. Since $E(t) = 1\frac{1}{c}\dot{A}(t)$, in Fourier components we rewrite the perturbation as $-\frac{1}{c}\hat{j}\left(-\frac{ic}{\omega}\right)E(\omega) =$

$\frac{i}{\omega}\hat{j}E(\omega)$. This is the correct coupling, since we want to find the system's response to the external electric field, not the vector potential. The Green's function thus will include equilibrium averages like $\left\langle \hat{j}(t)\hat{j}(t')\right\rangle_0$. I will not dwell on the detailed form of this specific Green's function. A general statement is, though, to be made here. The linear response of the system to the external electric field—the *nonequilibrium* current—turns out to be determined by the *equilibrium* correlators of the current itself. The external field, in a sense, simply reveals these equilibrium fluctuations. In the next subsection we will see that this is indeed the case, and we will give this vague statement an exact form in the *fluctuation-dissipation theorem*.

3.3.2 Fluctuation-Dissipation Theorem

First we introduce some basic apparatus of the mathematical theory of fluctuations.

The *autocorrelation function*, or correlator, of an observable $\mathcal{A}$ is defined as follows:

$$K_A(t, t') = \frac{1}{2}\left\langle\{\mathcal{A}(t), \mathcal{A}(t')\}\right\rangle. \tag{3.78}$$

In this definition we take into account that the operator taken at different moments of time may not commute with itself.

We will limit our consideration to the stationary case. This means that the average values of all observables are time independent, and the autocorrelation function depends only on the difference of its arguments:

$$K_A(t) = \frac{1}{2}\left\langle\{\mathcal{A}(t), \mathcal{A}(0)\}\right\rangle. \tag{3.79}$$

Often it is simpler to deal with the *autocovariation function*,

$$\begin{aligned}
K_{\delta A}(t) &= \frac{1}{2}\langle\{\delta\mathcal{A}(t), \delta\mathcal{A}(0)\}\rangle \\
&\equiv \frac{1}{2}\langle\{\mathcal{A}(t) - \langle\mathcal{A}\rangle, \mathcal{A}(0) - \langle\mathcal{A}\rangle\}\rangle \\
&= K_A(t) - \langle\mathcal{A}\rangle^2.
\end{aligned} \tag{3.80}$$

In this way we explicitly consider the fluctuations around the average value.

The Fourier transform of the correlator is called the *spectral density* of fluctuations:

$$(\mathcal{A}^2)_\omega = \int\limits_{-\infty}^{\infty} dt\, e^{i\omega t}\, K_A(t). \tag{3.81}$$

(It is often denoted by $S(\omega)$, but we will not use this notation here.)

Fig. 3.5 Noise measurement
in a resistor

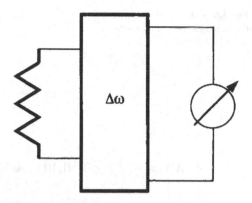

The *Wiener–Khintchin* theorem of the theory of random processes states that the
spectral density (3.81) gives the average power of the fluctuations of A in the fre-
quency interval $[\omega, \omega + \Delta\omega)$ through $(A^2)_\omega \Delta\omega$, and thus can be directly measured.
For example, if we are measuring the voltage fluctuations on a resistor (Fig. 3.5), this
is what the wattmeter shows.

For the equilibrium state average of the product of two operators A, B we have
the *Kubo–Martin–Schwinger identity*:

$$\langle A(t)B(0)\rangle = tr\left(e^{\beta(\Omega - \mathcal{H})}e^{i\mathcal{H}t}A(0)e^{-i\mathcal{H}t}B(0)\right) = \langle B(0)A(t + i\beta)\rangle. \quad (3.82)$$

It is an evident result of the cyclic invariance of trace and a special form of the
equilibrium statistical operator. This allows us to rewrite the spectral density as

$$\left(A^2\right)_\omega = \frac{1}{2}\left(1 + e^{\beta\omega}\right) \int\limits_{-\infty}^{\infty} dt\, e^{i\omega t}\langle A(0)A(t)\rangle. \quad (3.83)$$

In this expression enters the Fourier transform of the anticommutator, $\{A(t), A(0)\}$.
It is easy to find a like formula for the commutator:

$$([A, A])_\omega \equiv \int\limits_{-\infty}^{\infty} dt\, e^{i\omega t}\langle[A(t), A(0)]\rangle = \left(e^{\beta\omega} - 1\right)\int\limits_{-\infty}^{\infty} dt\, e^{i\omega t}\langle A(0)A(t)\rangle.$$

$$(3.84)$$

Then

$$(A^2)_\omega = \frac{1}{2}\coth\frac{\beta\omega}{2}([A, A])_\omega. \quad (3.85)$$

On the other hand, we can rewrite Eq. (3.84) in the form

Fig. 3.6 Fluctuation-dissipation
theorem in a curcuit: Johnson–
Nyquist noise

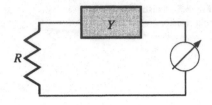

$$([A,\ A])_\omega = - \int\limits_0^\infty dt e^{-i\omega t}\{[\mathcal{A}(t),\ \mathcal{A}(0)]\} + \int\limits_0^\infty dt e^{i\omega t}\,\langle[\mathcal{A}(t),\ \mathcal{A}(0)]\rangle,$$

to find that

$$([A,\ A])_\omega = -2\Im \ll \mathcal{A}\mathcal{A} \gg_\omega^R.\tag{3.86}$$

We have proved the *fluctuation-dissipation theorem* (alias Callen–Welton formula): the spectral density of fluctuations of an observable $\mathcal{A}$ in equilibrium is proportional to the imaginary part of the generalized susceptibility of this system to a weak external perturbation coupled to this very observable:

$$\left(\mathcal{A}^2\right)_\omega = \coth\frac{\beta\omega}{2}\Im\chi(\omega) \equiv -\coth\frac{\beta\omega}{2}\Im \ll \mathcal{A}\mathcal{A} \gg_\omega^R.\tag{3.87}$$

It is the imaginary part of susceptibility that determines the energy dissipation rate in the system, hence the name of the theorem.

3.3.2.1 Current and Voltage Fluctuations in Linear Circuits: Nyquist's Theorem

Let us consider one of the most important applications of the theorem. Take the current operator, $\mathcal{J}$, in a simple electric circuit (Fig. 3.6). At frequencies low enough ($\omega \ll c/L$, where L is the size of the circuit) the current is the same throughout the circuit and depends only on time:

$$\langle\mathcal{J}(t)\rangle = J(t).$$

If there were an external emf in the circuit, $W(t)$, the energy dissipation per unit time would be

$$Q = JW = \langle\mathcal{J}\rangle W.$$

According to our previous considerations, the generalized force is then given by

$$\dot{f} = -W;\quad i\omega f(\omega) = W(\omega).$$

Then we see that

$$J(\omega) = \chi(\omega) f(\omega),$$

and

$$J(\omega) = W(\omega)/Z(\omega) = i\omega f(\omega)/Z(\omega),$$

where $Z(\omega)$ is the circuit impedance. Therefore,

$$\chi(\omega) = \frac{i\omega}{Z(\omega)}$$

$$= \frac{i\omega}{R(\omega) + iY(\omega)}; \tag{3.88}$$

$$\Im_\chi(\omega) = \frac{\omega}{|Z(\omega)|^2} R(\omega). \tag{3.89}$$

Using the fluctuation-dissipation theorem, we immediately find for the current fluctuations

$$\left(J^2\right)_\omega = \frac{\omega}{|Z(\omega)|^2} R(\omega) \coth \frac{\beta\omega}{2} = \Re\left(Z(\omega)^{-1}\right) \omega \coth \frac{\beta\omega}{2}. \tag{3.90}$$

(As it should be, the imaginary part of the susceptibility corresponds to the real part of the impedance: the reactance).

The corresponding voltage fluctuations, $W(\omega) = Z(\omega)J(\omega)$, are given by

$$\left(W^2\right)_\omega = \Re Z(\omega)\omega \coth \frac{\beta\omega}{2}. \tag{3.91}$$

In the classical limit ($T \gg \omega$) we obtain the famous *Nyquist theorem*:

$$\left(J^2\right)_\omega = 2TG(\omega); \tag{3.92}$$

$$\left(W^2\right)_\omega = 2TR(\omega). \tag{3.93}$$

($G = 1/R$ is the circuit conductance.) It relates the equilibrium thermal noise in an electrical circuit to its resistance. This beautiful relation was discovered experimentally by Johnson and theoretically by Nyquist, hence the often-used term "Johnson–Nyquist noise".

3.4 Nonequilibrium Green's Functions

3.4.1 Nonequilibrium Causal Green's Function: Definition

In the general definition of the causal many-particle Green's function (2.4),

$$G_{\alpha\beta}(\mathbf{x}_1, t_1; \mathbf{x}_2, t_2) = -i \left\langle T \psi_\alpha(\mathbf{x}_1, t_1) \psi_\beta^\dagger(\mathbf{x}_2, t_2) \right\rangle$$
$$\equiv -i \operatorname{tr}\left(\hat{\varrho} T \psi_\alpha(\mathbf{x}_1, t_1) \psi_\beta^\dagger(\mathbf{x}_2, t_2) \right), \qquad (3.94)$$

we imposed no specific limitations upon the quantum state of the system, i.e., on its statistical operator $\hat{\varrho}$. If later we had to deal with the system in its ground state (at $T = 0$) or in equilibrium (at $T \neq 0$), it was because our elegant diagram technique based on perturbation theory essentially used specific properties: the uniqueness of the ground state (up to the phase factor) in the first case, and formal equivalence of the equilibrium statistical operator and the (analytically continued) evolution operator in the second.

Extremely powerful with regard to the thermodynamic properties, these approaches are evidently unable to cope with the kinetic problems, which are very important in condensed matter theory. For example, they cannot describe the response of the system to time-dependent external perturbation, even at zero temperature (since such a perturbation will lead to energy pumping into the system, which can be transferred to some excited state). Linear response theory can answer this sort of question, but only in the first order, while there is no convenient way to write the higher-order terms in this method.

Nevertheless there exists a possibility to develop a diagram technique for the nonequilibrium Green's function, if we take into account several types of Green's functions simultaneously [7]. Then the Green's function becomes a *matrix*. This is the price we pay for universality, and the reason why this technique introduced by Keldysh did not replace the other two in their respective fields.

We define the causal nonequilibrium Green's function as follows:

$$G_{\alpha\beta}^{--}(\mathbf{x}_1, t_1; \mathbf{x}_2, t_2) = -i \left\langle \Phi | T \psi_\alpha(\mathbf{x}_1, t_1) \psi_\beta^\dagger(\mathbf{x}_2, t_2) | \Phi \right\rangle. \qquad (3.95)$$

Here $|\Phi\rangle$ denotes an *arbitrary quantum state* (in Heisenberg representation) of the system under consideration.[2]

---

[2] Statistical averaging can be included at any stage of the calculations without any problem, since

$$\operatorname{tr}(\hat{\varrho}\mathcal{A}) \equiv \sum_\Phi W_\Phi \langle \Phi | \mathcal{A} | \Phi \rangle$$

is a linear operation.

This is essentially the same expression as the one we introduced in Eq. (2.10), except that now $|\Phi\rangle$ generally is not the ground state. What are the consequences of this difference?

Let us try to repeat the steps that have led us to the generating formula (2.55) of the perturbation theory for Green's function. We express the Heisenberg field operators, ψ, through the interaction representation operators, Ψ:

$$\psi(\mathbf{x}, t) = \mathcal{U}^\dagger(t) e^{-i'\mathcal{H}_0 t} \Psi(\mathbf{x}, t) e^{i\mathcal{H}_0 t} \mathcal{U}(t).$$

Then note that the Heisenberg state $|\Phi\rangle$ is related to the corresponding state vector in the interaction representation, $|\Phi(t)\rangle_I$, by

$$e^{i\mathcal{H}_0 t} \mathcal{U}(t)|\Phi\rangle = |\Phi(t)\rangle_I \equiv S(t, -\infty)|\Phi_0\rangle. \tag{3.96}$$

Here the S-matrix (in the interaction representation given by Dyson's expansion, see Chap. 2) relates the actual state $|\Phi(t)\rangle_I$ to the presumably unperturbed one $|\Phi(-\infty)\rangle_I = |\Phi_0\rangle$, i.e., to the state of the system of free Particles.

This allows us to rewrite Green's function as[3]

$$
\begin{aligned}
G^{--}&(\mathbf{x}_1, t_1; \mathbf{x}_2, t_2) \\
&= -i\langle \Phi(\infty)|T S(\infty, -\infty)\Psi(\mathbf{x}_1, t_1)\Psi^\dagger(\mathbf{x}_2, t_2)|\Phi_0\rangle.
\end{aligned} \tag{3.97}
$$

The fundamental difference between this expression and the corresponding result for the ground state average (2.55) is that the state at $t = \infty$, $|\Phi(\infty)\rangle$, is by no means simply related to the unperturbed state at $t = -\infty$, $|\Phi_0\rangle$. The only way to bring it back is to use a straightforward formula,

$$|\Phi(\infty)\rangle = S(\infty, -\infty)|\Phi_0\rangle. \tag{3.98}$$

Then we obtain the basic formula for the nonequilibrium Green's function,

$$
\begin{aligned}
G^{--}&(\mathbf{x}_1, t_1; \mathbf{x}_2, t_2) \\
&= \langle \Phi_0|S^\dagger(\infty, -\infty)T S(\infty, -\infty)\Psi(\mathbf{x}_1, t_1)\Psi^\dagger(\mathbf{x}_2, t_2)|\Phi_0\rangle.
\end{aligned} \tag{3.99}
$$

Now we substitute in this expression the Dyson expansion for the S-matrix (1.90, 1.91),

---

[3] We omit the spin indices for clarity.

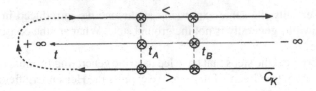

Fig. 3.7 Time ordering along the Keldysh contour

$$S(\infty, -\infty) = \mathcal{T} \exp \left\{ -\frac{i}{\hbar} \int_{-\infty}^{\infty} d\tau \mathcal{W}(\tau) \right\} ; \qquad (3.100)$$

$$S^{\dagger}(\infty, -\infty) = \tilde{\mathcal{T}} \exp \left\{ \frac{i}{\hbar} \int_{-\infty}^{\infty} d\tau \mathcal{W}(\tau) \right\} . \qquad (3.101)$$

The *anti* time ordering operator $\tilde{\mathcal{T}}$ arranges the operators in the opposite order to that of the $\mathcal{T}$-operator.

It can be shown that the main features of theory are kept intact; namely, (1) Wick's theorem is still valid, so that we can express anything in terms of pairings in the unperturbed state, and (2) the vacuum (disconnected) diagrams are canceled.

3.4.2 Contour Ordering and Three More Nonequilibrium Green's Functions

We see from (3.99) that the main difference between the present case and the zero temperature technique is the presence of two time orderings in the same formula:

$$G^{--}(\mathbf{x}_1, t_1; \mathbf{x}_2, t_2)$$
$$= (\Phi_0 | \tilde{\mathcal{T}} e^{\frac{i}{\hbar} \int_{-\infty}^{\infty} d\tau \mathcal{W}(\tau)} \mathcal{T} e^{-\frac{i}{\hbar} \int_{-\infty}^{\infty} d\tau \mathcal{W}(\tau)} \Psi(\mathbf{x}_1, t_1) \Psi^{\dagger}(\mathbf{x}_2, t_2) | \Phi_0). \quad (3.102)$$

It appears due to the necessity to return back in time, to the initial unperturbed state, before the interaction was turned on, since we don't know what the state will be like *after* it is finally turned off. Formally this can be presented as a single time ordering along the contour running from $-\infty$ to ∞ and back again (Fig. 3.7). (The time ordering along a contour that returns to $-\infty$ instead of running to ∞ was first suggested by Schwinger.) The operators standing to the right of the $\mathcal{T}$-operator in (3.102) belong to the right-going $(-)$, the other to the left-going $(+)$ branch of the contour. The $+$ operators always stand to the left of the—ones.

Evidently, if we use Wick's theorem, we obtain *four* types of pairings, namely ($\pm$ denotes the branch)

$$< \mathcal{T}\Psi_-\Psi_-^{\dagger} >, < \tilde{\mathcal{T}}\Psi_+\Psi_+^{\dagger} >, < \Psi_+\Psi_-^{\dagger} >, < \Psi_+^{\dagger}\Psi_- > . \qquad (3.103)$$

The first of these gives the causal Green's function; the rest we have not met before. The diagram technique in nonequilibrium thus includes *four* Green's functions, which we will define as follows[4]:

$$G^{--}(1,2) = -i \left\langle T\psi(1)\psi^\dagger(2) \right\rangle; \qquad (3.104)$$

$$G^{+-}(1,2) = -i \cdot \left\langle \psi(1)\psi^\dagger(2) \right\rangle; \qquad (3.105)$$

$$G^{-+}(1,2) = \pm i \left\langle \psi^\dagger(2)\psi(1) \right\rangle; \qquad (3.106)$$

$$G^{++}(1,2) = -i \left\langle \tilde{T}\psi(1)\psi^\dagger(2) \right\rangle. \qquad (3.107)$$

The $(-+)$ function is directly proportional to the density of real particles in the system.

3.4.2.1 Relations Between Different Nonequilibrium Green's Functions

The Green's functions defined above are not independent, since as follows from their definition,

$$G^{--}(1,2) + G^{++}(1,2) = G^{-+}(1,2) + G^{+-}(1,2). \qquad (3.108)$$

Then, we have

$$G^{--}(1,2) = - \left(G^{++}(2,1) \right)^*;$$

$$G^{-+}(1,2) = - \left(G^{-+}(2,1) \right)^*; \qquad (3.109)$$

$$G^{+-}(1,2) = - \left(G^{+-}(2,1) \right)^*. \qquad (3.110)$$

If we define the retarded and advanced Green's functions as before, they can be expressed as follows:

$$G^R(1,2) = G^{--}(1,2) - G^{-+}(1,2) = G^{+-}(1,2) - G^{++}(1,2); \quad (3.111)$$

$$G^A(1,2) = G^{--}(1,2) - G^{+-}(1,2) = G^{-+}(1,2) - G^{++}(1,2). \quad (3.112)$$

3.4.2.2 Nonequilibrium Green's Function for the System of Noninteracting Particles

The unperturbed Green's functions satisfy the following linear differential equations:

---

[4] The upper sign for the Fermi system.

$$\left(i\hbar\frac{\partial}{\partial t_1} - \mathcal{E}(\mathbf{x}_1)\right) G_0^{--}(1,2) \equiv (\mathcal{G}_0)^{-1}(1) G_0^{--}(1,2)$$

$$= \hbar\delta(1-2); \tag{3.113}$$

$$(\mathcal{G}_0)^{-1}(1) G_0^{-+}(1,2) = 0; \tag{3.114}$$

$$(\mathcal{G}_0)^{-1}(1) G_0^{+-}(1,2) = 0; \tag{3.115}$$

$$(\mathcal{G}_0)^{-1}(1) G_0^{++}(1;2) = -\hbar\delta(1-2). \tag{3.116}$$

Suppose that the system of noninteracting particles (ideal quantum gas) is in a stationary uniform state. Then it can be characterized by the (nonequilibrium) distribution function in the momentum space, $n_\mathbf{p}$. Then we easily get useful expressions for all Green's functions:

$$G_0^{--}(\mathbf{p},\omega) = \frac{1}{\omega - (\epsilon_\mathbf{p} - \mu) + i0} \pm 2\pi i n_\mathbf{p}\delta\left(\omega - (\epsilon_\mathbf{p} - \mu)\right); \tag{3.117}$$

$$G_0^{-+}(\mathbf{p},\omega) = \pm 2\pi i n_\mathbf{p}\delta\left(\omega - (\epsilon_\mathbf{p} - \mu)\right); \tag{3.118}$$

$$G_0^{+-}(\mathbf{p},\omega) = -2\pi i(1 \mp n_\mathbf{p})\delta\left(\omega - (\epsilon_\mathbf{p} - \mu)\right); \tag{3.119}$$

$$G_0^{++}(\mathbf{p},\omega) = -\frac{1}{\omega - (\epsilon_\mathbf{p} - \mu) - i0} \pm 2\pi i n_\mathbf{p}\delta\left(\omega - (\epsilon_\mathbf{p} - \mu)\right); \tag{3.120}$$

$$G_0^{R}(\mathbf{p},\omega) = \frac{1}{\omega - (\epsilon_\mathbf{p} - \mu) + i0}; \tag{3.121}$$

$$G_0^{A}(\mathbf{p},\omega) = \frac{1}{\omega - (\epsilon_\mathbf{p} - \mu) - i0}; \tag{3.122}$$

$$G_0^{K}(\mathbf{p},\omega) = -2\pi i(1 \mp 2n_\mathbf{p})\delta\left(\omega - (\epsilon_\mathbf{p} - \mu)\right). \tag{3.123}$$

Here we have introduced one more linear combination of $G^{\pm\pm}$, the so-called Keldysh Green's function:

$$G^{K}(1,2) = G^{-+}(1,2) + G^{+-}(1,2) = G^{--}(1,2) + G^{++}(1,2). \tag{3.124}$$

Note that the retarded and advanced Green's functions don't contain any information on the state of the system, which is given solely by the Keldysh Green's function. Since they are given by linear combinations of $G^{\pm\pm}$, we can use them as three *independent* functions, instead of four dependent $G^{\pm\pm}$. As we will see later, in many cases the set (G^R, G^A, G^K) indeed is simpler to use than the initial set $(G^{--}, G^{-+}, G^{+-}, G^{++})$.

3.4.3 The Keldysh Formalism

The rules of the Keldysh diagram technique directly follow from the expansion of S-matrices in Eq. (3.99).[5] First consider the rules for the scalar interaction with an external field $W(x, t)$.

The only difference from the zero-temperature case will be that each electron or interaction line bears at its ends $\pm$-indices, which show to which branch of the Keldysh contour the corresponding operator belongs; besides, the "$+$"-vertices are multiplied by $+i$ instead of $-i$, since they originate from the $S^\dagger$-operator. This can be taken into account in an elegant way, if we introduce the *matrix Green's function* in the Keldysh space:

$$\hat{G}(1, 2) = \begin{pmatrix} G^{--}(1, 2) & G^{-+}(1, 2) \\ G^{+-}(1, 2) & G^{++}(1, 2) \end{pmatrix} \tag{3.125}$$

and the matrix of the external potential

$$-i\hat{W}(1) = \begin{pmatrix} -iW(1) & 0 \\ 0 & iW(1) \end{pmatrix} = -iW(1)\hat{\tau}_3, \tag{3.126}$$

where $\hat{\tau}_3$ is one of the Pauli matrices.

This allows us to gather all the diagrams that differ only by the arrangement of the $\pm$-indices into a single one, which is understood as a single matrix expression, for example that shown in Fig. 3.8 (the integration over intermediate coordinates and times is implied):

$$i\hat{G}_1(1, 2) = i\hat{G}_0 \cdot i\hat{W} \cdot i\hat{G}_0(1, 2); \tag{3.127}$$

$$i\begin{pmatrix} G_1^{--}(1, 2) & G_1^{-+}(1, 2) \\ G_1^{+-}(1, 2) & G_1^{++}(1, 2) \end{pmatrix}$$

$$= \begin{pmatrix} iG_0^{--}(-iW)iG_0^{--}(1, 2) \ iG_0^{--}(-iW)iG_0^{-+}(1, 2) \\ +iG_0^{-+}(+iW)iG_0^{+-}(1, 2) \ +iG_0^{-+}(+iW)iG_0^{++}(1, 2) \\ \\ iG_0^{+-}(-iW)iG_0^{--}(1, 2) \ iG_0^{+-}(-iW)iG_0^{-+}(1, 2) \\ +iG_0^{++}(+iW)iG_0^{+-}(1, 2) \ +iG_0^{++}(+iW)iG_0^{++}(1, 2) \end{pmatrix}. \tag{3.128}$$

The Feynman rules for some cases of interest are given in Table 3.2. Note that four differential equations for the unperturbed Green's functions (3.113)–(3.116) are now gathered in one elegant expression:

$$(\mathcal{G}_0)^{-1}(1)\hat{G}_0(1, 2) = \hat{\tau}_3\delta(1 - 2). \tag{3.129}$$

---

[5] We follow the notation of [10].

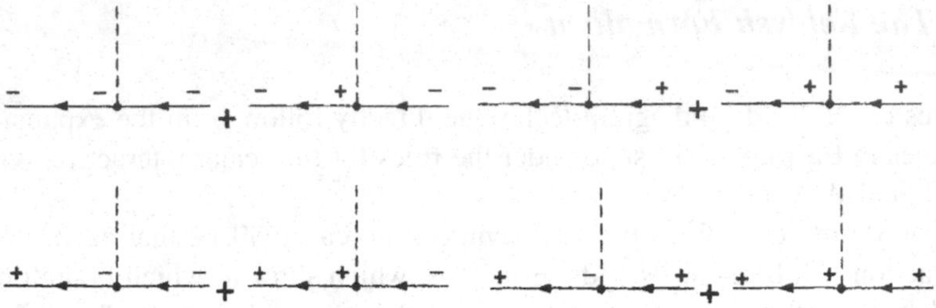

Fig. 3.8 First-order diagrams for the nonequilibrium Green's function

Table 3.2 Feynman rules for the matrix $\hat{G}$ (after [10])

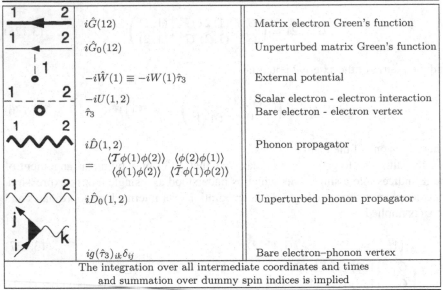

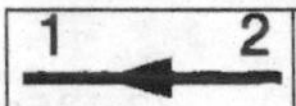	$i\hat{G}(12)$	Matrix electron Green's function
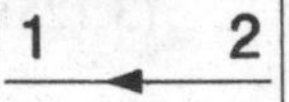	$i\hat{G}_0(12)$	Unperturbed matrix Green's function
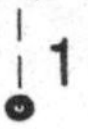	$-i\hat{W}(1) \equiv -iW(1)\hat{\tau}_3$	External potential
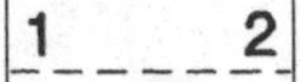	$-iU(1,2)$	Scalar electron - electron interaction
	$\hat{\tau}_3$	Bare electron - electron vertex
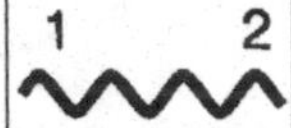	$i\hat{D}(1,2) = \begin{pmatrix} \langle T\phi(1)\phi(2)\rangle & \langle\phi(2)\phi(1)\rangle \\ \langle\phi(1)\phi(2)\rangle & \langle\tilde{T}\phi(1)\phi(2)\rangle \end{pmatrix}$	Phonon propagator
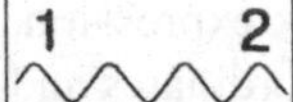	$i\hat{D}_0(1,2)$	Unperturbed phonon propagator
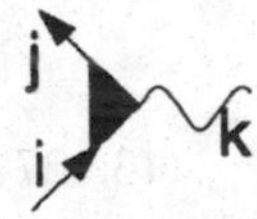	$ig(\hat{\tau}_3)_{ik}\delta_{ij}$	Bare electron–phonon vertex
The integration over all intermediate coordinates and times and summation over dummy spin indices is implied		

We can make use of the relations between the $G^{\pm\pm}$ functions to obtain another representation of the same formalism. If we perform the transformation

$$\hat{G} \to \bar{G} = \frac{1}{2}(\hat{\tau}_0 - i\hat{\tau}_2)\hat{\tau}_3\hat{G}(\hat{\tau}_0 - i\hat{\tau}_2)^{\dagger}, \tag{3.130}$$

we come to Green's function in the following form (check this!):

$$\bar{G} = \begin{pmatrix} G^R & G^K \\ 0 & G^A \end{pmatrix}. \tag{3.131}$$

The equation of motion for this matrix is

Table 3.3 Feynman rules for the matrix $\bar{G}$ (after [10])

1 ◄─── 2	$i\bar{G}(12)$	Matrix electron Green's function
1 ○─│1	$-i\bar{W}(1) \equiv -iW(1)\hat{\tau}_0$	External potential
1 ─ ─ ◄ ─ ─ 2	$-iU(1,2)$	Scalar electron - electron interaction
j ►─ i ─ k	γ_{ij}^k	Bare electron - electron vertex (absorption)
j ►─ i ─► k	$\tilde{\gamma}_{ij}^k$	Bare electron - electron vertex (emission)
1 ∿∿∿ 2	$i\bar{D}(1,2)$	Phonon progator
i ►∿ k	$-ig\gamma_{ij}^k$	Bare phonon vertex (absorption)
i ►∿► k	$-ig\tilde{\gamma}_{ij}^k$	Bare phonon vertex (emission)

$$\gamma_{ij}^1 = \tilde{\gamma}_{ij}^2/\sqrt{2}$$
$$\gamma_{ij}^2 = \tilde{\gamma}_{ij}^1 = (\tau_1)_{ij}/\sqrt{2}$$

The integration over all intermediate coordinates and times
and summation over dummy spin indices is implied

$$(\mathcal{G}_0)^{-1}(1)\bar{G}_0(1, 2) = \delta(1 - 2). \tag{3.132}$$

The Feynman rules for this representation can be obtained from the initial ones if we use the transformation inverse to (3.130). They are given in Table 3.3.

3.5 Quantum Kinetic Equation

The Keldysh Green's functions often contain more information than we need. Indeed, as we have seen, of the three independent components of the matrix $\bar{G}$, G^R and G^A contain only the information about the dispersion relation in the system; all information about the occupation of the states is contained in the component G^K (*plus* again the information about the dispersion relation). In many cases we are more interested in the kinetics, i.e., in how the states are occupied, than in what exactly these states are (after all, we can use some approximate relation $\epsilon(\mathbf{p})$ and forget about them). Since the matrix theory is rather awkward, this puts a premium on some sort of reduced description, which would let us get rid of nonessential information. In this way we will derive a *quantum kinetic equation* for the quantum analogue of the distribution function of classical statistical mechanics.

Defining the statistical distribution function in a quantum limit is a nontrivial problem. The uncertainty relations prohibit the use of the classical distribution function itself, $f(\mathbf{r}, \mathbf{p}, t)$. Instead, we can introduce the *Wigner function*,

$$f^W(\mathbf{r}, \mathbf{p}, t) \equiv \int d^3\xi e^{-i\mathbf{p}\cdot\xi/\hbar} \left\langle \psi^\dagger\left(\mathbf{r} - \frac{\xi}{2}, t\right) \psi\left(\mathbf{r} + \frac{\xi}{2}, t\right) \right\rangle \tag{3.133}$$

The Wigner function has many useful properties, but being positively determined is not one of them: $f^W(\mathbf{r}, \mathbf{p})$ can take negative values in some regions of the phase space. This is the price we have to pay for our wish to have some definite relation between momentum and coordinate in quantum mechanics! Nevertheless, if we average $f^W(\mathbf{r}, \mathbf{p})$ over the scale of h^d (d is the dimensionality of the system), this "roughened" distribution function coincides with the classical distribution function $f(\mathbf{r}, \mathbf{p})$ up to the terms of higher order in h:

$$\int_h \frac{d^d\mathbf{p}d^d\mathbf{r}}{h^d} f^W(\mathbf{r}, \mathbf{p}, t) = f(\mathbf{r}, \mathbf{p}, t) + o(h^d). \tag{3.134}$$

According to the principle of correspondence, this means that f^W is indeed the proper quantum analogue to the classical distribution function. A detailed discussion of the Wigner distribution function and related formalism can be found in [2].

Note that $f^W(\mathbf{r}, \mathbf{p}, t)$ is obtained from the $(-+)$-component of the nonequilibrium Green's function with coinciding temporal arguments by a specific Fourier transformation. Therefore, we can derive the equation for $f^W(\mathbf{r}, \mathbf{p}, t)$ (*quantum kinetic equation*) starting from the corresponding component of the matrix Dyson's equation (for G^{-+} or G^K). This can be done using *gradient expansion*. First, we present any Green's function $G(\mathbf{r}_1, \mathbf{r}_2) = \langle \psi(\mathbf{r}_1)\psi^\dagger(\mathbf{r}_2) \rangle$ as $G(\mathbf{R}, \mathbf{r})$, where $\mathbf{R} = \frac{\mathbf{r}_1 + \mathbf{r}_2}{2}$; $\mathbf{r} = \mathbf{r}_1 - \mathbf{r}_2$, and assume that the properties of the system "on a big scale," $\mathbf{R}$, change slowly compared to the characteristic quantum length, λ_B (e.g., Fermi wavelength, in the case of the Fermi system). The latter dominates the dependence on the small scale, given by $\mathbf{r}$. You see that Wigner functions are very well suited for such an approach, since they by definition depend on $\mathbf{R}$ ("slow" variable) and $\mathbf{p}$, the latter being the conjugate of $\mathbf{r}$ ("fast" variable). What we are going to do is to Taylor expand everything in gradients $\nabla_\mathbf{R}, \nabla_\mathbf{r}$ and derive a simpler equation. We will see how it works in a moment.

3.5.1 Dyson's Equations for Nonequilibrium Green's Functions

We start from the exact matrix Dyson's equation,

$$\hat{G} = \hat{G}_0 + \hat{G}_0\hat{\Sigma}\hat{G}; \tag{3.135}$$

$$(\bar{G} = \bar{G}_0 + \bar{G}_0 + \bar{\Sigma}\bar{G}); \tag{3.136}$$

or (the conjugated equation)

$$\hat{G} = \hat{G}_0 + \hat{G}\hat{\Sigma}\hat{G}_0; \tag{3.137}$$

$$(\bar{G} = \bar{G}_0 + \bar{G}\bar{\Sigma}\bar{G}_0). \tag{3.138}$$

Here we introduce the self energy matrix,

$$\hat{\Sigma} = \begin{pmatrix} \Sigma^{--} & \Sigma^{-+} \\ \Sigma^{+-} & \Sigma^{++} \end{pmatrix}; \tag{3.139}$$

$$\left(\bar{\Sigma} = \begin{pmatrix} \Sigma^R & \Sigma^K \\ 0 & \Sigma^A \end{pmatrix}\right). \tag{3.140}$$

Only three components of the former matrix are independent, since as is easy to see,

$$\Sigma^{--} + \Sigma^{++} = -\left(\Sigma^{-+} + \Sigma^{+-}\right). \tag{3.141}$$

The independent combinations are given by

$$\begin{aligned} \Sigma^K &= \Sigma^{--} + \Sigma^{++}; \\ \Sigma^R &= \Sigma^{--} + \Sigma^{-+}; \\ \Sigma^A &= \Sigma^{--} + \Sigma^{+-}. \end{aligned} \tag{3.142}$$

Acting on Dyson's equation from the left (or from the right on its conjugate) by the operator $\mathcal{G}_0^{-1}$, we find the *Keldysh equations*, equivalent to the set of integro-differential equations for the component Green's functions:

$$\left(\mathcal{G}_0^{-1} - \hat{\tau}_3\hat{\Sigma}\right) \cdot \hat{G}(1, 2) = \hat{\tau}_3\delta(1 - 2); \tag{3.143}$$

$$\hat{G}(1, 2) \cdot \left(\mathcal{G}_0^{-1} - \hat{\Sigma}\hat{\tau}_3\right) = \hat{\tau}_3\delta(1 - 2), \tag{3.144}$$

or in the "barred" representation,

$$(\mathcal{G}_0 - \bar{\Sigma}) \cdot \bar{G}(1, 2) = \delta(1 - 2); \tag{3.145}$$

$$\bar{G}(1, 2) \cdot \left(\mathcal{G}_0^{-1} - \bar{\Sigma}\right) = \delta(1 - 2). \tag{3.146}$$

3.5.2 The Quantum Kinetic Equation

Now we can obtain the quantum kinetic equation. We will use the "hat" representation as more straightforward.

Set $T = \frac{t_1+t_2}{2}, \tau = t_1 - t_2;$ $\mathbf{R} = \frac{\mathbf{x}_1+\mathbf{x}_2}{2}, \xi = \mathbf{x}_1 - \mathbf{x}_2$. Then Wigner's function can be written as follows:

$$f^W(\mathbf{R}, \mathbf{p}, T) = -i \int\limits_{-\infty}^{\infty} \frac{d\omega}{2\pi} G^{-+}(\mathbf{R}, T; \mathbf{p}, \omega); \tag{3.147}$$

$$G^{-+}(\mathbf{R}, T; \mathbf{p}, \omega)$$
$$= \int d^3\xi d\tau e^{i\omega\tau - i\mathbf{p}\cdot\xi} G^{-+}\left(\mathbf{R} + \frac{\xi}{2}, T + \frac{\tau}{2}; \mathbf{R} - \frac{\xi}{2}, T - \frac{\tau}{2}\right). \tag{3.148}$$

Taking the $(-+)$-component of Eqs. (3.143), (3.144), we find:

$$\left(\mathcal{G}_0^{-1}\right)(1)G^{-+}(1, 2) = \int d^4 3 \left(\Sigma^{--}(1, 3)G^{-+}(3, 2) + \Sigma^{-+}(1, 3)G^{++}(3, 2)\right) \tag{3.149}$$

$$\left(\mathcal{G}_0^{-1}\right)^*(2)G^{-+}(1, 2) = -\int d^4 3 \left(G^{--}(1, 3)\Sigma^{-+}(3, 2) + G^{-+}(1, 3)\Sigma^{++}(3, 2)\right).$$

After subtracting the first line from the second and noticing that

$$\left(\mathcal{G}_0^{-1}\right)^*(2) - \left(\mathcal{G}_0^{-1}\right)(1) = -i \left(\frac{\partial}{\partial T} - \frac{i}{m}\nabla_{\mathbf{R}} \cdot \nabla_\xi\right), \tag{3.150}$$

we can integrate the equation over $\frac{d\omega}{2\pi}$ and see that the resulting equation takes the standard form[6]:

$$\left(\frac{\partial}{\partial T} + \frac{\mathbf{p}}{m} \cdot \nabla_{\mathbf{R}}\right) f^W(\mathbf{R}, \mathbf{p}, T) = I(\mathbf{R}, \mathbf{p}, T). \tag{3.151}$$

This is the quantum kinetic equation. Its right-hand side in the quasiclassical limit must yield the (quasli)classical collision integral, St $\left[f^W(\mathbf{R}, \mathbf{p}, T)\right]$. Generally, there appears one more term, which is responsible for the *renormalization of the energy spectrum* of the quasiparticles and can be merged with the dynamic left-hand side of (3.151). This is consistent with the fact that the imaginary part of the self energy (entering the right-hand side of (3.151)) determines the lifetimes of the quasiparticles

---

[6] In the most general case, we could introduce the distribution function of all four conjugated variables, $f(\mathbf{R}, \mathbf{p}, T, \epsilon)$, which in the presence of the external potential obeys the equation (Footnote 6 continued)

$$\left(\frac{\partial}{\partial T} + \frac{\mathbf{p}}{m} \cdot \nabla_{\mathbf{R}} - \nabla_{\mathbf{R}}U(\mathbf{R}, T) \cdot \nabla_{\mathbf{p}} + \frac{\partial U(\mathbf{R}, T)}{\partial T} \frac{\partial}{\partial \epsilon}\right) f(\mathbf{R}, \mathbf{p}, T, \epsilon)$$
$$= I[f(\mathbf{R}, \mathbf{p}, T, \epsilon)].$$

(in our case, through the collision integral governing the in- and outscattering rates), while its real part changes their dispersion law (thus modifying the dynamical part of the kinetic equation).

But in the quasiclassical limit the corrections to the dispersion law are negligible, and only the collision integral survives. To show this, we take into account that then we can write

$$
\int d^4 3\Sigma(1,3)G(3,2)
$$

$$
= \int d^4 3\Sigma \left(\frac{X_1+X_3}{2} + \frac{X_1-X_3}{2}, \frac{X_1+X_3}{2} - \frac{X_1-X_3}{2} \right)
$$

$$
\times G \left(\frac{X_3+X_2}{2} + \frac{X_3-X_2}{2}, \frac{X_3+X_2}{2} - \frac{X_3-X_2}{2} \right)
$$

$$
\approx \int d^4 3\Sigma \left(X + (X_1 - X_3)/2, X - (X_1 - X_3)/2 \right)
$$

$$
\times G \left(X + (X_3 - X_2)/2, X - (X_3 - X_2)/2 \right).
$$

Using the identities for different components of $\hat{G}$ and $\hat{\Sigma}$, we find that the collision integral takes the following form:

$$
\mathrm{St}\left[f^w(\mathbf{R}, \mathbf{p}, T) \right] = \int_{-\infty}^{\infty} \frac{d\omega}{2\pi} \left(-\Sigma^{-+}(\mathbf{R}, T; \mathbf{p}, \omega) G^{+-}(\mathbf{R}, T; \mathbf{p}, \omega) \right. \tag{3.152}
$$

$$
\left. + \Sigma^{+-}(\mathbf{R}, T; \mathbf{p}, \omega) G^{-+}(\mathbf{R}, T; \mathbf{p}, \omega) \right).
$$

In the quasiclassical limit, due to the slow variation of the distribution function, we can substitute into the previous expression the values of Green's function for the uniform stationary case [Eqs. (3.117)–(3.123)], changing there $n_\mathbf{p}$ to $f^W(\mathbf{R}, \mathbf{p}, T)$:

$$
\mathrm{St}\left[f^w(\mathbf{R}, \mathbf{p}, T) \right] = i\Sigma^{-+}(\mathbf{R}, T; \mathbf{p}, \epsilon_\mathbf{p} - \mu) \left[1 - f^W(\mathbf{R}, \mathbf{p}, T) \right] \tag{3.153}
$$

$$
+ i\Sigma^{+-}(\mathbf{R}, T; \mathbf{p}, \epsilon_\mathbf{p} - \mu) f^W(\mathbf{R}, \mathbf{p}, T).
$$

The specific form of the collision integral is determined by the interaction, which enters the self energy functions.

3.6 Application: Electrical Conductivity of Quantum Point Contacts

As an example of how the above formalism can be applied, we discuss here the quantum conductivity of quantum point contacts (QPC). Point contacts are the contacts between two conductors, whose dimension is much less than the inelastic scattering

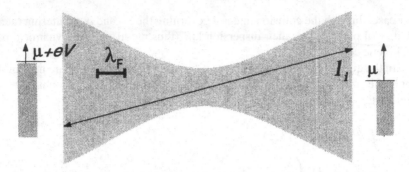

Fig. 3.9 Quantum point contact

length of carriers, l_j (Fig. 3.9). The size of the *quantum point contact*, moreover, is of the order of the Fermi wavelength or smaller.

In the last 10 years, quantum point contacts were realized in highly mobile 2-dimensional electron gas (2DEG) of semiconductor heterostructures by "squeezing" it from under the gate electrodes by applied negative voltage. In this case, there are two benefits: the size of the contact can be changed at will, and—because λ_F in 2DEG is large, about 400° A—they really are quantum point contacts. On the other hand, three-dimensional metallic QPC of atomic size are now being created using a variety of experimental techniques.

We begin with the three-dimensional case, following the analysis of [15]. The model of such a contact is presented in Fig. 3.10: it is the round opening with radius a in a thin planar dielectric barrier Σ separating two conducting half-spaces.

The matrix Green's function in this system satisfies the Keldysh equation (3.143), where the electron–phonon interaction enters the self energy operator. (We consider electron–phonon interactions as the only interactions present.)

The geometry of the system imposes specific boundary conditions. Far from the contact, the electron gas of either bank does not feel the presence of the contact at all and is in equilibrium, so that the Wigner's distribution function of the electron (related to the $(-+)$-component of the Keldysh matrix Green's function by (3.147)) must satisfy the boundary conditions

$$\lim_{r \to \infty} f^W(\mathbf{r}, \mathbf{p}) = n_{\mathbf{p},\sigma}, \sigma = \begin{cases} 1, z > 0, \\ 2, z < 0, \end{cases} \tag{3.154}$$

where

$$n_{\mathbf{p},\sigma} = \frac{1}{\exp \frac{\epsilon_\mathbf{p} - \mu_\sigma}{T} + 1} \tag{3.155}$$

is the equilibrium distribution. The chemical potentials in the banks of the contact are biased by the driving voltage

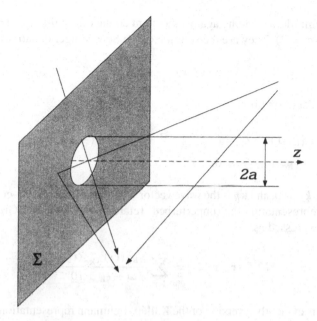

Fig. 3.10 Model of three-dimensional point contact

$$\mu_1 - \mu_2 = eV. \tag{3.156}$$

The impenetrability of the dielectric barrier Σ for the electrons is taken into account by setting

$$\hat{G}(1,2)|_{\mathbf{r}_1 \subset \Sigma} = \hat{G}(1,2)|_{\mathbf{r}_2 \in \Sigma} = 0. \tag{3.157}$$

The density of the electric current in the system is given by

$$\mathbf{j}(\mathbf{r}) = \frac{e\hbar}{m} \left(\frac{\partial}{\partial \mathbf{r}_2} - \frac{\partial}{\partial \mathbf{r}_1} \right) G^{-+}(1,2) \bigg|_{1=2}; \tag{3.158}$$

$$\mathbf{j}(\mathbf{r}) = 2e \int \frac{d^3 p}{(2\pi\hbar)^3} \mathbf{v} f^W(\mathbf{r}, \mathbf{p}). \tag{3.159}$$

Here $\mathbf{v} = \mathbf{p}/m$.

3.6.1 Quantum Electrical Conductivity in the Elastic Limit

In the absence of collisions ($l \to \infty$) the motion of the electrons is nondissipative. It can be described with a complete system of wave functions $\chi_{\mathbf{k}\sigma}(\mathbf{r})$ of electrons with momentum $\hbar\mathbf{k}$, impinging against the point contact from the right ($\sigma = 1$) and from the left ($\sigma = 2$). They are the solution of the Schrödinger equation

$$\Delta\chi_{\mathbf{k}\sigma}(\mathbf{r}) = \epsilon_{\mathbf{k}}\chi_{\mathbf{k}\sigma}(\mathbf{r}) \tag{3.160}$$

with the boundary conditions

$$\chi_{\mathbf{k}\sigma}(\mathbf{r})|_{r\to\infty} = (-1)^{\sigma}\left(e^{i\mathbf{kr}} - e^{i\mathbf{k}_R\mathbf{r}}\right)\theta\left(-(-1)^{\alpha}z\right), \tag{3.161}$$

$$\chi_{\mathbf{k}\sigma}(\mathbf{r})|_{\mathbf{r}\in\Sigma} = 0, \tag{3.162}$$

where we take $k_z > 0$, and $\mathbf{k}_R$ is the wave vector with the opposite to $\mathbf{k}$ z-component.[7]

In ($\mathbf{r}, \omega$)-representation the (unperturbed) retarded and advanced Green's functions can be expressed as

$$G^{R(A)}(\mathbf{r}_1, \mathbf{r}_2, \omega) = \sum_{\mathbf{k}\sigma} \frac{\chi_{\mathbf{k}\sigma}(\mathbf{r}_1)\chi_{\mathbf{k}\sigma}^*(\mathbf{r}_2)}{\omega - \epsilon_{\mathbf{k}} \pm i0}. \tag{3.163}$$

This formula is evidently a version of the Källén–Lehmann representation for unperturbed Green's functions. It already has correct analytic properties, and can be checked directly by substituting it in (2.19) and using the completeness of the set of one-particle eigenstates, $\sum_{\mathbf{k}\sigma} \chi_{\mathbf{k}\sigma}(\mathbf{r}_1)\chi_{\mathbf{k}\sigma}^*(\mathbf{r}_2) = \delta(\mathbf{r}_1 - \mathbf{r}_2)$.

Making use of the relations (3.117)–(3.123), we now build the solution of the Keldysh matrix equation (3.143) with zero self energy (i.e., in the elastic limit) with proper boundary conditions, in the following form:

$$\hat{G}_0(1, 2) = i\sum_{\sigma=1}^{2}\int\frac{d^3k}{(2\pi)^3}\theta(k_z)\hat{n}_{\mathbf{k}\sigma}(t_1 - t_2)e^{-i\epsilon_{\mathbf{k}}(t_1-t_2)}$$

$$\times \chi_{\mathbf{k}\sigma}(\mathbf{r}_1)\chi_{\mathbf{k}\sigma}^*(\mathbf{r}_2). \tag{3.164}$$

The matrix $\hat{n}_{\mathbf{k}\sigma}(t)$ has the form

$$\hat{n}_{\mathbf{k}\sigma}(t) = \begin{pmatrix} n_{\mathbf{k}\sigma} - \theta(t) & n_{\mathbf{k}\sigma} \\ n_{\mathbf{k}\sigma} - 1 & n_{\mathbf{k}\sigma} - \theta(-t) \end{pmatrix}. \tag{3.165}$$

Substituting (3.164) into the relation (3.158) and carrying out the integration over any surface enveloping the point contact from the right ($z > 0$) or the left ($z < 0$), we obtain the following expression for the total current in the elastic limit:

---

[7] We can neglect the effect of the electric field in our considerations due to the fact that the potential drop in the vicinity of the contact is (a/r_D) times smaller than the total potential drop, V. Here r_D is the screening length, which is large enough in semiconductors to provide the condition $r_D \gg a$.

$$I(V) = \frac{2em}{\hbar} \int \frac{d^3k}{(2\pi)^3} \theta(k_z)(n_{k2} - n_{k1}) \int_{S_1} dS\, (\chi_{k2}^* \nabla \chi_{k1}). \tag{3.166}$$

This expression can be rewritten in an alternative form, namely as

$$I(V) = \int \frac{d^3k}{(2\pi)^3} \theta(k_z)(n_{k2} - n_{k1}) J_k, \tag{3.167}$$

where $J_k = \frac{2em}{\hbar} \theta(k_z) \int_{S_1} dS(\chi_{k2}^* \nabla \chi_{k1})$ is the *partial current* carried through the contact by the electron incident from infinity with the momentum $\mathbf{k}$. This expression is very useful in the calculation of the electrical conductance of a mesoscopic system in the elastic limit.

3.6.1.1 Current in the Quasiclassical Limit

In the limit $k_F a \gg 1$ far from the contact, the wave function has the asymptotic form (for example, for $\sigma = 2$)

$$\chi_{k2}(\mathbf{r}) = \frac{iae^{ikr}}{2qr} \left(k_z + \frac{k_z}{r} \right) J_1(qa), \quad z > 0, \tag{3.168}$$

where $q = |\mathbf{k}_{||} - k\mathbf{r}_{||}/r|$, here $\mathbf{k}_{||}$ is the component of the vector parallel to the dielectric barrier; $J_1(x)$ is a Bessel function.

The contact current is thus given by

$$I(V) = I_1(V) - I_2(V), \tag{3.169}$$

$$I_\sigma(V) = I_\sigma^{(0)} \left[1 - \frac{(k_{F\sigma}a)^{-9/2}}{64\sqrt{2\pi}} \cos(4k_{F\sigma}a - \pi/4) \right]. \tag{3.170}$$

Here

$$I_\sigma^{(0)} = \frac{|e|ma^2\mu_\sigma^2}{4\pi\hbar^3} \tag{3.171}$$

is the point contact current in the classical limit. In the linear response regime the latter formula yields the expression for the classical point contact resistance (*Sharvin's resistance*) in the form

$$R_0^{-1} = \frac{e^2 S S_F}{h^3}, \tag{3.172}$$

where $S = \pi a^2$, $S_F = 4\pi p_F^2$ are the areas of the contact and Fermi surface respectively. Equations (3.169), (3.170) show that in the limit $k_F a \gg 1$, only small corrections to this resistance appear, oscillating with k_F (that is, with driving voltage).

A two-dimensional version of the above model is an isolating *line* (or, more exactly, two rays: say, $-\infty < x < -d/2$ and $d/2 < x < \infty$), separating two conducting half-planes ($y < 0, y > 0$) [22]. The 2D analogue of (3.167) is, evidently,

$$I(V) = \int \frac{d^2k}{4\pi^2} J_{\mathbf{k}}\theta(k_y)(n_{\mathbf{k}2} - n_{\mathbf{k}1}). \tag{3.173}$$

By standard methods of Green's functions for the classical wave equation, described, e.g., in Chap. 7 of [8], it can be expressed through the values of the wave function in the opening, $\chi_{\mathbf{k}}(x, 0)$:

$$J_{\mathbf{k}} = \frac{ek^2\hbar}{m^*} \int\limits_{-d/2}^{d/2} dx \int\limits_{-d/2}^{d/2} dx' \frac{\chi_{\mathbf{k}}(x, 0)\chi_{\mathbf{k}}(x, 0)^*}{k(x' - x)} J_1(k(x - x')). \tag{3.174}$$

Here $J_1(z)$ is the Bessel function. In the quasiclassical limit we can take $\chi_{\mathbf{k}}(x, 0) \approx e^{ik_x x}$, and find

$$I(V) = G(d)V = \frac{e^2}{\pi\hbar}V\left(\frac{k_F d}{\pi} + \frac{\sin 2k_F d}{2\pi k_F d} - \frac{1}{4}\right). \tag{3.175}$$

We again found an analogue to Sharvin resistance $\left(R_{0,2D}^{-1} = \frac{e^2}{\pi\hbar}\frac{k_F d}{\pi}\right)$ plus small oscillating corrections. As we will see shortly, this is not a purely academic exercise.

Now a natural question arises: What is Sharvin resistance?

3.6.2 Elastic Resistance of a Point Contact: Sharvin Resistance, the Landauer Formula, and Conductance Quantization

Let us return to the picture presented in Fig. 3.9. We apply a voltage across a point contact. The carriers flow through it, and since the contact has finite size, there will be finite current I at any finite voltage V, which means finite contact resistance, Sharvin resistance R_0. Then the energy dissipation rate is $IV = I^2R$. But the size of the contact is less than l_i, inelastic scattering length, therefore the carriers cannot lose energy in the contact!

This apparent paradox is resolved if we ask what happens at distances larger than l_i from the contact. At infinity we have equilibrium, at differing chemical potentials (Eq. 3.154) to the right and to the left of the contact. On the way to infinity from the contact, which is somewhat farther than l_j, electrons will relax to one of those equilibrium distributions, due to various inelastic processes. Thus, dissipation will occur far away from the contact; and its rate is nevertheless determined by the contact resistance, R_0, which we calculated in apparent neglect of any inelastic scattering!

This would be strange indeed, but actually we did take inelastic processes into account, when we postulated the boundary conditions at infinity (3.154). They may seem self-evident, but they imply that all electrons impinging at the contact have equilibrium distribution—are *completely thermalized* (and thus have no "memory" of their previous history). This thermalization is vital for the theory and can be achieved only if there is sufficiently strong inelastic scattering in the system. Its details are, though, irrelevant, since as long as l_i exceeds the contact size, the dissipation rate is determined by Sharvin resistance.

The point contact is thus a very fine example of the *Landauer formalism*, a powerful tool in transport theory of small systems. Landauer considered a one-dimensional (1D) wire a scatterer, its quantum-mechanical transition and reflection amplitudes being t and r respectively, $|t|^2 + |r|^2 = 1$, and asked a question:

What is its electrical resistance? To answer this question, he connected with this wire two equilibrium electron reservoirs (that is, systems containing vast numbers of electrons at equilibrium, and with effective energy and momentum relaxation mechanisms) at differing chemical potentials, $\mu_1 - \mu_2 = eV$ (Fig. 3.9). Assuming that once leaving the wire for a reservoir, an electron never returns (or, more exactly, immediately thermalizes)—which is an exact analogue to our boundary condition (3.154)—it is easy to write down the current (since it is the same in all of the wire, we can calculate it to the right of the scatterer):

$$I(V) = 2e \int \frac{dk}{2\pi} v(k) \left(n_1(k)|t|^2 - n_2(k)(1 - |r|^2) \right)$$

$$= |t|^2 \frac{2e}{2\pi\hbar} \int d\epsilon \frac{1}{v(\epsilon)} v(\epsilon)(n_F(\epsilon - \mu - eV) - n_F(\epsilon - \mu))$$

$$\approx \frac{2e^2}{h} |t|^2 V. \tag{3.176}$$

Here $v(\epsilon) = \frac{\partial \epsilon}{\partial \hbar k}$ is the velocity of an electron, and the factor of 2 comes from spins. Had we several 1D wires in parallel, the currents would simply add up. Thus we come to the *Landauer formula* for electrical conductance of a quantum wire (or contact):

$$G \equiv \frac{1}{R} = \frac{2e^2}{h} \sum_{a=1}^{N_\perp} |t_a|^2. \tag{3.177}$$

The quantum resistance unit $h/(2e^2)$ (which is, evidently, $137\pi/3 \times 10^{-10}$ s/cm, or approximately $13\,k\Omega$ in more convenient units) is the same as appears, e.g., in the quantum Hall effect. The sum is taken over so-called *quantum channels*. It is proper to say that $2e^2/h$ is a conductance of an ideal quantum channel.

How this is related to Sharvin resistance? We had previously

$$R_{0,3D}^{-1} = \frac{e^2 \pi a^2 4\pi p_F^2}{h^3}$$

$$= \frac{2e^2}{h}(2\pi a/\lambda_F)^2;$$

$$R_{0,2D}^{-1} = \frac{e^2}{\pi\hbar} \frac{k_F d}{\pi}$$
$$= \frac{2e^2}{h} \frac{2d}{\lambda_F}.$$

We see that in both cases the conductance is really given by $2e^2/h$ times approximately $N_\perp \approx (a/\lambda_F)^{\mathrm{dim}}$, where a is the size of the opening and dim its dimensionality. (There is no additional scattering, so $|t|^2 = 1$.) In this case the number of channels tells us how many electrons at the Fermi surface (because they are carrying current) can squeeze through the opening simultaneously, given the uncertainty relation which keeps them $\approx \lambda_F$ apart. But thus defined the number of channels is fuzzy: the conductance, as we see, changes continuously with the size of the opening.

The situation changes dramatically if instead of an opening in an infinitely thin wall, we consider a long channel that smoothly enters the reservoirs. This system is more like a quantum wire of Landauer formula, and we should expect that as we change the width of the wire, the number of modes changes by 1, so that the conductance is quantized in units of $2e^2/h$. This quantization was indeed observed in 2D quantum point contacts [19, 20], which we have mentioned above. The beauty of the thing is that in this system the size of the contact can be continually changed by changing the gate voltage, and it turns out that the conductance changes by $2e^2/h$-steps. (Unfortunately, the accuracy of these steps is far inferior to those of quantum Hall effect.)

To understand this, let us return to (3.167). The partial current $J_\mathbf{k}$ is the current carried through the opening by a particle incident from infinity. In a smooth (*adiabatic*) channel (with diameter $d(z)$ slowly varying with longitudinal coordinate az) we can approximately present such a wave function in factorized form, $\chi_{k,a}(\rho,z) = \phi_a(\rho; z)e^{ikz}$, where now k is the longitudinal momentum, ρ is the transverse coordinate, and a labels transverse eigenfunctions (which in the 2D case can be, e.g., $\approx \sin(\pi a\rho/d(z))$). The channel is effectively presented as a set of 1D "subbands", playing the role of Landauer quantum wires. It is almost self-evident that only if the energy of transverse motion in the narrowest pan of the channel, $\approx \hbar^2 a^2/(2m^* d_{\min}^2)$, is less than the Fermi energy, can the corresponding ath mode participate in conductivity. Otherwise the particle will be reflected back to the initial reservoir. Here indeed, the number of conducting modes is determined by how many of them can squeeze through the narrowest part of the channel [14].

This picture should naturally hold in the three-dimensional case as well, with the only complication that now there may occur multiple steps, $n \times 2e^2/h$, due to accidental degeneracy of different transverse modes [11]. Such conductance quantization was observed in 3D metal point contacts—mechanically controllable break-junctions [16] and scanning tunneling microscopy (STM) devices [18].

For further reading on the Landauer formalism and transport in point contacts and other mesoscopic devices I refer you to [3, 4, 9], and references therein.

3.6.3 The Electron–Phonon Collision Integral in 3D Quantum Point Contact

Now let us return to the interactions. As we have said above, generally the righthand side of the quantum kinetic equation (3.151) yields not only the collision integral, but also the renormalization effects. Therefore, we will call its right-hand side the "collision integral":

$$I_{ph}(\mathbf{r}, \mathbf{p}) = \int d^3\xi e^{-i\mathbf{p}\xi} I_{ph}(\mathbf{r} + \xi/2, t; \mathbf{r} - \xi/2, t), \qquad (3.178)$$

where

$$I_{ph}(1, 2) = -\int d^4 3\{\Sigma^{-+}(1, 3)G^{++}(3, 2) + \Sigma^{--}(1, 3)G^{-+}(3, 2)$$
$$+ G^{--}(1, 3)\Sigma^{-+}(3, 2) + G^{-+}(1, 3)\Sigma^{++}(3, 2)\}. \quad (3.179)$$

In the lowest order in the electron–phonon interaction the self energy components are given by the expressions (see Table 3.2)

$$\hat{\Sigma} = \hat{\Sigma} = \underset{}{\overbrace{}} \qquad (3.180)$$

$$\Sigma^{-+}(1, 3) = -iG_0^{-+}(1, 3)D_0^{+-}(3, 1);$$
$$\Sigma^{+-}(1, 3) = -iG_0^{+-}(1, 3)D_0^{-+}(3, 1);$$
$$\Sigma^{--}(1, 3) = iG_0^{--}(1, 3)D_0^{--}(3, 1);$$
$$\Sigma^{++}(1, 3) = iG_0^{++}(1, 3)D_0^{++}(3, 1). \qquad (3.181)$$

(We can take $g = 1$ by redefining the phonon operators.)

It is more convenient to present $I_{ph}(1, 2)$ as

$$I_{ph}(1, 2) = -\int d^3r_3 \int \frac{d\omega}{2\pi} \{\Sigma^{-+}(\mathbf{r}_1, \mathbf{r}_3; \omega)G^{++}(\mathbf{r}_3, \mathbf{r}_2; \omega)$$
$$+ \Sigma^{--}(\mathbf{r}_1, \mathbf{r}_3; \omega)G^{-+}(\mathbf{r}_3, \mathbf{r}_2; \omega)$$
$$+ G^{--}(\mathbf{r}_1, \mathbf{r}_3; \omega)\Sigma^{-+}(\mathbf{r}_3, \mathbf{r}_2; \omega)$$
$$+ G^{-+}(\mathbf{r}_1, \mathbf{r}_3; \omega)\Sigma^{++}(\mathbf{r}_3, \mathbf{r}_2; \omega)\}. \qquad (3.182)$$

The phonon field operator can be written as follows:

$$\phi(\mathbf{r}, t) = \sum_q \left(\Phi_q(\mathbf{r}) e^{-i\omega_q t} b_q + \Phi_q^*(\mathbf{r}) e^{i\omega_q t} b_q^\dagger \right). \tag{3.183}$$

The phonon distribution function N_q (not necessarily equilibrium) is defined by

$$\left\langle b_q^\dagger b_q \right\rangle \equiv N_q; \left\langle b_q b_q^\dagger \right\rangle \equiv N_q + 1. \tag{3.184}$$

With the use of Eqs. (3.164), (3.183) we find the following expressions for the self energy components (to spare space, we will later on denote the combination $(\mathbf{k}, \sigma)(k_z > 0)$ by K):

$$\Sigma^{--}(\mathbf{r}_1, \mathbf{r}_3; \omega) = \sum_{qK} \chi_K(\mathbf{r}_1) \chi_K^*(\mathbf{r}_3)$$

$$\times \left\{ \Phi_q(\mathbf{r}_3) \Phi_q^*(\mathbf{r}_1) \left(\frac{n_K(N_q+1)}{\omega + \omega_q - \epsilon_K - i0} + \frac{(1-n_K)N_q}{\omega + \omega_q - \epsilon_K + i0} \right) \right.$$
$$\left. + \Phi_q^*(\mathbf{r}_3) \Phi_q(\mathbf{r}_1) \left(\frac{n_K N_q}{\omega - \omega_q - \epsilon_K - i0} + \frac{(1-n_K)(N_q+1)}{\omega - \omega_q - \epsilon_K + i0} \right) \right\};$$

$$\tag{3.185}$$

$$\Sigma^{++}(\mathbf{r}_1, \mathbf{r}_3; \omega) = -\sum_{qK} \chi_K(\mathbf{r}_1) \chi_K^*(\mathbf{r}_3)$$

$$\times \left\{ \Phi_q(\mathbf{r}_3) \Phi_q^*(\mathbf{r}_1) \left(\frac{n_K(N_q+1)}{\omega + \omega_q - \epsilon_K + i0} + \frac{(1-n_K)N_q}{\omega + \omega_q - \epsilon_K - i0} \right) \right.$$
$$\left. + \Phi_q^*(\mathbf{r}_3) \Phi_q(\mathbf{r}_1) \left(\frac{n_K N_q}{\omega - \omega_q - \epsilon_K + i0} + \frac{(1-n_K)(N_q+1)}{\omega - \omega_q - \epsilon_K - i0} \right) \right\};$$

$$\Sigma^{-+}(\mathbf{r}_1, \mathbf{r}_3; \omega) = -2\pi i \sum_{qK} \chi_K(\mathbf{r}_1) \chi_K^*(\mathbf{r}_3)$$

$$\times \left\{ \Phi_q(\mathbf{r}_3) \Phi_q^*(\mathbf{r}_1) n_K (N_q+1) \delta(\omega + \omega_q - \epsilon_K) \right.$$
$$\left. + \Phi_q^*(\mathbf{r}_3) \Phi_q(\mathbf{r}_1) n_K N_q \delta(\omega - \omega_q - \epsilon_K) \right\}; \tag{3.186}$$

$$\Sigma^{+-}(\mathbf{r}_1, \mathbf{r}_3; \omega) = -2\pi i \sum_{qK} \chi_K(\mathbf{r}_1) \chi_K^*(\mathbf{r}_3)$$

$$\times \left\{ \Phi_q(\mathbf{r}_3) \Phi_q^*(\mathbf{r}_1) (n_K - 1) N_q \delta(\omega + \omega_q - \epsilon_K) \right.$$
$$\left. + \Phi_q^*(\mathbf{r}_3) \Phi_q(\mathbf{r}_1) (n_K - 1)(N_q+1) \delta(\omega - \omega_q - \epsilon_K) \right\}.$$

Substituting these expressions into (3.182) and (3.178), and gathering like terms, we finally obtain the following expression for the electron–phonon collision integral:

$$I_{ph}(\mathbf{r}, \mathbf{p}) = -2 \sum_{K,K'} \sum_{q} C^{q}_{K'K} \Im \left[S^{q}_{K'K}(\mathbf{r}, \mathbf{p}) \right.$$
$$\left. \times \frac{n_K(1 - n_{K'})(N_q + 1) - n_{K'}(1 - n_K)N_q}{\epsilon_K - \epsilon_{K'} - \omega_q - i0} \right].$$
(3.187)

Here the function

$$C^{q}_{K'K} \equiv \int d^3 r_3 \chi^*_{K'}(\mathbf{r}_3) \Phi^*(\mathbf{r}_3) \chi_K(\mathbf{r}_3)$$
(3.188)

in the uniform case would yield the momentum conservation law, $C^{q}_{K'K} \propto \delta(\mathbf{k} - \mathbf{k'} - \mathbf{q})$. The function $S^{q}_{K'K}(\mathbf{r}, \mathbf{p})$ is defined as

$$S^{q}_{K'K}(\mathbf{r}, \mathbf{p}) = \int d^3 \xi e^{-i \mathbf{p}\xi} \chi_{K'}(\mathbf{r} + \xi/2) \chi^*_K(\mathbf{r} - \xi/2)$$
$$\times \left(\Phi_q(\mathbf{r} + \xi/2) - \Phi_q(\mathbf{r} - \xi/2) \right).$$
(3.189)

It is clearly seen that Eq. (3.187) yields two different terms. One would contain the energy-conserving delta functions $\delta(\epsilon_K - \epsilon_{K'} - \omega_q)$ multiplied by the usual statistical in–out factors for the electronic scattering with emission or absorption of phonons. The other (expressed through the main value integrals) is responsible for the spectrum renormalization.

3.6.4 *Calculation of the Inelastic Component of the Point Contact Current

The kinetic equation (3.151) in the point contact,

$$\mathbf{v}\nabla_r f^W(\mathbf{r}, \mathbf{p}) = I_{ph}(\mathbf{r}, \mathbf{p}),$$
(3.190)

must be supplemented by appropriate boundary conditions. They are conveniently written after introducing the inverse Fourier transform of Wigner's function,

$$f^W(\mathbf{r}_1, \mathbf{r}_2) = \int \frac{d^3 p}{(2\pi)^3} e^{i\mathbf{p}(\mathbf{r}_1 - \mathbf{r}_2)} f^W\left(\frac{\mathbf{r}_1 + \mathbf{r}_2}{2}, \mathbf{p}\right),$$
(3.191)

as follows:

$$f^W(\mathbf{r}_1, \mathbf{r}_2)|_{\substack{r_1, r_2 \to \infty \\ z_1, z_2 > 0}} = 0;$$

$$f^W(\mathbf{r}_1, \mathbf{r}_2)|_{\mathbf{r}_1 \in \Sigma} = f^W(\mathbf{r}_1, \mathbf{r}_2)|_{\mathbf{r}_2 \in \Sigma}$$

$$= 0. \tag{3.192}$$

Then the solution to the boundary value problem (3.190), (3.192) has the form

$$f^W(\mathbf{r}, \mathbf{p}) = \int d^3 p' \int d^3 r' g_{\mathbf{pp}'}(\mathbf{r}, \mathbf{r}') I_{ph}(\mathbf{r}', \mathbf{p}'), \tag{3.193}$$

where the function $g_{\mathbf{pp}'}(\mathbf{r}, \mathbf{r}') = -g_{\mathbf{p}'\mathbf{p}}(\mathbf{r}', \mathbf{r})$ satisfies the linear equation

$$\mathbf{v} \nabla_{\mathbf{r}} g_{\mathbf{pp}'}(\mathbf{r}, \mathbf{r}') = \delta(\mathbf{r} - \mathbf{r}') \delta(\mathbf{p} - \mathbf{p}') \tag{3.194}$$

with boundary conditions for its inverse Fourier transform analogous to (3.192).

This allows us to express the inelastic correction to the point contact current as follows (we used Eq. (3.159)):

$$I_{ph} = \frac{2e}{(2\pi\hbar)^3} \int d^3 p \int d^3 r F_{\mathbf{p}}(\mathbf{r}) I_{ph}(\mathbf{r}, \mathbf{p}). \tag{3.195}$$

The function F is defined by

$$F_{\mathbf{p}}(\mathbf{r}) \equiv \int_0^{} dS' \int d^3 p' v_z' g_{\mathbf{p}'\mathbf{p}}(\mathbf{r}', \mathbf{r}) \tag{3.196}$$

(the integration is taken over the opening $\mathbf{O}$) and is the solution of the following boundary value problem:

$$\mathbf{v} \nabla_{\mathbf{r}} F_{\mathbf{p}}(\mathbf{r}) = -v_z \delta(z);$$

$$F(\mathbf{r}_1, \mathbf{r}_2)|_{\substack{r_1, r_2 \to \infty \\ z_1, z_2 > 0}} = 0;$$

$$F(\mathbf{r}_1, \mathbf{r}_2)|_{\mathbf{r}_1 \in \Sigma} = F(\mathbf{r}_1, \mathbf{r}_2)|_{\mathbf{r}_2 \in \Sigma}$$

$$= 0. \tag{3.197}$$

It can be shown that this quantity can be written as [15]

$$F_{\mathbf{p}}(\mathbf{r}) = \alpha_{-\mathbf{p}}(\mathbf{r}) - \theta(z). \tag{3.198}$$

The quantity $\alpha_{\mathbf{p}}(\mathbf{r})$ in (3.198), defined as

$$\alpha_{\mathbf{p}}(\mathbf{r}) = \int d^3 r' e^{-i\mathbf{pr}'} \int_{k_z > 0} \frac{d^3 k}{(2\pi)^3} \chi_{\mathbf{k}1}\left(\mathbf{r} + \frac{\mathbf{r}'}{2}\right) \chi_{\mathbf{k}2}^*\left(\mathbf{r} - \frac{\mathbf{r}'}{2}\right), \tag{3.199}$$

is the quantum analogue of the classical probability for an electron moving from infinity with momentum $\mathbf{p}$ to be located at the point $\mathbf{r}$ to the right of the contact.

The above results allow us to explain the experimentally observed nonlinear current–voltage dependence, $I(V)$, in point contacts, the origins of the nonlinearity being (1) electron–phonon interaction and (2) renormalization of the electron mass. It turns out that the peaks of the function $d^2V/dI^2(eV)$ are situated at the maxima of the phonon density of states. Qualitatively this is understandable, since on the one hand, the distribution functions of the electrons injected through the point contact is shifted by eV with respect to the surrounding electrons, and on the other hand, its relaxation to the distribution function of the surroundings will be accompanied by emission of phonons with energy $\hbar\omega = eV$.

This effect [17, 21] is a basis for *point contact spectroscopy*, the method of restoration of the phonon density of states from the nonlinear current–voltage characteristics of a point contact (for a review see [5]). It was first developed in metals, where due to both screening length and Fermi wavelength being very small, the quasiclassical theory is already sufficient.

3.7 Method of Tunneling Hamiltonian

The tunneling Hamiltonian approximation (THA) is most often used in the theory of the Josephson effect—superconducting current flow between two weakly coupled superconductors (separated by a potential barrier), which we will address later in the book. But it can be successfully employed in a variety of problems where electron transfer between the conductors is realized through a "weak link," be it a tunneling barrier or a point contact, and it is natural to discuss this method here.

The idea of the method consists in presenting the total Hamiltonian of the system as a sum:

$$\mathcal{H} = \mathcal{H}_L + \mathcal{H}_R + \mathcal{H}_T. \tag{3.200}$$

Here $\mathcal{H}_{L,R}$ are the Hamiltonians of isolated (i.e., unperturbed) (left/right) conductors, while the *tunneling term* $\mathcal{H}_T$ describes the effect of electron transitions between them:

$$\mathcal{H}_L = \sum_{\mathbf{k},\sigma} \epsilon_{\mathbf{k},\sigma} c^\dagger_{\mathbf{k},\sigma} c_{\mathbf{k},\sigma}; \tag{3.201}$$

$$\mathcal{H}_R = \sum_{\mathbf{q},\sigma} \epsilon_{\mathbf{q},\sigma} d^\dagger_{\mathbf{q},\sigma} d_{\mathbf{q},\sigma}; \tag{3.202}$$

$$\mathcal{H}_T = \sum_{\mathbf{k},\mathbf{q},\sigma} \left(T_{\mathbf{kq}} c^\dagger_{\mathbf{k},\sigma} d_{\mathbf{q},\sigma} + T^*_{\mathbf{kq}} d^\dagger_{\mathbf{q},\sigma} c_{\mathbf{k},\sigma} \right). \tag{3.203}$$

As you see, $\mathcal{H}_{L,R}$ are written as though the banks of the contact were infinite, and as though quasiparticle states could be characterized by their momenta (which in reality are not good quantum numbers).

The tunneling matrix element in the case of time reversal symmetry satisfies the relation

$$T_{\mathbf{kq}}^* = T_{-\mathbf{k},-\mathbf{q}}. \tag{3.204}$$

In the case of planar barrier of amplitude U and thickness d in WKB approximation we would obtain

$$T_{\mathbf{kq}} \propto k_x q_x e^{-\frac{1}{\hbar}\sqrt{2mUd}} \delta(\mathbf{k}_\perp - \mathbf{q}_\perp),$$

but the detailed behavior of $T_{\mathbf{kq}}$ is not relevant. It will suffice to note that the component of the momentum parallel to the interface is conserved, and that in many cases the energy dependence of the tunneling matrix element can be neglected in a rather wide interval around the Fermi energy.

If we apply to the left bank the voltage V, the Hamiltonian will acquire the form

$$\mathcal{H}_L(V) = \mathcal{H}_L(0) - |e|V\mathcal{N}_L. \tag{3.205}$$

The particle number operator in the left/right bank can be defined as

$$\mathcal{N}_L = \sum_{\mathbf{k},\sigma} c_{\mathbf{k},\sigma}^\dagger c_{\mathbf{k},\sigma}; \quad \mathcal{N}_R = \sum_{\mathbf{q},\sigma} d_{\mathbf{q},\sigma}^\dagger d_{\mathbf{q},\sigma}, \tag{3.206}$$

and it commutes with the corresponding unperturbed Hamiltonian. The only term changing particle numbers in each separate bank is, of course, $\mathcal{H}_T$.

The Heisenberg equation of motion for the electron annihilation operator in the left bank for nonzero bias, $\tilde{c}(t)$, is then

$$i\hbar\dot{\tilde{c}}_{\mathbf{k},\sigma}(t) = [\tilde{c}_{\mathbf{k},\sigma}(t),\ \mathcal{H}_L(0)] - |e|V\tilde{c}_{\mathbf{k},\sigma}(t). \tag{3.207}$$

The tunneling current can be written, e.g., as

$$I(V,t) = -|e|\left\langle \dot{\mathcal{N}}_R(t)\right\rangle. \tag{3.208}$$

Commuting $\mathcal{N}_R$ with $\mathcal{H}_T$, we find that

$$I(V,t) = \frac{2|e|}{\hbar}\Im \sum_{\mathbf{kq}\sigma} T_{\mathbf{kq}}\left\langle \tilde{c}_{\mathbf{k}\sigma}^\dagger(t) d_{\mathbf{q}\sigma}(t)\right\rangle. \tag{3.209}$$

We have thus reduced the problem of calculating the current to one of calculating an average of four field operators over a nonequilibrium quantum state—which, as we know well, can be expressed as an infinite series in the perturbation, $\mathcal{H}_T$. Here we will use the Keldysh formalism and consider the simplest case of a constant tunneling

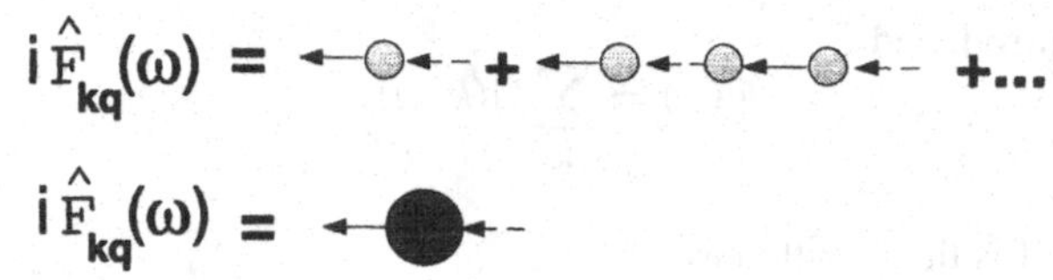

$$i\hat{F}_{kq}(\omega) = \text{—○—} + \text{—○—○—○—} + \dots$$

$$i\hat{F}_{kq}(\omega) = \text{—●—}$$

Fig. 3.11 "Left–right" Green's function in the Keldysh formalism for the tunneling contact

matrix element,

$$T_{kq} \equiv T,$$

in which case the whole matrix series can be summed explicitly.

Then we can write the current as

$$I = -\frac{4|e|T}{\hbar}\Im\sum_{k,q}\left\langle d_q^\dagger \tilde{c}_k\right\rangle = \frac{4|e|T}{\hbar}\Re\sum_{k,q}F_{kq}^{-+}(0). \tag{3.210}$$

We have introduced Keldysh Green's functions, mixing the states in different banks of the contact:

$$\hat{F}_{kq}(t) = \begin{pmatrix} F_{kq}^{--}(t) & F_{kq}^{-+}(t) \\ F_{kq}^{+-}(t) & F_{kq}^{++}(t) \end{pmatrix};$$

$$F_{kq}^{--}(t) = \frac{1}{i}\left\langle T\tilde{c}_k(t)d_q^\dagger(0)\right\rangle;$$

$$F_{kq}^{++}(t) = \frac{1}{i}\left\langle \tilde{T}\tilde{c}_k(t)d_q^\dagger(0)\right\rangle; \tag{3.211}$$

$$F_{kq}^{-+}(t) = -\frac{1}{i}\left\langle d_q^\dagger(0)\tilde{c}_k(t)\right\rangle;$$

$$F_{kq}^{+-}(t) = \frac{1}{i}\left\langle \tilde{c}_k(t)d_q^\dagger(0)\right\rangle.$$

The diagram series for $\hat{F}_{kq}(\omega)$ is shown in Fig. 3.11 (here solid and broken lines correspond to the unperturbed Keldysh matrix in the left and right bank, $\hat{G}_{l,r}$).[8]

Due to T being momentum independent, we can easily integrate over internal momenta. For example,

$$\text{—○—} = \frac{T}{\hbar}\sum_{k,q}i\hat{G}_l(k,\ \omega-\omega_0)\cdot(-i\hat{\tau}_3)\cdot i\hat{G}_r(q,\ \omega)$$

$$= iT\hat{g}_l(\omega-\omega_0)\cdot\hat{\tau}_3\cdot\hat{g}_r(\omega).$$

---

[8] All left-bank Green's functions (labeled by k, k', k'', …) depend on shifted frequency, $\omega-\omega_0$, where $\omega_0 = |e|V/\hbar$. This answers to the difference between $\tilde{c}_k$ and c_k.

Here we have introduced

$$\hat{g}(\omega) = \sum_{\mathbf{k}} \hat{G}(k, \omega). \tag{3.212}$$

The current itself is thus written as

$$I(V) = \frac{4|e|\mathrm{T}}{\hbar} \Re \int\limits_{-\infty}^{\infty} \frac{d\omega}{2\pi} \left[\hat{g}_l(\omega - \omega_0) \cdot \hat{\Sigma}(\omega, \omega_0) \cdot \hat{g}_r(\omega) \right]^{-+} \tag{3.213}$$

where

$$-i\hat{\Sigma}(\omega, \omega_0) = \bigcirc + \bigcirc\!\!\leftarrow\!\!\bigcirc\!\!\leftarrow\!\!\cdot\!\bigcirc + \ldots$$

$$= -\frac{i}{\hbar}\mathrm{T}\hat{\tau}_3 + \left(-\frac{i}{\hbar}\mathrm{T}\hat{\tau}_3\right) \cdot i\hat{g}_r(\omega) \cdot \left(-\frac{i}{\hbar}\mathrm{T}\hat{\tau}_3\right) \cdot i\hat{g}_l(\omega - \omega_0) \cdot \left(-\frac{i}{\hbar}\mathrm{T}\hat{\tau}_3\right)$$

$$= +\cdots -i\frac{\mathrm{T}}{\hbar}\hat{\tau}_3 \cdot \left(\hat{1} - \left(\frac{\mathrm{T}}{\hbar}\right)^2 \hat{\eta}(\omega, \omega_0)\right)^{-1} \tag{3.214}$$

and

$$\hat{\eta}(\omega, \omega_0) = \hat{g}_r(\omega) \cdot \hat{\tau}_3 \cdot \hat{g}_l(\omega - \omega_0) \cdot \hat{\tau}_3.$$

Calculations are easier in the "rotated" representation:

$$\hat{g} \to \check{g} = \mathrm{L}\hat{g}\mathrm{L}^{\dagger}; \ \mathrm{L} = \left(\mathrm{L}^{\dagger}\right)^{-1} = \frac{1}{\sqrt{2}} \begin{pmatrix} 1 & -1 \\ 1 & 1 \end{pmatrix}.$$

Now we are left only with independent components of the Keldysh matrix:

$$\hat{g} = \begin{pmatrix} g^{--} & g^{-+} \\ g^{+-} & g^{++} \end{pmatrix} \to \check{g} = \begin{pmatrix} 0 & g^A \\ g^R & g^K \end{pmatrix}. \tag{3.215}$$

The components of the $\check{g}$-matrix are easily found:

$$g^{R,A}(\omega) = \sum_{\mathbf{k}} \frac{1}{\omega - \xi_{\mathbf{k}} \pm i0} = \mathrm{P} \sum_{\mathbf{k}} \frac{1}{\omega - \xi_{\mathbf{k}}} \mp i\pi \sum_{\mathbf{k}} \delta(\omega - \xi_{\mathbf{k}});$$

$$\mathrm{P} \sum_{\mathbf{k}} \frac{1}{\omega - \xi_{\mathbf{k}}} = \mathrm{P} \int \frac{N(\xi)}{\omega - \xi} d\xi \approx N(\omega)\mathrm{P} \int\limits_{-\infty}^{\infty} \frac{d\xi}{\omega - \xi} = 0.$$

Therefore,

$$g^{R,A}(\omega) \approx \mp i\pi \sum_{\mathbf{k}} \delta(\omega - \xi_{\mathbf{k}}) \equiv \mp i\pi N(\omega), \tag{3.216}$$

while the Keldysh component is, by definition,

$$g^K(\omega) = \sum_{\mathbf{k}} G^K(k, \omega)$$

$$= \sum_{\mathbf{k}} \frac{1}{i} (1 - 2n_F(\omega)) \cdot 2\pi\delta(\omega - \xi_{\mathbf{k}})$$

$$= -2\pi i (1 - 2n_F(\omega)) N(\omega). \tag{3.217}$$

After performing the inverse rotation, we find the expression for the normal tunneling current in the first nonvanishing order:

$$I^{(1)}(V) = \frac{4|e|T}{\hbar} \Re \int\limits_{-\infty}^{\infty} \frac{d\omega}{2\pi} \frac{2T}{\hbar} \pi N_l(\omega - \omega_0) \cdot \pi N_r(\omega) [n_F(\omega) - n_F(\omega - \omega_0)]$$

$$= V \cdot \frac{e^2}{\pi h} \cdot \left(\frac{2\pi T}{\hbar} N_l(0)\right) \left(\frac{2\pi T}{\hbar} N_r(0)\right) \tag{3.218}$$

in the linear response limit ($V \to 0$). The conductance is thus

$$G^{(1)} = \left(\frac{2\pi T}{\hbar} N_l(0)\right) \left(\frac{2\pi T}{\hbar} N_r(0)\right), \tag{3.219}$$

and the effective barrier transparency in the sense of the Landauer formula is

$$T_{\text{eff}} = \left(\frac{2\pi T}{\hbar} N_l(0)\right) \left(\frac{2\pi T}{\hbar} N_r(0)\right) \approx \left(\frac{2\pi}{\hbar} N(0)\right)^2 T^2. \tag{3.220}$$

The evaluation of $\hat{\Sigma}$ leads to the following correction: the integrand of the expression for $I(V)$ (3.218) acquires the factor

$$\left|1 - \frac{T^2}{\hbar^2} g_l^A(\omega - \omega_0) g_r^A(\omega)\right|^{-2} \approx \left[1 + \left(\frac{\pi T}{\hbar}\right)^2 N_l(0) N_r(0)\right]^{-2},$$

which (for $V \to 0$) leads to

$$I(V) = V \cdot \frac{e^2}{\pi\hbar} \cdot \frac{T_{\text{eff}}}{(1 + T_{\text{eff}}/4)^2}. \tag{3.221}$$

This result was first obtained by [13] using the Matsubara formalism. Note that the actual small parameter of the problem is not T, but rather $T_{\text{eff}}/4$.

A rather disturbing property of the above result is that it agrees with the Landauer formula only in the lowest order; moreover, for large T conductance tends to zero as transparency grows! Nevertheless, neither of the approaches is at fault. Indeed, the tenet of the Landauer formalism is that once leaving the contact, the particle never comes back (or immediately loses the phase memory, which is the same). On the other hand, from the first look at the diagram series that we draw, it becomes clear that the higher-order terms represent exactly these processes of multiple coherent reentrances. Then Landauer transparency should be *defined* as

$$T_{\text{Landauer}} = \frac{T_{\text{eff}}}{(1 + T_{\text{eff}}/4)^2}; \tag{3.222}$$

that is all: after the many happy returns to the contact, an electron finally goes to infinity, and the Landauer approach becomes legitimate.

The question of why conductance goes to zero as T grows may seem irrelevant, because the THA itself cannot possibly apply in this limit. We could, though, consider a formally equivalent case of a one-dimensional tight-binding Hamiltonian:

$$\mathcal{H}_L = -t_0 \sum_{i=-\infty}^{-1} \left(c_{i-1}^\dagger c_i + c_i^\dagger c_{i-1} \right); \tag{3.223}$$

$$\mathcal{H}_R = -t_0 \sum_{i=1}^{\infty} \left(d_i^\dagger d_{i+1} + d_{i+1}^\dagger d_i \right); \tag{3.224}$$

$$\mathcal{H}_T = -T \left(c_{-1}^\dagger d_1 + d_1^\dagger c_{-1} \right). \tag{3.225}$$

It is evident that if $T = t_0$, we have an ideal 1D chain, and there will be no reflection at the "contact" between the sites -1 and 1. On the other hand, both $T < t_0$ and $T > t_0$ would disrupt the chain and eventually break it in two [12], in agreement with our result.

3.8 Problems

- *Problem 1*
  Check whether the approximation for the retarded Green's function

$$G^R(\mathbf{p}, \omega) = \frac{Z}{\omega - (\epsilon_\mathbf{p} - \mu) - \Sigma(\mathbf{p}, \omega)}$$

satisfies

(1) the Kramers–Kronig relations;
(2) the sum rule;
(3) asymptotics at $|\omega| \to \infty$, if the approximation for the self energy is given by:

   (A) $\Sigma(\mathbf{p}, \omega) = \Sigma' - i\Sigma''$;

(B) $\Sigma(\mathbf{p}, \omega) = A\omega - i\Sigma''$;

(C) $\Sigma(\mathbf{p}, \omega) = A\omega - iB\omega^2$

   (where Z, Σ', Σ'', A, B are positive constants).

- *Problem 2*

Write the analytical expression for the polarization operator at finite temperature in the lowest order and evaluate it.

- *Problem 3*

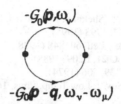

Using the above result, find the large wavelength ($q \to 0$) screening of Coulomb potential in the nondegencrate limit ($e^{\mu\beta} \ll 1$). Calculate the (Debye–Hückel) screening length and compare it to the Thomas–Fermi screening length.

References

Books and Reviews

1. Abrikosov, A.A., Gorkov, L.P., Dzyaloshinski, I.E.: Methods of Quantum Field Theory in Statistical Physics. Dover Publications, New York (1975) (Matsubara formalism)
2. Balescu, R.: Equilibrium and Nonequilibrium Statistical Mechanics. Wiley, New York (1975) (Definition and properties of Wigner's functions quantum distribution functions)
3. Datta, S.: Electronic Transport in Mesoscopic Systems. Cambridge University Press, Cambridge (1995) (Very detailed and pedagogical presentation of transport theory in normal mesoscopic systems)
4. Imry, Y.: Physics of mesoscopic systems. In: Grinstein, G., Mazenko, G. (eds.) Directions in Condensed Matter Physics: Memorial Volume in Honor of Shang-Keng Ma. World Scientific, Singapore (1986)
5. Jansen, A.G.M., van Gelder, A.P., Wyder, P.: Point-contact spectroscopy in metals. J. Phys. C **13**, 6073 (1980)
6. Lifshitz, E.M., Pitaevskii, L.P.: Statistical Physics pt. II. (Landau and Lifshitz Course of Theoretical Physics, v. IX.). Pergamon Press, Oxford (1980) (Matsubara formalism)
7. Lifshitz, E.M., Pitaevskii, L.P.: Physical Kinetics. (Landau and Lifshitz Course of Theoretical Physics, v.X.). Pergamon Press, Oxford (1981) (Keldysh formalism)
8. Morse, P.M., Feshbach, H.: Methods of Theoretical Physics. McGraw-Hill, New York (1953)
9. Washburn, S.,Webb, R.A.: Quantum transport in small disordered samples from the diffusive to the ballistic regime. Rep. Progr. Phys. **55**, 1311 (1992) (A review of theoretical and experimental results on mesoscopic transport)
10. Rammer, J., Smith, H.: Quantum field-theoretical methods in transport theory of metals. Rev. Mod. Phys. **58**, 323 (1986)

Articles

11. Bogachek, E.N., Zagoskin, A.M., Kulik, I.O.: Sov. J. Low Temp. Phys. **16**, 796 (1990) (An emphasis is made on the Keldysh formalism and the method of quantum kinetic equations)
12. Cuevas, J.C., Martín-Rodero, A.: Phys. Rev. B **54**, 7366 (1996)
13. Genenko, Yu.A., Ivanchenko, Yu.M.: Theor. Math. Physics **69**, 1056 (1986)
14. Glazman, L.I., Lesovik, G.B., Khmelnitskii, D.E., Shekhter, R.I.: JETP Lett. **48**, 239 (1988)
15. Itskovich, I.F., Shekhter, R.I.: Sov. J. Low Temp. Phys. **11**, 202 (1985)
16. Krans, J.M., van Ruitenbeek, J.M., Fisun, V.V., Yanson, I.K., de Jongh, L.J.: Nature **375**, 767 (1995)
17. Kulik, I.O., Omelyanchuk, A.N., Shekhter, R.I.: Sov. J. Low Temp. Phys. **3**, 1543 (1977)
18. Pascual, J.I., et al.: Science **267**, 1793 (1995)
19. van Wees, B.J., et al.: Phys. Rev. Lett. **60**, 848 (1988)
20. Wharam, D.A., et al.: J. Phys. C **21**, L209 (1988)
21. Yanson, I.K.: Sov. Phys. JETP **39**, 506 (1974)
22. Zagoskin, A.M., Kulik, I.O.: Sov. J. Low Temp. Phys. **16**, 533 (1990)

Chapter 4
Methods of the Many-Body Theory in Superconductivity

There's a fallacy somewhere: he murmured drowsily, as he stretched his long legs upon the sofa. "I must think it over again." He closed his eyes, in order to concentrate his attention more perfectly, and for the next hour or so his slow and regular breathing bore witness to the careful deliberation with which he was investigating this new and perplexing view of the subject.

Lewis Carroll
"A Tangled Tale"

Abstract Physical origins of superconductivity and peculiarity of the superconducting state. Cooper pairing. Instability of the normal state of a system of fermions. BCS Hamiltonian. Elementary excitations in a superconductor. Nambu-Gor'kov formalism for matrix Green's functions in a superconductor. Andreev reflection and Josephson effect. Coulomb blockade.

4.1 Introduction: General Picture of the Superconducting State

The discovery of superconductivity by Kamerlingh Onnes in 1911 was a real challenge for contemporary physical theory. The theory of metals developed by Drude on the basis of classical statistical physics, while giving a very good explanation of their normal properties, was absolutely unable to deal with superconducting ones. This was also true later for Sommerfeld's and Bloch's theories, based on quantum mechanics. It took almost a half century of studies for the microscopic theory of this phenomenon to appear. (One of the boldest sci-fi writers, R. Heinlein, in a novel written in the mid-40s, predicted the creation of such a theory only by the middle of the next millennium.)

We will not give a detailed account on the properties of superconductors. On the one hand, they are well known. On the another, we are not giving a course on superconductivity, or even on the theory of superconductivity, but on the applications of the many-body theory methods to condensed matter, including the superconducting

A. Zagoskin, *Quantum Theory of Many-Body Systems*,
Graduate Texts in Physics, DOI: 10.1007/978-3-319-07049-0_4,
© Springer International Publishing Switzerland 2014

state. For our purposes, thus, it is enough to recollect two deciding consequences of the experimental data.

The ground state of the superconductor is unusual

At sufficiently low temperature below the superconducting transition temperature T_c there exists *macroscopic phase coherence* of the electrons throughout the sample. You can imagine the superconductor as a single giant molecule in the sense that its electrons are described rather by the wave function than by the density matrix (as in any decent macroscopic system).

The elementary excitations in the superconductor

Elementary excitations separated from the ground state by a *finite energy gap*. This means that the system opposes attempts to excite it, staying in its ground state if the perturbation is not sufficiently strong.

These key properties can be reproduced by the theory, provided it accounts for the following things:

(1) Degeneracy of the electron gas (exclusion principle): The existence of the Fermi surface is essential.
(2) Attraction between electrons. This property seems incredible due to the inevitable Coulomb repulsion between electrons. Nevertheless, in the previous chapter you were shown the possibility of such an attraction due to the electron–phonon interaction (EPI). The role of EPI was understood by Frölich even before the appearance of BCS theory and is directly confirmed by an isotope effect.
(3) The characteristic interaction energy, i.e., the range of electrons involved in the interaction (the order of the Debye energy, $\hbar\omega_D$), must be much less than the Fermi energy:

$$\hbar\omega_D \ll E_F.$$

If these demands are met, the normal zero-temperature ground state of the metal-filled Fermi sphere becomes unstable. The qualitatively different ground state appears instead, with all the strange features that we observe.

This instability of the ground state, or of the *vacuum* (you remember this terminology), means that we can no longer use the perturbation techniques starting from the normal ground state. As in the search for the root of a function by iteration, we must start not too far from the root we want to find, lest we get a wrong answer, or no answer at all. Then we have no regular way to build the new vacuum; we need an insight.

But analogies can help to bring the insight closer. The normal ground state has no gaps. The superconducting one has. In the usual quantum mechanics we also find the situation when the gap appears: when the bound state exists. For an electron in the isolated atom, e.g., the gap between the level and the continuous spectrum is the ionization potential.

Note that the bound state cannot be obtained by the perturbation theory from the propagating one. Indeed, the bound state does not carry a current. So, no matter how weak the attractive potential is, there is a finite, qualitative difference between these states.

"No matter how weak" is a slightly inaccurate statement, for in the three-dimensional (3D) case too weak an attractive potential cannot create the bound state. But in the two- and one-dimensional cases the bound state can be created by an arbitrarily weak attractive potential.

Here the *dimensionality* is physically significant. We guess that the attraction between the electrons is fairly weak. (Experimental measurements of the gap confirm this supposition.) Thus we must have an effectively low-dimensional situation to get a bound state.

What takes place in the one-body case? Assuming that the attractive potential is of the simplest form,

$$U(\mathbf{r}) = -u\delta(\mathbf{r}), u > 0,$$

we have the Schrödinger equation

$$-\nabla^2 \Psi - u\delta(\mathbf{r})\Psi = E\Psi.$$

Fourier-expanding the wave function,

$$\Psi(\mathbf{r}) = \sum_{\mathbf{k}} \Psi_{\mathbf{k}} \exp(i\mathbf{k}\mathbf{r}),$$

we find

$$\sum_{\mathbf{k}} (k^2 - E)\Psi_{\mathbf{k}} e^{i\mathbf{k}\mathbf{r}} = \sum_{\mathbf{k}} u\Psi(0)e^{i\mathbf{k}\mathbf{r}},$$

or

$$\Psi_{\mathbf{k}} = \frac{u}{k^2 - E} \Psi(0) \equiv \frac{u}{k^2 - E} \sum_{\mathbf{k}'} \Psi_{\mathbf{k}'}.$$

Summing over wave vectors $\mathbf{k}$, we find the relation

$$\frac{1}{u} = \sum_{k^2 > 0} \frac{1}{k^2 - E}. \tag{4.1}$$

In the propagating states we have $E, k^2 > 0$, in the bound ones $E, k^2 < 0$. First of all, we see that for *repulsive* potential ($u < 0$) only the propagating states are possible (the r.h.s. of 4.1 is strictly positive for negative E).

Then, for weak attraction (the case we are interested in) the l.h.s. of this equation is a large positive number. For positive E it can be easily matched by one of the terms in the r.h.s. sum. This demands only a slight shift of the energy levels relative to their

Fig. 4.1 Formation of the
bound state

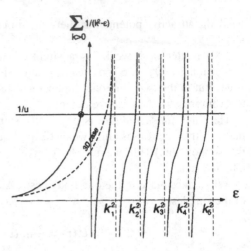

positions without a potential (see Fig. 4.1). But for *negative E* it can be matched
only by the whole sum, since none of its terms is any longer arbitrarily large. If the
attraction is infinitely weak, the excitation energy, $|E|$, of the possible bound state
tends to zero, and can be neglected in (4.1).

Thus, for the bound state to be created by an infinitesimal attractive potential,
the sum $\sum_{k^2>0} k^{-2}$ must diverge. Here the dimensionality role becomes absolutely
clear, for

$$\sum_k \frac{1}{k^2} = \begin{cases} 4\pi \int_0^\infty (k^2 dk)\frac{1}{k^2} \propto k|_0^\infty, & (3D) \\ 2\pi \int_0^\infty (k dk)\frac{1}{k^2} \propto \ln k|_0^\infty, & (2D) \\ 2\int_0^\infty dk \frac{1}{k^2} \propto \frac{1}{k}|_0^\infty. & (1D) \end{cases} \qquad (4.2)$$

The divergence on the upper limit is an artifact due to our choice of δ-like potential;
we should renormalize it, which means introducing a cutoff at some $k_{max} \approx 1/a$,
where a is the scattering length, related to scattering cross-section $\sigma = 4\pi a^2$. The
physically significant divergence, at small momenta (i.e., large distances), occurs
only in the 1D and 2D cases.

These academic speculations become practically important as soon as we recollect
that in the metal at low temperature we have effectively a 2D situation.

Indeed, a filled Fermi sphere does not allow the electrons to be scattered inside
it. On other hand, the energy transfer during the electron–electron collision due to
our attractive potential is of the order of its energy scale, $\omega_D \ll E_F$. Thus the
electrons under consideration are confined to an effectively 2D layer around the
Fermi surface. It was Cooper who first understood this fact, and now we shall sketch
his considerations of the famous *Cooper pairing.*

Consider two electrons (the "pair") above the filled Fermi sphere (at $T = 0$)
(Fig. 4.2a). We neglect all the electrons inside in any other relation. We then assume
the translation invariance of the whole system, and neglect all spin-dependent forces.

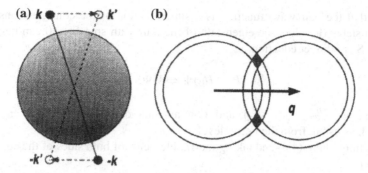

Fig. 4.2 **a** Cooper pair scattering, **b** Scattering phase volume

Fig. 4.3 Pair formation in the superconductor

Then the momentum of the pair mass center, $\hbar\mathbf{q}$, and its total spin, $\mathbf{S}$, are the motion constants. The pair orbital wave function is then

$$\Psi(\mathbf{r}_1, \mathbf{r}_2) = \phi_{\mathbf{q}}(\varrho)e^{i\mathbf{q}\mathbf{R}}, \tag{4.3}$$

where $\varrho = \mathbf{r}_1 - \mathbf{r}_2$, $\mathbf{R} = \frac{\mathbf{r}+\mathbf{r}}{2}$.

The problem is greatly simplified if the pair is at rest ($\mathbf{q} = 0$). (It is easy to see (Fig. 4.2b) that indeed in this case the scattering phase volume is largest.) In this case the problem is spherically symmetric; $\phi(\varrho)$ is an eigenfunction of the moment $\mathbf{L}$. For the singlet spin state of the pair ($\mathbf{S} = 0$) the orbital wave function is symmetric (Fig. 4.3).

We can write

$$\Psi(\mathbf{r}_1, \mathbf{r}_2) = \phi(\varrho) = \sum_{k>k_F} a_{\mathbf{k}}e^{i\mathbf{k}\varrho} = \sum_{k>k_F} a_{\mathbf{k}}e^{i\mathbf{k}\mathbf{r}_1}e^{-i\mathbf{k}\mathbf{r}_2}, \tag{4.4}$$

and see that the pair wave function is a superposition of the configurations with occupied states $(\mathbf{k}, -\mathbf{k})$. The eigenstate of the pair with spin $\mathbf{S} = 0$ can be found from the Schrödinger equation:

$$(E - H_0)\Psi = V\Psi, \tag{4.5}$$

where $H_0 = -\frac{\hbar^2}{2m}(\nabla_{\mathbf{r}1}^2 + \nabla_{\mathbf{r}2}^2)$, and V is an interaction potential. The energy is measured, as usual, from the Fermi level.

Substituting here (4.4) and taking matrix elements of both sides of the equation, we obtain

$$(E - 2E_{\mathbf{k}})a_{\mathbf{k}} = \sum_{k' > k_F} \langle \mathbf{k}, -\mathbf{k}|V|\mathbf{k}', -\mathbf{k}'\rangle a_{\mathbf{k}'}. \tag{4.6}$$

This equation can be solved if we take V in the factorized form,

$$\langle \mathbf{k}, -\mathbf{k}|V|\mathbf{k}', -\mathbf{k}'\rangle = \lambda w_{\mathbf{k}}^* w_{\mathbf{k}'}. \tag{4.7}$$

Then we obtain the same sort of equation as in our quantum-mechanical example:

$$-\frac{1}{\lambda} = \sum_{k > k_F} \frac{|w_{\mathbf{k}}|^2}{2E_{\mathbf{k}} - E}. \tag{4.8}$$

Now we see that in the case of attraction $(\lambda < 0)$ the bound state appears if the interaction is confined to the vicinity of the Fermi surface, i.e., if

$$|w_{\mathbf{k}}|^2 = \begin{cases} 1, & 0 < E_{\mathbf{k}} < \omega_c, \ (\omega_c \ll E_F) \\ 0. & \text{otherwise} \end{cases} \tag{4.9}$$

Indeed, in this case

$$-\frac{1}{\lambda} = \int_0^{\omega_c} dE' \frac{N(E')}{2E' - E} \approx \frac{N(0)}{2} \ln \left| \frac{2\omega_c + |E|}{E} \right|, \tag{4.10}$$

and

$$|E| = \frac{2\omega_c}{1 - \exp(-2/|\lambda|N(0))} \exp(-2/|\lambda|N(0)). \tag{4.11}$$

Of course, if the attraction is strong, there is a bound state with $|E| \approx \omega_c |\lambda| N(0)$. But in the weak coupling limit it still exists, and its energy is given by

$$|E| \approx 2\omega_c \exp(-2/|\lambda|N(0)). \tag{4.12}$$

Thus, even an infinitesimal electron–electron attraction near the Fermi surface creates a bound state of two electrons. This is greatly due to the effectively 2 D situation, which allowed us to write instead of the 3 D D.O.S., $N(E) \propto E^{1/2}$, the 2 D one, $N(E) \approx N(0) = \text{const.}$

The binding energy is enormously sensitive to the interaction strength, $|\lambda|$ (4.12) and is an essentially nonanalytic function of $|\lambda|$ when $|\lambda| \to 0$ (all terms in its Taylor expansion around this point are zeros). This indicates the instability of the normal ground state and unavoidable failure of any perturbation scheme starting from it.

The above calculations have a serious flaw: they are essentially "few-body" and deal rather with a single pair of real electrons injected into metal than with quasiparticles. Thus, only the possibility to leaving the $(\mathbf{k} \uparrow, -\mathbf{k} \downarrow)$ state due to interaction is considered, while the inverse process is omitted (when *another* pair is scattered from $(\mathbf{k'} \uparrow, -\mathbf{k'} \downarrow)$ to $(\mathbf{k} \uparrow, -\mathbf{k} \downarrow)$). The quasiparticles *below* the Fermi surface (quasiholes) also were not considered, leading to a wrong exponent in the expression for the binding energy (4.12): in BCS theory we have $\exp(-1/|\lambda|N(0))$ instead of $\exp(-2/|\lambda|N(0))$ (an enormous difference, since $\exp(-1/|\lambda|N(0))$ is already a small number). Then, the quasiparticle approach allows us to explain how the pairs (binding energy $\approx k_B T_c \approx 10^{-4}$ eV) survive the Coulomb electron–electron and electron–phonon interactions, with a scale of some 1 eV per particle. (In the quasiparticle picture all these interactions are included into background, renormalizing the quasiparticle mass, etc.; what is left (if any) is just the effective weak (quasi)electron-(quasi)electron attraction.)

The existence of many pairs means that we no longer can be sure whether a given $(\mathbf{k} \uparrow, -\mathbf{k} \downarrow)$-state is occupied or empty; we have instead to introduce the *probability amplitudes* that it is occupied, $v_{\mathbf{k}}$, or free, $u_{\mathbf{k}}$;

$$|v_{\mathbf{k}}|^2 + |u_{\mathbf{k}}|^2 = 1. \tag{4.13}$$

The fact that we use the amplitudes, not the probabilities, shows that this uncertainty is *not* due to the usual scattering: we have a *coherent state*, preserving the quantum-mechanical phase effects.

As usual, the price to be paid for the advantages of the many-body approach is high. The calculations, though, could be greatly simplified if some sort of a mean field approximation (MFA) can be applied. For this, the pair concentration must be large enough. Let us make an estimate. The size of a Cooper pair may be defined as

$$
\begin{aligned}
(\Delta\varrho)^2 &= \frac{\int |\phi(\varrho)|^2 \varrho^2 d^3\varrho}{\int |\phi(\varrho)|^2 d^3\varrho} = \frac{\sum_{\mathbf{k}} |\nabla_{\mathbf{k}} a_{\mathbf{k}}|^2}{\sum_{\mathbf{k}} |a_{\mathbf{k}}|^2} \\
&\approx \frac{N(0)(\partial\xi/\partial k)^2_{\xi=0} \int_0^\infty (\partial a/\partial\xi)^2 d\xi}{N(0)\int_0^\infty a^2 d\xi}
\end{aligned} \tag{4.14}
$$

(ξ is the energy measured from the Fermi level).

Because of

$$(E - 2E_{\mathbf{k}})a_{\mathbf{k}} = \sum_{\mathbf{k}'} \langle \mathbf{k}\uparrow, -\mathbf{k}\downarrow |V|\mathbf{k}'\uparrow, -\mathbf{k}'\downarrow\rangle a_{\mathbf{k}'}$$

$$= \lambda w_{\mathbf{k}}^* \sum_{\mathbf{k}'} w_{\mathbf{k}'} a_{\mathbf{k}'} = \text{const}$$

(if $\mathbf{k}$ is near the Fermi surface), we find that

$$a(\xi) = \frac{\text{const}}{E - 2\xi}; \qquad \frac{\partial a}{\partial \xi} = \frac{2\,\text{const}}{(E - 2\xi)^2}.$$

Therefore,

$$\Delta\varrho \approx \sqrt{\frac{4}{3}\frac{\hbar v_F}{E}}. \tag{4.15}$$

If we take $E = k_B T_c$, we get quite a macroscopic length, $\Delta\varrho \approx 10^{-4}$ cm. The density of pairs at sufficiently low temperature $T \ll T_c$ must be, on the other hand, of the order of the density of the electrons themselves, $n_e \approx 10^{22}$ cm^{-3}. Thus in the volume of a single pair there exist at least 10^{10} more pairs. This provides us with perfect conditions for MFA application.

To use this advantage, note that the term in the Hamiltonian for the electron–electron attraction described above is of the form

$$H_i = \sum_{\mathbf{k}\mathbf{k}'} V_{\mathbf{k}\mathbf{k}'} a_{\mathbf{k}\uparrow}^\dagger a_{-\mathbf{k}\downarrow}^\dagger a_{\mathbf{k}'\downarrow} a_{-\mathbf{k}'\uparrow}.$$

This term contains four operators and will lead to the four-legged vertex in the diagram techniques. Along with the sharp momentum dependence of the matrix element, this is a rather unpleasant thing to deal with. The MFA immediately allows us to get rid of two of these operators, writing instead of them the average value (and then obtaining for it the self-consistent equation). But the usual choice in accordance with Wick's theorem, say, $a^\dagger a^\dagger aa \to \langle a^\dagger a\rangle a^\dagger a$, does not give rise to any superconductivity: this term is for usual electron scattering and has nothing to do with pairing instability. Anyway, it could have been included in other terms of the Hamiltonian or/and in the renormalization of quasiparticle characteristics.

To keep the pairs in our approximation, we must make a crazy step and write

$$H_i \to \mathcal{H}_{i,\text{MFA}} = \sum_{\mathbf{k}\mathbf{k}'} V_{\mathbf{k}\mathbf{k}'} a_{\mathbf{k}\uparrow}^\dagger a_{-\mathbf{k}\downarrow}^\dagger \langle a_{\mathbf{k}'\downarrow} a_{-\mathbf{k}'\uparrow}\rangle + \text{Hermitian conjugate}$$

$$= \sum_{\mathbf{k}} \left\{ \Delta_{\mathbf{k}} a_{\mathbf{k}\uparrow}^\dagger a_{-\mathbf{k}\downarrow}^\dagger + \Delta_{\mathbf{k}}^* a_{-\mathbf{k}\downarrow} a_{\mathbf{k}\uparrow} \right\}. \tag{4.16}$$

The quantity $\Delta_{\mathbf{k}}$ is, naturally, called the *pairing potential*:

$$\Delta_{\mathbf{k}} = \sum_{\mathbf{k}'} V_{\mathbf{k}\mathbf{k}'} \langle a_{\mathbf{k}'\downarrow} a_{-\mathbf{k}'\uparrow} \rangle. \qquad (4.17)$$

The craziness of this step consists in the fact that we have introduced *anomalous averages* of two creation or annihilation operators, which must be zero in any state with a fixed number of particles! But there is a method (or at least self-consistency) in this madness. The Hamiltonian $\mathcal{H}_{i,\mathrm{MFA}}$ also violates the law of conservation of the particles' number, creating and annihilating them in pairs. Then an average like $\langle aa \rangle$ or $\langle a^{\dagger} a^{\dagger} \rangle$ calculated using this Hamiltonian will *not* be zero, and the ends are met. Moreover, as you certainly know, the nonzero anomalous averages are the fundamental feature of the superconducting state. They describe the *off-diagonal long-range order* (ODLRO) of this state (the concept introduced by Yang [27])—the sort of symmetry that makes it qualitatively different from the normal state.

ODLRO is so called because the anomalous averages are related to off-diagonal terms of the density matrix of the system (which, as you know, are responsible for the quantum coherence phenomena). The long-range order can be understood as follows. The *pair-pair correlation function,*

$$S_{\uparrow\downarrow}(\mathbf{r}_1, \mathbf{r}_2) = \langle \Psi_{\uparrow}^{\dagger}(\mathbf{r}_1) \Psi_{\downarrow}^{\dagger}(\mathbf{r}_1) \Psi_{\downarrow}(\mathbf{r}_2) \Psi_{\uparrow}(\mathbf{r}_2) \rangle, \qquad (4.18)$$

describes the correlation between the pairs at $\mathbf{r}_1$ and at $\mathbf{r}_2$. When the distance between them grows, this function must factorize (because of the general principle of the correlations' extinction):

$$S_{\uparrow\downarrow}(\mathbf{r}_1, \mathbf{r}_2) \underset{|\mathbf{r}_1 - \mathbf{r}_2| \to \infty}{\sim} \langle \Psi_{\uparrow}^{\dagger}(\mathbf{r}_1) \Psi_{\downarrow}^{\dagger}(\mathbf{r}_1) \rangle \times \langle \Psi_{\downarrow}(\mathbf{r}_2) \Psi_{\uparrow}(\mathbf{r}_2) \rangle. \qquad (4.19)$$

In the normal state it factorizes trivially, for anomalous averages are zero. But in the superconducting state we get a nontrivial factorization

$$S_{\uparrow\downarrow}(\mathbf{r}_1, \mathbf{r}_2) \underset{|\mathbf{r}_1 - \mathbf{r}_2| \to \infty}{\propto} \Delta^*(\mathbf{r}_1) \Delta(\mathbf{r}_2), \qquad (4.20)$$

which means that there exists a long-range correlation between the electronic states (*macroscopic quantum coherence*). The superconductor is in an *ordered* state. Therefore, the pairing potential Δ is also called the *order parameter.*

But if these anomalous averages are so important, then the superconductivity should not appear in a system with a fixed number of particles. That is, the super-conducting ground state seems to be forbidden in a closed system.

The most trivial answer (and the correct one) is that there are no isolated objects in this Universe; the electrons are, in principle, delocalized, etc., so that two electrons more or less don't play any role in practice.

Another answer, also true, is that, O.K., anomalous averages are important. But let us think: what are the *observable* consequences of nonzero Δ, Δ^*? We shall see that the superconductor is described in terms of $|\Delta|$ and $\mathrm{Arg}\Delta$.

Fig. 4.4 Are the anomalous
averages real?

We notice, then, that for an isolated system *anything* (thermodynamic properties, for example) is defined only by $|\Delta|^2 \equiv \Delta^*\Delta \propto \langle a^\dagger a^\dagger a a\rangle$, which is nonzero even if the number of particles is fixed. And the phase of Δ in this situation is physically senseless.

Of course, there exist situations when the phase *is* significant (the Josephson effect, for instance). But then we have an *essentially* open system-where electrons can tunnel this way and that way, so their number in either superconductor is fluctuating.

We see, then, that this "nonconservation" is rather an artifact related to a certain way of describing the superconductivity, MFA Hamiltonian, than some mystical physical phenomenon (see Fig. 4.4).[1]

Let us recall that in the second quantization method we also had violated all the conservation laws, by introduction of the creation/annihilation operators. The only difference is that then we restored justice immediately, in the very term (writing something like $a^\dagger a$), and now we wait until the end of the calculations.

Now let us look at the problem from another point of view.

As you know, if we calculate the average magnetic moment of the magnetic system in zero magnetic field, the result is zero even below the Curie point. To get a proper result—finite spontaneous magnetization **M**—you have to perform a trick: turn on an infinitesimal magnetic field **h**, find **M**, and then set **h** = 0.

Why? It is simple. The magnetic moment is a vector. Then the calculation of its average value includes an averaging over directions. But without the magnetic field, all directions of **M** are equally probable, and the average is zero! An arbitrarily small field, though, orients **M** along itself, and it is no longer averaged to zero. This sort of average was first introduced by Bogoliubov and is called *quasiaverages*.

---

[1] Of course, if you deal with a *very small* system, when one or two extra electrons significantly change the total energy, certain precautions must be taken; an example-the *parity effect*—will be discussed later.

Fig. 4.5 Spontaneous symmetry breaking

This is an example of a very general and very important situation. The symmetry of the Hamiltonian is higher, than the symmetry of the ground state: that is, the symmetry is *spontaneously broken*.

The role played by the field **h** is just to reduce the symmetry of the Hamiltonian to that of the ground state, through the term**-Mh**.

The fact that below the transition temperature an infinitesimal external field leads to finite magnetization in our calculations means that the previous ground state, with zero **M**, has lost stability and spontaneously acquired a finite magnetic moment. (Therefore we speak of a *spontaneous symmetry breaking*; see Fig. 4.5.)

We encounter a like situation in a superconductor. If we introduce the pair sources into the Hamiltonian, say $f\, a^\dagger a^\dagger$ and $f^* a a$, calculate $\langle a^\dagger a^\dagger \rangle$ and $\langle a a \rangle$, and then put $f = f^* = 0$, above T_c we get zero. But below T_c we find a finite result even for zero pair sources, and it is possible to show that all the averages of observables calculated with the help of our "pair" Hamiltonian $\mathcal{H}_{\mathrm{MFA}}$ are the same as quasiaverages obtained by more complicated procedure explained above. The normal state is then unstable, but the real particle non-conservation is not necessary.

The principle "push the one who is falling" is never violated. Thus, *any* instability will fully develop, and below T_c we never find the normal state. We find instead the qualitatively different, superconducting one.

Conclusions

The normal ground state of the superconductor is unstable below the transition temperature with regard to the creation of Cooper pairs of electrons (in the case of arbitrarily weak electron–electron attraction near the Fermi surface).

The superconducting ground state is qualitatively different from the normal one and possesses the special kind of symmetry that is revealed in the existence of nonzero

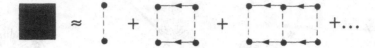

Fig. 4.6 Ladder series for the vertex function

anomalous averages (related to the superconducting order parameter) and leads to the macroscopic quantum coherence in the system.

The standard many-body perturbation theory starting from the normal state is unable to describe the superconducting one; nevertheless, it allows us to find the transition point, where the instability arises.

The apparent violation of the particle conservation law due to the existence of nonzero anomalous averages does not lead to any physically observable consequences in massive superconductors.

4.2 Instability of the Normal State

The poles of the two-particle Green's function correspond to the bound states of two quasiparticles (such as plasmons). This is me for the temperature Green's functions as well. We will show that if there exists an arbitrarily weak attraction between the quasiparticles on the Fermi surface, the pole appears in the vertex function (see Lifshits and Pitaevski [5]). As we know, this is equivalent to the pole in the two-particle Green's function itself.

Let us consider the temperature vertex function,

$$\Gamma(\mathbf{p}_1 \varpi_1, \mathbf{p}_2 \varpi_2; \mathbf{p}_1' \varpi_1', \mathbf{p}_2' \varpi_2'), \tag{4.21}$$

under the conditions

$$\mathbf{p}_1' + \mathbf{p}_2' = 0; \quad |\mathbf{p}_1'| = p_F; \quad \varpi_1' = \varpi_2' = 0. \tag{4.22}$$

This means that the pairing occurs on the Fermi surface, with zero binding energy (this must be so at the very transition point), and zero total momentum (this was discussed earlier).

The pole arises due to the ladder diagrams of Fig. 4.6. (We do not have to consider the diagrams with interchanged ends, for the pole arises in both series simultaneously.) The Bethe-Salpeter equation can be thus written as follows (Fig. 4.7):

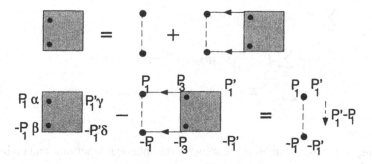

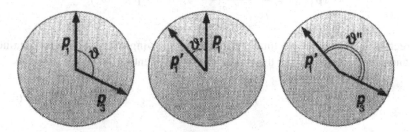

Fig. 4.7 Bethe-Salpeter equation in the ladder approximation

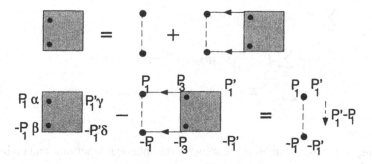

Fig. 4.8 Parametrization of momenta vectors on the Fermi sphere

$$\Gamma(\mathbf{p}_1, \ -\mathbf{p}_1; \mathbf{p}'_1, -\mathbf{p}'_1)\delta_{\alpha\gamma}\delta_{\beta\delta}$$
$$+\frac{1}{\beta}\sum_{s=-\infty}^{\infty}\int\frac{d^3\mathbf{p}_3}{(2\pi)^3}\delta_{\alpha\eta}\mathcal{G}^0(\mathbf{p}_3\varpi_s)\delta_{\beta\mu}\mathcal{G}^0(-\mathbf{p}_3,-\varpi_s)$$
$$\times U(\mathbf{p}_3-\mathbf{p}_1)\Gamma(\mathbf{p}_3,-\mathbf{p}_3;\mathbf{p}'_1,-\mathbf{p}'_1)\delta_{\eta\gamma}\delta_{\mu\delta} \tag{4.23}$$
$$= -U(\mathbf{p}'_1-\mathbf{p}_1)\delta_{\alpha\gamma}\delta_{\beta\delta}.$$

In the sum and integral, only $\mathbf{p}_3$ close to the Fermi surface and small Matsubara frequencies ϖ are important. Therefore, we can set $|\mathbf{p}_3| = p_F, \varpi_s = 0$ in the arguments of both Γ and U under the summation and integration.

Now all the vector arguments lie on the Fermi sphere, and Γ, U depend each on a *single* variable (the angle between the corresponding momenta). Then they can be expanded in Legendre polynomials:

$$U(\theta) = \sum_{l=0}^{\infty}(2l+1)u_l P_l(\cos\theta); \tag{4.24}$$

$$\Gamma(\theta) = \sum_{l=0}^{\infty}(2l+1)\gamma_l P_l(\cos\theta). \tag{4.25}$$

Then the equation takes the form

$$\sum_l (2l + 1)(-u_l - \gamma_l) P_l(\cos \theta)$$

$$= \frac{1}{\beta} \sum_s \int \frac{d^3 \mathbf{p}_3}{(2\pi)^3} \mathcal{G}^0(\mathbf{p}_3 \varpi_s) \mathcal{G}^0(-\mathbf{p}_3, -\varpi_s)$$

$$\times \sum_{l'} \sum_{l''} (2l' + 1)(2l'' + 1) u_{l'} \gamma_{l''} P_{l'}(\cos \theta') P_{l''}(\cos \theta'') \qquad (4.26)$$

(see Fig. 4.8).

Making use of the summation theorem for the spherical functions, and taking into account that

$$\mathcal{G}^0(\mathbf{p}_3 \varpi_s) = \frac{1}{i\varpi_s - \frac{p_3^2}{2m} + \mu} = \mathcal{G}^0(-\mathbf{p}_3, -\varpi_s)^*$$

is direction independent, we finally obtain the following expression, first obtained by Landau and Pitaevskii, for the lth angular component of the vertex function:

$$\gamma_l = \frac{-u_l}{1 + u_l \Pi}. \qquad (4.27)$$

In this expression

$$\Pi = \frac{1}{\beta} \sum_s \int \frac{d^3 \mathbf{p}_3}{(2\pi)^3} \left| \mathcal{G}^0(\mathbf{p}_3 \varpi_s) \right|^2 = \int \frac{d^3 \mathbf{p}_3}{(2\pi)^3} \frac{1}{\beta} \sum_s \frac{1}{\varpi_s^2 + \xi_p^2}. \qquad (4.28)$$

The sum quickly converges. Using the formula for summation over fermionic Matsubara frequencies, we find that

$$\Pi = \int \frac{d^3 \mathbf{p}}{2(2\pi)^3} \frac{\tanh \frac{\beta \xi_p}{2}}{\xi_p}. \qquad (4.29)$$

The high-momentum divergence is not a physical one, and as we have mentioned earlier, it is to be cut off at some $p_{\max} \approx 1/a$. Thus we have obtained the condition for the appearance of the pole in the vertex function:

$$\frac{-u_l}{2(2\pi)^3} \int d^3 \mathbf{p} \frac{\tanh \frac{\beta \xi_p}{2}}{\xi_p} = 1. \qquad (4.30)$$

This very equation (with $l = 0$) is obtained from BCS theory if we put the order parameter $\Delta = 0$ (i.e., in the transition point). The critical temperature in the lth channel is thus, up to a factor of order unity,

$$T_c^{(l)} \approx \epsilon_{\max} \exp\{-\frac{1}{N(0)|u_l|}\}. \qquad (4.31)$$

Here ϵ_{max} is the energy cutoff parameter. We see that the bound state appears if at least one of the coefficients u_l in the potential's angular expansion is negative (attraction). The transition takes place at the largest of the $T_c^{(l)}$, and the nonanalytic dependence of $T_c^{(l)}$ on the interaction parameter is properly restored.

In conventional superconductors the electron–electron attraction appears already in the s-wave channel ($l = 0$) due to the phonon-mediated electron–electron coupling (see Sect. 4.4.2). But for $l > 0$ negative coefficients may arise from the bare repulsive electron–electron interaction without such an intermediary. This so-called *Kohn–Luttinger pairing mechanism* [21] has the same origin as Friedel oscillations of screened electrostatic potential at large distances (see Appendix A). Namely, due to the sharp edge of Fermi distribution the polarization operator, as a function of momentum, is non-analytic when the momentum transfer is close to $2p_F$. As a result, even starting from a purely repulsive bare potential, *all* angular components u_l of which are positive, one obtains the screened potential, where at least some components u_l^{eff} are negative. Substituting such a u_l^{eff} in Eq. (4.27) instead of u_l, we obtain the instability of the normal state and the superconducting transition. Note that the existence of a sharp Fermi surface and the resulting effective reduction of the dimensionality of the momentum space, which was critically important for the formation of Cooper pairs, is also crucial here.

The Kohn–Luttinger mechanism offers a possibility to explain the superconducting pairing with $l \neq 0$ in systems, where the phonon-mediated coupling is too weak (like in high-T_c cuprates). While the original version of the Kohn–Luttinger argument does not directly apply to the systems without spherical symmetry, its recent extensions to lattice models provide interesting insight in the possible nature of superconducting state in the cuprates, the Fe- graphene (see Maiti and Chubukov [6]). In the following, though, we will mainly concern ourselves with conventional, s-wave, phonon-mediated superconductivity.

Finally, note also an important difference of Eq. (4.31) from the result we obtained earlier: $-1/(N(0)|u_l|)$ in the exponent, instead of twice this value, as followed from Cooper's initial argument. This is because now we have properly taken into account the many-body character of the problem. Unfortunately, we cannot proceed any further: the fact of *some* transition due to electron–electron coupling, and the temperature of this transition are the *only* things which can be obtained by the usual technique, which starts from the normal ground state. In fact, no real bound state (in terms of quasiparticles above the normal vacuum) appear, but rather instability arises, which cannot be dealt with.

Therefore, we need the modified formalism, which will be built with the help of the so-called pairing Hamiltonian.

4.3 Pairing (BCS) Hamiltonian

4.3.1 Derivation of the BCS Hamiltonian

We start from the Hamiltonian with a four-fermion attractive term

$$\mathcal{H} = \mathcal{H}_0 + \mathcal{H}_{\text{int}} = \mathcal{H}_0 + g \int d^3\mathbf{r} \psi^\dagger_\uparrow(\mathbf{r})\psi^\dagger_\downarrow(\mathbf{r})\psi_\downarrow(\mathbf{r})\psi_\uparrow(\mathbf{r}), \qquad g < 0 \qquad (4.32)$$

(an appropriate momentum cutoff confining the interaction into the narrow layer near the Fermi surface is implied). From the naïve point of view we have followed earlier, the MFA pairing Hamiltonian is obtained by selective averaging of operators in the previous expression:

$$\mathcal{H}_{\text{MFA}} = \mathcal{H}_0 + g \int d^3\mathbf{r} \left\{ \psi^\dagger_\uparrow(\mathbf{r})\psi^\dagger_\downarrow(\mathbf{r})\langle\psi_\downarrow(\mathbf{r})\psi_\uparrow(\mathbf{r})\rangle + \left\langle \psi^\dagger_\uparrow(\mathbf{r})\psi^\dagger_\downarrow(\mathbf{r}) \right\rangle \psi_\downarrow(\mathbf{r})\psi_\uparrow(\mathbf{r}) \right\}$$

$$\equiv \mathcal{H}_0 - \int d^3\mathbf{r} \left\{ \psi^\dagger_\uparrow(\mathbf{r})\psi^\dagger_\downarrow(\mathbf{r})\Delta(\mathbf{r}) + \Delta^*(\mathbf{r})\psi_\downarrow(\mathbf{r})\psi_\uparrow(\mathbf{r}) \right\}; \qquad (4.33)$$

$$\Delta(\mathbf{r}) = |g|\langle\psi_\downarrow(\mathbf{r})\psi_\uparrow(\mathbf{r})\rangle. \qquad (4.34)$$

The result is correct, but its foundation seems to be a bit shaky. In the following we pursue two objectives: to show that the pairing Hamiltonian is more reliable than one may guess, and how the corrections can be introduced in a regular way (following Svidzinskii [8]). The equilibrium properties of the system can be derived from its grand partition function,

$$\Xi = \text{tr} e^{-\beta(\mathcal{H}_0 + \mathcal{H}_{\text{int}})}, \qquad (4.35)$$

which, as we know from the Matsubara formalism, can be written as

$$\Xi = \text{tr} \left\{ e^{-\beta\mathcal{H}_0} T_\tau e^{-\int_0^\beta d\tau \mathcal{H}_{\text{int}}(\tau)} \right\}, \qquad (4.36)$$

where $\mathcal{H}_{\text{int}}$ is taken in Matsubara interaction representation. Rewriting Eq. 4.36 in more detail, we have

$$\Xi = \text{tr} \left\{ e^{-\beta\mathcal{H}_0} T_\tau e^{+\int_0^\beta d\tau \int d^3\mathbf{r} \overline{\mathcal{A}}(\mathbf{r},\tau)\mathcal{A}(\mathbf{r},\tau)} \right\}, \qquad (4.37)$$

where

$$\overline{\mathcal{A}}(\mathbf{r}, \tau) = \sqrt{|g|}\overline{\psi}_\uparrow(\mathbf{r}, \tau)\overline{\psi}_\downarrow(\mathbf{r}, \tau); \qquad (4.38)$$

$$\mathcal{A}(\mathbf{r}, \tau) = \sqrt{|g|}\psi_\downarrow(\mathbf{r}, \tau)\psi_\uparrow(\mathbf{r}, \tau). \qquad (4.39)$$

Now we use one of the functional integration formulae, which is a direct generalization of the standard formula for Gaussian integrals

$$e^{A^2} = \frac{1}{\sqrt{\pi}} \int_{-\infty}^{\infty} dx e^{-x^2+2Ax} = \frac{\int_{-\infty}^{\infty} dx e^{-x^2+2Ax}}{\int_{-\infty}^{\infty} dx e^{-x^2}}. \tag{4.40}$$

Let us write

$$\mathcal{A} = \mathcal{P} + i\mathcal{Q}, \quad \overline{\mathcal{A}} = \mathcal{P} - i\mathcal{Q}, \tag{4.41}$$

where $\mathcal{P}$, $\mathcal{Q}$ are Hermitian operators. Then reordering the operators under the sign of the $\mathcal{T}_\tau$-operator, we can obtain the expression

$$\mathcal{T}_\tau e^{\int_0^\beta d\tau \int d^3 \mathbf{r} \overline{\mathcal{A}}(\mathbf{r},\tau)\mathcal{A}(\mathbf{r},\tau)} = \mathcal{T}_\tau e^{\int_0^\beta d\tau \int d^3 \mathbf{r} \overline{\mathcal{P}}^2(\mathbf{r},\tau)} e^{\int_0^\beta d\tau \int d^3 \mathbf{r} \overline{\mathcal{Q}}^2(\mathbf{r},\tau)}, \tag{4.42}$$

and then, introducing two real auxiliary fields $\xi(\mathbf{r}, \tau)$, $\eta(\mathbf{r}, \tau)$:

$$\mathcal{T}_\tau e^{\int_0^\beta d\tau \int d^3 \mathbf{r} \overline{\mathcal{A}}(\mathbf{r},\tau)\mathcal{A}(\mathbf{r},\tau)} = \int \mathcal{D}\xi(\mathbf{r},\tau)\mathcal{D}\eta(\mathbf{r},\tau) \left\{ e^{-\int_0^\beta d\tau \int d^3 \mathbf{r}[\xi^2(\mathbf{r},\tau)+\eta^2(\mathbf{r},\tau)]} \right.$$

$$\times \mathcal{T}_\tau e^{\int_0^\beta d\tau \int d^3 \mathbf{r} 2[\mathcal{P}(\mathbf{r},\tau)\xi(\mathbf{r},\tau)+\mathcal{Q}(\mathbf{r},\tau)\eta(\mathbf{r},\tau)]} \Big\}$$

$$\times \left\{ \int \mathcal{D}\xi(\mathbf{r},\tau)\mathcal{D}\eta(\mathbf{r},\tau) e^{-\int_0^\beta d\tau \int d^3 \mathbf{r}[\xi^2(\mathbf{r},\tau)+\eta^2(\mathbf{r},\tau)]} \right\}^{-1}. \tag{4.43}$$

Introducing then instead of $\xi(\mathbf{r}, \tau)$, $\eta(\mathbf{r}, \tau)$ a *complex* field

$$\varsigma(\mathbf{r}, \tau) = \xi(\mathbf{r}, \tau) + i\eta(\mathbf{r}, \tau), \tag{4.44}$$

noticing that

$$-[\xi^2 + \eta^2] + 2[\mathcal{P}\xi + \mathcal{Q}\eta] = -|\varsigma|^2 + \mathcal{A}\varsigma^* + \overline{\mathcal{A}}\varsigma, \tag{4.45}$$

and finally setting

$$\varsigma(\mathbf{r}, \tau) = \frac{1}{\sqrt{|g|}} \Delta(\mathbf{r}, \tau), \tag{4.46}$$

we can write the partition function as follows:

$$\Xi = \Xi_0 \int \mathcal{D}\Delta(\mathbf{r},\tau)\mathcal{D}\Delta^*(\mathbf{r},\tau) e^{-\frac{1}{|g|}\int_0^\beta d\tau \int d^3 \mathbf{r}|\Delta(\mathbf{r},\tau)|^2}$$

$$\times \left\langle \mathcal{T}_\tau e^{+\int_0^\beta d\tau \int d^3 \mathbf{r}[\Delta(\mathbf{r},\tau)\overline{\psi}_\uparrow(\mathbf{r},\tau)\overline{\psi}_\downarrow(\mathbf{r},\tau)+\Delta(\mathbf{r},\tau)^*\psi_\downarrow(\mathbf{r},\tau)\psi_\uparrow(\mathbf{r},\tau)]} \right\rangle_0 \tag{4.47}$$

$$\times \left\{ \int \mathcal{D}\Delta(\mathbf{r},\tau)\mathcal{D}\Delta^*(\mathbf{r},\tau) e^{-\frac{1}{|g|}\int_0^\beta d\tau \int d^3 \mathbf{r}|\Delta(\mathbf{r},\tau)|^2} \right\}^{-1}.$$

The partition function contains a Gaussian average over the pair sources' fields, Δ, Δ^*, of the *Bogoliubov functional*

$$\Xi_B[\Delta, \Delta^*] \equiv e^{-\beta \Omega_B[\Delta, \Delta^*]} = \left\langle T_\tau e^{-\int_0^\beta d\tau \mathcal{H}_B(\tau)} \right\rangle_0, \tag{4.48}$$

with the pairing Hamiltonian $\mathcal{H}_B$ (in Matsubara representation). Then the approximation of the pairing Hamiltonian corresponds to the main order in the expansion of the functional integral over the pair sources' fields in (4.48). The corresponding values of these fields are given by the extremum condition:

$$e^{-\frac{1}{|g|} \int_0^\beta d\tau \int d^3 r |\Delta(\mathbf{r}, \tau)|^2 - \beta \Omega_B[\Delta, \Delta^*]} = \max, \tag{4.49}$$

that is[2]

$$\Delta^*(\mathbf{r}, \tau) = -\beta|g| \frac{\delta \Omega_B[\Delta, \Delta^*]}{\delta \Delta(\mathbf{r}, \tau)}. \tag{4.50}$$

By our definition, $\Xi = \Xi_0 \Xi_B = \Xi_0 e^{-\beta \Omega_B}$. Then

$$\frac{\delta \Omega_B}{\delta \Delta} = -\beta^{-1} \frac{\delta \ln(\Xi/\Xi_0)}{\delta \Delta} = -\beta^{-1} \frac{\delta \ln \Xi}{\delta \Delta}. \tag{4.51}$$

Substituting here the partition function (4.48), we obtain

$$\frac{\delta \Omega_B}{\delta \Delta(\mathbf{r}, \tau)} = -\frac{1}{\beta \Xi} \text{tr} \left\{ e^{-\beta \mathcal{H}_0} T_\tau \overline{\psi}_\uparrow(\mathbf{r}, \tau) \overline{\psi}_\downarrow(\mathbf{r}, \tau) \right.$$
$$\times \left. e^{+\int_0^\beta d\tau \int d^3 r [\Delta(\mathbf{r}, \tau)_{-\uparrow}(\mathbf{r}, \tau)_{-\downarrow}(\mathbf{r}, \tau) + \Delta(\mathbf{r}, \tau)^* \psi_\downarrow(\mathbf{r}, \tau) \psi_\downarrow(\mathbf{r}, \tau)]} \right\}$$
$$= -\beta^{-1} \langle T_\tau \overline{\psi}_\uparrow(\mathbf{r}, \tau) \overline{\psi}_\downarrow(\mathbf{r}, \tau) \rangle, \tag{4.52}$$

which finally yields the *self-consistency relation*

$$\Delta^*(\mathbf{r}, \tau) = |g| \langle T_\tau \overline{\psi}_\uparrow(\mathbf{r}, \tau) \overline{\psi}_\downarrow(\mathbf{r}, \tau) \rangle. \tag{4.53}$$

Its conjugate coincides with the relation (4.34) obtained from the "naïve" point of view.

---

[2] The functional, or variational, derivative of a functional $F[f]$ of a function $f(x)$ is denoted by $\delta F/\delta f(x)$. It describes the variation of the functional when $f(x)$ is replaced by $f(x) + \delta f(x)$, in (Footnote 2 continued) the first order in $\delta f(x)$,

$$F[f + \delta f] - F[f] \equiv \int dx \left(\frac{\delta F}{\delta f(x)} \right) \delta f(x) + \cdots.$$

Higher-order functional derivatives are introduced in the similar way.

The advantage of our approach is that we not only have established the validity of the "MFA" Hamiltonian in the superconducting case, but also found its limits and way of introducing the necessary corrections: there may occur situation in which not only the extremal value (4.53), but its vicinity as well is to be taken into account.[3]

4.3.2 Diagonalization of the BCS Hamiltonian: The Bogoliubov Transformation—Bogoliubov-de Gennes Equations

The BCS Hamiltonian can be diagonalized by a canonical *Bogoliubov transformation*, which expresses the electron field operators in terms of new creation/annihilation Fermi operators,

$$\alpha^{\dagger}_{q,\downarrow\uparrow}, \alpha_{q,\uparrow\downarrow}$$

(now we use a usual, time-dependent representation; q labels the eigenstates). The direct transformation is given by

$$\alpha_{q,\uparrow} = \int d^3\mathbf{r} \left[u^*_q(\mathbf{r})\psi_{\uparrow}(\mathbf{r},t) - v^*_q(\mathbf{r})\psi^{\dagger}_{\downarrow}(\mathbf{r},t) \right]; \qquad (4.54)$$

$$\alpha^{\dagger}_{q,\downarrow} = \int d^3\mathbf{r} \left[u_q(\mathbf{r})\psi^{\dagger}_{\downarrow}(\mathbf{r},t) + v_q(\mathbf{r})\psi_{\uparrow}(\mathbf{r},t) \right], \qquad (4.55)$$

while the inverse transformation is

$$\psi_{\uparrow}(\mathbf{r},t) = \sum_q \left[u_q(\mathbf{r})\alpha_{q,\uparrow} + v^*_q(\mathbf{r})\alpha^{\dagger}_{q,\downarrow} \right]; \qquad (4.56)$$

$$\psi^{\dagger}_{\downarrow}(\mathbf{r},t) = \sum_q \left[u^*_q(\mathbf{r})\alpha^{\dagger}_{q,\downarrow} - v_q(\mathbf{r})\alpha_{q,\uparrow} \right]. \qquad (4.57)$$

As you see, the Bogoliubov transformation mixes electron and hole operators with opposite spins—this is the only way to get rid of nondiagonal pairing terms! The physical significance of this is that the quasiparticles in the superconductor are rather like centaurs: "part electrons, part holes." For obvious reasons they are called *bogolons*.

The coefficients of the Bogoliubov transformation must satisfy the following canonical relations,

---

[3] You can see from the structure of the functional integral that the overall sign (or, more generally, the initial phase) of the complex field Δ is of no importance. In different books you thus can find the same equation with opposite signs of Δ. This is a matter of convention.

$$\sum_q \left[u_q(\mathbf{r}) u_q^*(\mathbf{r'}) + v_q(\mathbf{r'}) v_q^*(\mathbf{r}) \right] = \delta(\mathbf{r} - \mathbf{r'}); \tag{4.58}$$

$$\sum_q \left[u_q(\mathbf{r}) v_q^*(\mathbf{r'}) - u_q(\mathbf{r'}) v_q^*(\mathbf{r}) \right] = 0; \tag{4.59}$$

$$\int d^3\mathbf{r} \left[u_q(\mathbf{r}) u_{q'}^*(\mathbf{r}) + v_q(\mathbf{r}) v_{q'}^*(\mathbf{r}) \right] = \delta_{qq'}; \tag{4.60}$$

$$\int d^3\mathbf{r} \left[u_q(\mathbf{r}) v_{q'}(\mathbf{r}) - u_{q'}(\mathbf{r}) v_q(\mathbf{r}) \right] = 0, \tag{4.61}$$

in order to comply with Fermi statistics of both old and new creation/annihilation operators. (To check this would be *a* really useful exercise, even in the simplest case of the plane wave basis.)

In new operators the Hamiltonian takes the simple form

$$\mathcal{H}_B = U_0 + \sum_q E_q \left(\alpha_{q,\uparrow}^\dagger \alpha_{q,\uparrow} + \alpha_{q,\downarrow}^\dagger \alpha_{q,\downarrow} \right). \tag{4.62}$$

The first term here is the ground state energy of the superconductor. The second one is the quasiparticle term, which describes the elementary excitations above the ground state.

The excitation energies along with the transformation coefficients are given by the solution of the following system of *Bogoliubov-de Gennes equations*:

$$\left[\frac{1}{2m} \left(\frac{1}{i} \nabla - e\mathbf{A}/c \right)^2 - \mu + V(\mathbf{r}) \right] u_q(\mathbf{r}) + \Delta(\mathbf{r}) v_q(\mathbf{r}) = E_q u_q(\mathbf{r});$$

$$\left[\frac{1}{2m} \left(\frac{1}{i} \nabla + e\mathbf{A}/c \right)^2 - \mu + V(\mathbf{r}) \right] v_q(\mathbf{r}) - \Delta^*(\mathbf{r}) u_q(\mathbf{r}) = -E_q v_q(\mathbf{r}),$$

or denoting the kinetic energy operator and its conjugate by $\hat{\xi}, \hat{\xi}_c$ (it is convenient to measure energies from the Fermi level),

$$\hat{\xi} = \frac{1}{2m} \left(\frac{1}{i} \nabla - e\mathbf{A}/c \right)^2 - \mu,$$

$$\hat{\xi}_c = \frac{1}{2m} \left(\frac{1}{i} \nabla + e\mathbf{A}/c \right)^2 - \mu,$$

we can write the Bogoliubov-de Gennes equations in matrix form:

$$\begin{bmatrix} \hat{\xi} + V(\mathbf{r}) & \Delta(\mathbf{r}) \\ \Delta^*(\mathbf{r}) & -\hat{\xi}_c - V(\mathbf{r}) \end{bmatrix} \begin{bmatrix} u_q \\ v_q \end{bmatrix} = E_q \begin{bmatrix} u_q \\ v_q \end{bmatrix} \tag{4.63}$$

These equations can be easily derived if we write down the Heisenberg equations of motion for the field operators, $i\dot{\psi} = [\psi, \mathcal{H}_B]$; that is,

$$
i\dot{\psi}_\uparrow(\mathbf{r}, t) = \left[\hat{\xi} + V(\mathbf{r})\right]\psi_\uparrow(\mathbf{r}, t) - \Delta(\mathbf{r})\psi_\downarrow^\dagger(\mathbf{r}, t);
$$
$$
i\dot{\psi}_\downarrow^\dagger(\mathbf{r}, t) = -\left[\hat{\xi}_c + V(\mathbf{r})\right]\psi_\downarrow^\dagger(\mathbf{r}, t) - \Delta^*(\mathbf{r})\psi_\uparrow(\mathbf{r}, t)
$$

(4.64)

or in matrix form,

$$
i\frac{\partial}{\partial l}\begin{bmatrix}\psi_\uparrow(\mathbf{r}, t) \\ \psi_\downarrow^\dagger(\mathbf{r}, t)\end{bmatrix} = \begin{bmatrix}\hat{\xi} + V(\mathbf{r}) & -\Delta(\mathbf{r}) \\ -\Delta^*(\mathbf{r}) & -\hat{\xi}_c - V(\mathbf{r})\end{bmatrix}\begin{bmatrix}\psi_\uparrow(\mathbf{r}, t) \\ \psi_\downarrow^\dagger(\mathbf{r}, t)\end{bmatrix}.
$$

(4.65)

The change of sign before eA/c in the conjugate operator appears in the process of integration by parts, which we must perform while calculating the commutator $[\psi^\dagger, \mathcal{H}_B]$. This change of sign of the electrical charge should be expected, since $\psi^\dagger$ is the *hole* annihilation operator.

The ψ-operators are *not* the eigenvectors of the Hamiltonian and have no definite frequency. We can, though, express them through the eigenoperators of the Hamiltonian, $\alpha, \alpha^\dagger$:

$$
\alpha_{q,\uparrow}(t) = \alpha_{q,\uparrow}e^{-iE_q t}; \quad \alpha_{q,\downarrow}^\dagger(t) = \alpha_{q,\downarrow}^\dagger e^{iE_q t}.
$$

Gathering the terms, say, with $\alpha_{q,\uparrow}$, we get the Bogoliubov-de Gennes equations.

Here we will derive a useful general relation between the excitation energies, order parameter, and coherence factors, which follows directly from the Bogoliubov-de Gennes equations (4.63). Multiplying the first line of (4.63) from the left by $u_q^*(\mathbf{r})$ and the second by $v_q^*(\mathbf{r})$, adding them, and integrating over the whole space, we find (with the help of (4.60) and after an integration by parts)

$$
E_q = \int d^3\mathbf{r}\left\{u_q^*(\mathbf{r})(\hat{\xi} + V(\mathbf{r}))u_q(\mathbf{r}) - v_q(\mathbf{r})(\hat{\xi} + V(\mathbf{r}))v_q^*(\mathbf{r})\right.
$$
$$
\left. + 2\Re(\Delta(\mathbf{r})u_q^*(\mathbf{r})v_q(\mathbf{r}))\right\}.
$$

(4.66)

4.3.3 Bogolons

The elementary excitations above the ground state of the superconductor, bogolons, are created and annihilated by the $\alpha, \alpha^\dagger$-operators.

It follows from the relation (4.55) that this quasiparticle is a *coherent combination of an electron-like and hole-like excitations* with opposite spins. The coefficients u_q, v_q (they, or some their bilinear combinations-depending on the book you read - are called *coherence factors*) give the probability amplitudes of these states in the actual mixture and are defined by the Bogoliubov-de Gennes equations. Bogolons should not be mixed up with Cooper pairs, which are not excitations, but form the ground state of the superconductor. Two bogolons appear when a Cooper pair is torn apart (by thermal fluctuations, for example); of course, they keep some information about the superconducting phase coherence, which is contained in the coefficients u, v, as we will see later.

The *electric charge* of such a quasiparticle is no longer an integer multiple of e, but rather equals

$$Q = |u_q|^2 \cdot e + |v_q|^2 \cdot (-e). \tag{4.67}$$

Of course, the charge conservation law is here violated no more than the mass conservation law was by the fact that quasielectrons can have a mass different from m_0. The extra charge is taken or supplied by the condensate.

In the spatially uniform case and in the absence of external fields we can use the momentum representation to simplify equations (4.63). In the absence of the supercurrent, we can choose the order parameter Δ to be real:

$$\begin{bmatrix} \xi_{\mathbf{p}} - E_{\mathbf{p}} & \Delta \\ \Delta & -\xi_{\mathbf{p}} - E_{\mathbf{p}} \end{bmatrix} \begin{bmatrix} u_{\mathbf{p}} \\ v_{\mathbf{p}} \end{bmatrix} = 0. \tag{4.68}$$

The solvability condition gives the *dispersion law of bogolons* (see Fig. 4.9):

$$E_{\mathbf{p}} = \sqrt{\xi_{\mathbf{p}}^2 + \Delta^2}. \tag{4.69}$$

The solutions are thus

$$|u_{\mathbf{p}}| = \frac{1}{\sqrt{2}} \sqrt{1 + \frac{\xi_{\mathbf{p}}}{E_{\mathbf{p}}}}; \quad |v_{\mathbf{p}}| = \frac{1}{\sqrt{2}} \sqrt{1 - \frac{\xi_{\mathbf{p}}}{E_{\mathbf{p}}}}. \tag{4.70}$$

The self-consistency relation for the gap follows from the extremum condition $\Delta^* = |g|\langle \psi^\dagger \psi^\dagger \rangle$ and reads in the general case as follows:

$$\Delta^*(\mathbf{r}, T) = |g| \sum_q u_q^*(\mathbf{r}) v_q(\mathbf{r}) \tanh \frac{E_q(\Delta^*)}{2T}. \tag{4.71}$$

In the uniform case this is reduced to the famous BCS equation for the energy gap:

$$1 = |g| \int \frac{d^3 p}{(2\pi\hbar)^3 2} \frac{\tanh \frac{E_p(\Delta^*)}{2T}}{E_p(\Delta^*)}. \tag{4.72}$$

Finally, since in equilibrium bogolons satisfy the Fermi distribution, an *equilibrium average of a single-electron operator* $\mathcal{O}$ in the superconducting state can be easily calculated. Using the Bogoliubov transformation, we find

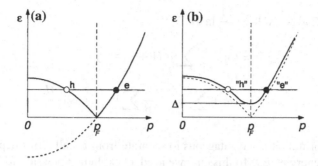

Fig. 4.9 Quasiparticle dispersion law: **a** normal metal (electrons and holes), **b** superconductor (electron-like and hole-like bogolons)

$$\langle \mathcal{O} \rangle = \sum_{\sigma=\uparrow,\downarrow} \left\langle \psi_\sigma^\dagger \mathcal{O} \psi_\sigma \right\rangle$$

$$= \sum_q \Big\langle (u_q^* \alpha_{q\uparrow}^\dagger + v_q \alpha_{q\downarrow}) \mathcal{O} (u_q \alpha_{q\uparrow} + v_q^* \alpha_{q\downarrow}^\dagger)$$

$$| (u_q^* \alpha_{q\downarrow}^\dagger - v_q \alpha_{q\uparrow}) \mathcal{O} (u_q \alpha_{q\downarrow} - v_q^* \alpha_{q\uparrow}^\dagger) \Big\rangle$$

$$- 2 \sum_{E_q > 0} \int d^3\mathbf{r} \left\{ u_q^* \mathcal{O} u_q n_F(E_q) + v_q \mathcal{O} v_q^* (1 - n_F(E_q)) \right\}. \quad (4.73)$$

Here we have assumed for simplicity that $\mathcal{O}$ is spin independent, but the generalization is trivial.

For example, the *equilibrium current density* is given by

$$\mathbf{j} = \frac{2e}{m} \sum_{E>0} \Re \left\{ u_v^* (\mathbf{p} + \frac{e}{c}\mathbf{A}) u_v n_F(E_v) + v_v (\mathbf{p} + \frac{e}{c}\mathbf{A}) v_v^* (1 - n_F(E_v)) \right\}. \quad (4.74)$$

4.3.4 Thermodynamic Potential of a Superconductor

Returning to the extremum condition (4.49) that we used earlier to derive the self-consistency relation for the order parameter, we see that the role of the thermodynamic potential of the superconductor is played by

$$\Omega[\Delta, \Delta^*] = \frac{1}{|g|} \int d^3\mathbf{r} |\Delta(\mathbf{r})|^2 + \Omega_B[\Delta, \Delta^*], \quad (4.75)$$

where the fluctuation corrections are thus neglected. The Bogoliubov functional Ω_B is easily calculated from the diagonalized form of the BCS Hamiltonian, (4.62):

$$\Omega_B[\Delta, \Delta^*] = -\frac{1}{\beta} \ln \mathrm{tr}\, e^{-\beta \mathcal{H}_B}$$

$$= U_0 - \frac{2}{\beta} \sum_q \ln(1 + e^{-\beta E_q})$$

$$= U_0 + \sum_q E_q - \frac{2}{\beta} \sum_q \ln\left(2\cosh\frac{\beta E_q}{2}\right).$$

In the latter formula it is advantageous to separate from U_0 all terms dependent on quasiparticle energies E_q. To this end, we need an explicit expression for U_0. Since U_0 originates from all terms in the BCS Hamiltonian that are not normally ordered after Bogoliubov transformation (that is, contain $\alpha\alpha^\dagger$), we can easily see that

$$U_0 = 2 \sum_q \int d^3\mathbf{r} \left\{ v_q(\mathbf{r})(\hat{\xi} + V(\mathbf{r}))v_q^*(\mathbf{r}) - \Re(\Delta(\mathbf{r})u_q^*(\mathbf{r})v_q(\mathbf{r})) \right\}.$$

From (4.66) we find

$$-\int d^3\mathbf{r}\, 2\Re(\Delta(\mathbf{r})u_q^*(\mathbf{r})v_q(\mathbf{r}))$$

$$= -E_q + \int d^3\mathbf{r} \left(u_q^*(\mathbf{r})(\hat{\xi} + V(\mathbf{r}))u_q(\mathbf{r}) - v_q(\mathbf{r})(\hat{\xi} + V(\mathbf{r}))v_q^*(\mathbf{r}) \right),$$

and therefore

$$U_0 = -\sum_q E_q + \sum_q \int d^3\mathbf{r} \left\{ v_q(\mathbf{r})(\hat{\xi} + V(\mathbf{r}))v_q^*(\mathbf{r}) + u_q^*(\mathbf{r})(\hat{\xi} + V(\mathbf{r}))u_q(\mathbf{r}) \right\}.$$

$$(4.76)$$

Finally, the (infinite) sums of excitation energies in Ω_B exactly cancel, and we obtain an important formula

$$\Omega[\Delta, \Delta^*] = \frac{1}{|g|} \int d^3\mathbf{r} |\Delta(\mathbf{r})|^2 + \sum_q \int d^3\mathbf{r} \left\{ v_q(\mathbf{r})(\hat{\xi} + V(\mathbf{r}))v_q^*(\mathbf{r}) \right.$$

$$\left. + u_q^*(\mathbf{r})(\hat{\xi} + V(\mathbf{r}))u_q(\mathbf{r}) \right\} - \frac{2}{\beta} \sum_q \ln\left(2\cosh\frac{\beta E_q}{2}\right). \quad (4.77)$$

4.4 Green's Functions of a Superconductor: The Nambu-Gor'kov Formalism

4.4.1 Matrix Structure of the Theory

We have already seen how the pairing Hamiltonian can be derived from the one with two-particle point interactions, and how anomalous averages appear if accept such a Hamiltonian. Here we will arrive at anomalous averages from a different direction. Namely, we will develop a special Green's function technique, where both normal and anomalous averages appear in a natural way.

We have seen from the Bogoliubov transformation that the superconducting state somehow mixes electrons and holes. It is then natural to introduce two-component field operators (*Nambu* operators)

$$\Psi(\mathbf{r}) = \begin{bmatrix} \psi_\uparrow(\mathbf{r}) \\ \psi_\downarrow^\dagger(\mathbf{r}) \end{bmatrix}, \qquad \Psi^\dagger(\mathbf{r}) = \begin{bmatrix} \psi_\uparrow^\dagger(\mathbf{r}), \psi_\downarrow(\mathbf{r}) \end{bmatrix}. \tag{4.78}$$

In the momentum representation, Nambu operators are given by

$$\Psi_\mathbf{p}(E) = \begin{bmatrix} \psi_{\mathbf{p}\uparrow}(E) \\ \psi_{-\mathbf{p}\downarrow}^\dagger(-E) \end{bmatrix}, \qquad \Psi_\mathbf{p}^\dagger(E) = \begin{bmatrix} \psi_{\mathbf{p}\uparrow}^\dagger(E), \psi_{-\mathbf{p}\downarrow}(-E) \end{bmatrix}. \tag{4.79}$$

Using them, we can rewrite the pairing Hamiltonian as

$$\mathcal{H}_B = \int d^3\mathbf{r}\Psi^\dagger(\mathbf{r}) \cdot \hat{\mathbf{H}}(\mathbf{r}) \cdot \Psi(\mathbf{r}), \tag{4.80}$$

where the matrix $\hat{\mathbf{H}}$ is

$$\hat{\mathbf{H}}(\mathbf{r}) = \begin{bmatrix} \hat{\xi} + V(\mathbf{r}) & -\Delta(\mathbf{r}) \\ -\Delta^*(\mathbf{r}) & -(\hat{\xi}_c + V(\mathbf{r})) \end{bmatrix}. \tag{4.81}$$

Then we quite naturally can introduce a matrix (*Gor'kov*) Green's function,

$$\hat{\mathbf{G}}_{jl} = \frac{1}{i}\left\langle \mathcal{T}\Psi_j(X)\Psi_l^\dagger(X')\right\rangle; \tag{4.82}$$

$$\hat{\mathbf{G}} = \begin{bmatrix} \frac{1}{i}\left\langle \mathcal{T}\psi_\uparrow(X)\psi_\uparrow^\dagger(X')\right\rangle & \frac{1}{i}\left\langle \mathcal{T}\psi_\uparrow(X)\psi_\downarrow(X')\right\rangle \\ \frac{1}{i}\left\langle \mathcal{T}\psi_\downarrow^\dagger(X)\psi_\uparrow^\dagger(X')\right\rangle & \frac{1}{i}\left\langle \mathcal{T}\psi_\downarrow^\dagger(X)\psi_\downarrow(X')\right\rangle \end{bmatrix}$$

$$\equiv \begin{bmatrix} G(X, X') & F(X, X') \\ F^+(X, X') & -G(X', X) \end{bmatrix}. \tag{4.83}$$

We see that the relevant terms arise from the pairings of Nambu operators of the type $\langle \Psi \Psi^\dagger \rangle$, while the pairings $\langle \Psi^\dagger \Psi^\dagger \rangle$, $\langle \Psi \Psi \rangle$ contain terms like $\langle \psi_\uparrow \psi_\uparrow \rangle$ or $\langle \psi_\uparrow^\dagger \psi_\downarrow \rangle$ (which would correspond to triplet pairing or magnetic ordering respectively) and are equal to zero in our case. (Of course, we could not rule them out a priori: here we use our knowledge of the properties of the superconducting state, based on experimental data, to narrow the field of search.) Then Wick's theorem for the Nambu operators looks the same as for the usual ones, and we can at once build the diagram technique. It is done, of course, along the same lines as before, and we need not repeat all the calculations.

4.4.2 Elements of the Strong Coupling Theory

We start from the unperturbed Hamiltonian *without* the pairing potentials:

$$\mathcal{H}_0 = \int d^3 r \Psi^\dagger \cdot \begin{bmatrix} \hat{\xi} & 0 \\ 0 & -\hat{\xi}_c \end{bmatrix} \cdot \Psi. \tag{4.84}$$

This is an important point: neither the unperturbed Hamiltonian nor the unperturbed Green's function contains anomalous (i.e., off-diagonal) terms (see Table 4.1). Nevertheless we will see that they naturally appear in Nambu-Gor'kov picture after interactions are taken into account [12].

As before, we use the Pauli matrices as a basis in the space of 2×2 matrices. Denoting them by

$$\hat{\tau}_0 = \begin{bmatrix} 1 & 0 \\ 0 & 1 \end{bmatrix}; \, \hat{\tau}_1 = \begin{bmatrix} 0 & 1 \\ 1 & 0 \end{bmatrix}; \, \hat{\tau}_2 = \begin{bmatrix} 0 & i \\ -i & 0 \end{bmatrix}; \, \hat{\tau}_3 = \begin{bmatrix} 1 & 0 \\ 0 & -1 \end{bmatrix},$$

we can write the perturbation terms (due to electron–phonon and electron–electron interactions) as follows:

$$\mathcal{H}_{EPI} = \sum_{p-p'=q} \sum_j g_j(\mathbf{p}, \mathbf{p}') \Psi_{\mathbf{p}}^\dagger \cdot \hat{\tau}_3 \cdot \Psi_{\mathbf{p}'} \left(b_{qj} + b_{-qj}^\dagger \right); \tag{4.85}$$

$$\mathcal{H}_C = \sum_{p_1+p_2=p_1'+p_2'} \langle \mathbf{p}_1 \mathbf{p}_2 | V_C | \mathbf{p}_1' \mathbf{p}_2' \rangle$$

$$\times \left(\Psi_{\mathbf{p}1}^\dagger \cdot \hat{\tau}_3 \cdot \Psi_{\mathbf{p}2} \right) \left(\Psi_{\mathbf{p}1'}^\dagger \cdot \hat{\tau}_3 \cdot \Psi_{\mathbf{p}2'} \right). \tag{4.86}$$

The Feynman rules are given in Table 4.1.

Now we can write the Dyson equation at once (Fig. 4.10):

$$\hat{G}^{-1} = \hat{G}_0^{-1} - \hat{\Sigma} \equiv E\hat{\tau}_0 - \xi_{\mathbf{p}}\hat{\tau}_3 - \hat{\Sigma}. \tag{4.87}$$

Table 4.1 Feynman rules for Nambu-Gor'kov Green's function (Momentum space)

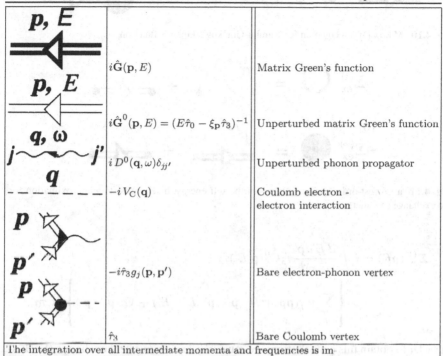

$i\hat{\mathbf{G}}(\mathbf{p}, E)$		Matrix Green's function
$i\hat{\mathbf{G}}^0(\mathbf{p}, E) = (E\hat{\tau}_0 - \xi_{\mathbf{p}}\hat{\tau}_3)^{-1}$		Unperturbed matrix Green's function
$i D^0(\mathbf{q}, \omega)\delta_{jj'}$		Unperturbed phonon propagator
$-i V_C(\mathbf{q})$		Coulomb electron - electron interaction
$-i\hat{\tau}_3 g_j(\mathbf{p}, \mathbf{p}')$		Bare electron-phonon vertex
$\hat{\tau}_3$		Bare Coulomb vertex

The integration over all intermediate momenta and frequencies is implied, taking into account energy/momentum conservation in every vertex and the matrix structure of Nambu-Gor'kov Green's function and vertices

Now we can present the self energy in the form

$$\hat{\Sigma}(\mathbf{p}, E) = [1 - Z(\mathbf{p}, E)]E\hat{\tau}_0 + Z(\mathbf{p}, E)\Delta(\mathbf{p}, E)\hat{\tau}_1 + \delta\epsilon(\mathbf{p})\hat{\tau}_3. \qquad (4.88)$$

The function $\Delta(\mathbf{p}, E)$ here is not the "initial" pairing potential, which we set to zero. In the absence of the magnetic field, in a stationary state, we can choose the phase in such a way as to eliminate the $\hat{\tau}_2$-component. Then

$$\hat{\mathbf{G}}(\mathbf{p}, E) = \frac{Z(\mathbf{p}, E)E\hat{\tau}_0 + Z(\mathbf{p}, E)\Delta(\mathbf{p}, E)\hat{\tau}_1 + (\xi_{\mathbf{p}} + \delta\epsilon_{\mathbf{p}})\hat{\tau}_3}{Z^2(\mathbf{p}, E)E^2 - Z^2(\mathbf{p}, E)\Delta^2(\mathbf{p}, E) - (\xi_{\mathbf{p}} + \delta\epsilon_{\mathbf{p}})^2}. \qquad (4.89)$$

The off-diagonal contribution to $\hat{\Sigma}$ should appear due to exchange terms, but it is absent in any finite approximation (see Fig. 4.11a):

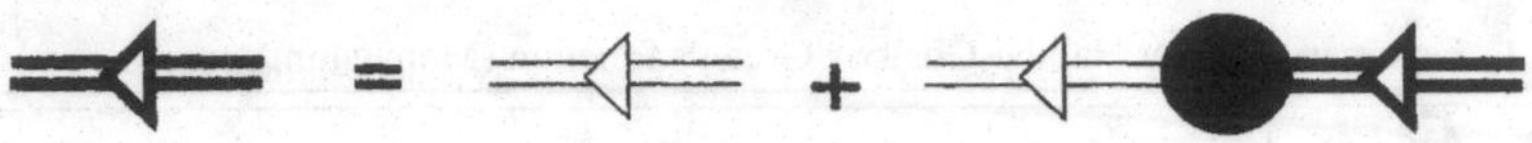

Fig. 4.10 Matrix Dyson equation for Nambu-Gor'kov's Green's function

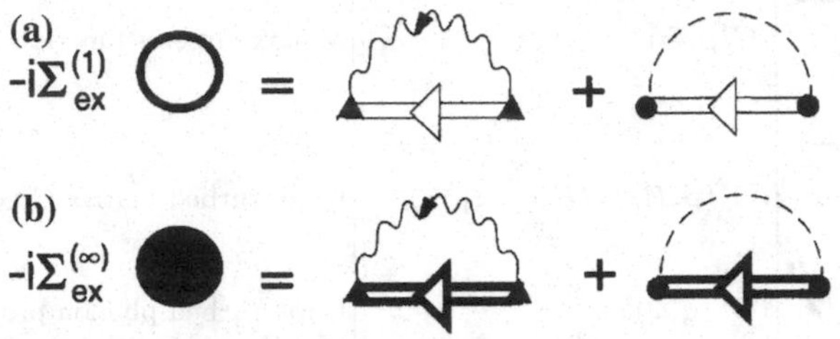

Fig. 4.11 **a** Lowest-order exchange terms in the self energy, **b** self-consistent approximation for the exchange self energy

$$\hat{\Sigma}_{ex}^{(1)}(\mathbf{p}E) = i \int \frac{d\,E'd\mathbf{p}'}{(2\pi)^4} \hat{\tau}_3 \hat{G}^0(\mathbf{p}'E')\hat{\tau}_3$$

$$\times \left\{ \sum_j |g_j(\mathbf{pp}')|^2 D_j^0(\mathbf{p}-\mathbf{p}', E-E') + V_C(\mathbf{p}-\mathbf{p}') \right\} \propto \hat{\tau}_0.$$

In order to obtain this contribution, we must use a self-consistent exchange (Fock's) approximation, effectively summing up an infinite subsequence of diagrams (Fig. 4.11b):

$$\hat{\Sigma}_{ex}^{(\infty)}(\mathbf{p}E) = i \int \frac{d\,E'd\mathbf{p}'}{(2\pi)^4} \hat{\tau}_3 \hat{G}(\mathbf{p}'E')\hat{\tau}_3$$

$$\times \left\{ \sum_j |g_j(\mathbf{pp}')|^2 D_j^0(\mathbf{p}-\mathbf{p}', E-E') + V_C(\mathbf{p}-\mathbf{p}') \right\}, \quad (4.90)$$

because of the identity

$$\hat{\tau}_3\hat{\tau}_1\hat{\tau}_3 = -\hat{\tau}_1.$$

In this way self energy and Green's function acquire an off-diagonal $\hat{\tau}_1$-term:

$$\hat{\tau}_3\hat{G}(\mathbf{p}E)\hat{\tau}_3 = \frac{Z(\mathbf{p},E)E\hat{\tau}_0 - Z(\mathbf{p},E)\Delta(\mathbf{p},E)\hat{\tau}_1 + (\xi_{\mathbf{p}}+\delta\epsilon_{\mathbf{p}})\hat{\tau}_3}{Z^2(\mathbf{p},E)E^2 - Z^2(\mathbf{p},E)\Delta^2(\mathbf{p},E) - (\xi_{\mathbf{p}}+\delta\epsilon_{\mathbf{p}})^2}. \quad (4.91)$$

Thus we arrive at a matrix equation:

$$[1 - Z(\mathbf{p}, E)]E\hat{\tau}_0 + Z(\mathbf{p}, E)\Delta(\mathbf{p}, E)\hat{\tau}_1 + \delta\epsilon(\mathbf{p})\hat{\tau}_3$$

$$= i \int \frac{d\,E'd\mathbf{p}'}{(2\pi)^4} \frac{Z(\mathbf{p}', E')E'\hat{\tau}_0 - Z(\mathbf{p}', E')\Delta(\mathbf{p}', E')\hat{\tau}_1 + (\xi_{\mathbf{p}'} + \delta\epsilon_{\mathbf{p}'})\hat{\tau}_3}{Z^2(\mathbf{p}', E')(E')^2 - Z^2(\mathbf{p}', E')\Delta^2(\mathbf{p}', E') - (\xi_{\mathbf{p}'} + \delta\epsilon_{\mathbf{p}'})^2}$$

$$\times \left\{ \sum_j |g_j(\mathbf{p}\mathbf{p}')|^2 D_j^0(\mathbf{p} - \mathbf{p}', E - E') + V_C(\mathbf{p} - \mathbf{p}') \right\}. \tag{4.92}$$

From this matrix relation the set of two nonlinear integral equations for Z, Δ follows, the so-called *Eliashberg equations*. They are central to the theory of superconductors with strong coupling and contain the expression for the intensity of electron–phonon interaction, traditionally written as

$$\alpha^2(\omega)F(\omega) \equiv \frac{\int_{SF} \frac{d^2\hat{p}}{v_p} \int_{SF} \frac{d^2\hat{p}'}{v_{p'}} \sum_j |g_j(\mathbf{p}\mathbf{p}')|^2 \delta(\omega - \omega_j(\mathbf{p} - \mathbf{p}'))}{\int_{SF} \frac{d^2\hat{p}}{v_p}}.$$

In the weak interaction limit these equations reduce to the BCS theory. For example, the transition temperature is given by the same formula:

$$T_c = 1.14\omega_D \exp\left\{ -\frac{1}{\lambda - \mu^*} \right\}. \tag{4.93}$$

Here μ^* is the Coulomb pseudopotential, which would appear in the BCS theory if Coulomb repulsion were explicitly taken into account.[4] Usually $\mu^* \ll \lambda$. The *electron–phonon coupling constant*, λ, is here defined as

$$\lambda = 2 \int_0^\infty d\omega \frac{\alpha^2(\omega)F(\omega)}{\omega}. \tag{4.94}$$

4.4.3 Gorkov's Equations for the Green's Functions

The matrix Green's function contains only two independent components: normal and anomalous Green's functions. The set of equations for these functions (*Gor'kov equations*) immediately follows from their definitions and the equation of motion for the field operators (4.64):

$$\left(i\frac{\partial}{\partial t} - \left(\hat{\xi} + V\right)_X\right) G(X, X') + \Delta(\mathbf{r})F^+(X, X') = \delta(X - X');$$
$$\left(i\frac{\partial}{\partial t} + \left(\hat{\xi}_c + V\right)_X\right) F^+(X, X') + \Delta^*(\mathbf{r})G(X, X') = 0. \tag{4.95}$$

---

[4] See Vonsovsky et al. [12] for more details and references.

In the stationary homogeneous case these equations take the following form (in the momentum representation):

$$(\omega - \xi_{\mathbf{p}})G(\mathbf{p}, \omega) + \Delta F^+(\mathbf{p}, \omega) = 1; \\ (\omega + \xi_{\mathbf{p}})F^+(\mathbf{p}, \omega) + \Delta^* G(\mathbf{p}, \omega) = 0. \tag{4.96}$$

At zero temperature

the solution to this set is given by

$$F^+(\mathbf{p}, \omega) = -\frac{\Delta^*}{\omega + \xi_{\mathbf{p}}} G(\mathbf{p}, \omega);$$

$$G(\mathbf{p}, \omega) = \frac{\omega + \xi_{\mathbf{p}}}{\omega^2 - (\xi_{\mathbf{p}}^2 + |\Delta|^2)}.$$

The infinitesimal imaginary term in the denominator of $G(\mathbf{p}, \omega)$ is determined by comparing it with the Källén-Lehmann representation:

$$G(\mathbf{p}, \omega) = \frac{|u_p|^2}{\omega - E_p + i0} + \frac{|v_p|^2}{\omega + E_p - i0}. \tag{4.97}$$

Here u, v are the parameters of the Bogoliubov transformation.

The Gor'kov equations must be completed with the self-consistency relation between Δ^* and F^+.

$$\Delta^*(\mathbf{rt}) = |g| \left\langle \psi_\uparrow^\dagger(\mathbf{rt}) \psi_\downarrow^\dagger(\mathbf{rt}) \right\rangle = -i|g| F^+(\mathbf{rt}^+, \mathbf{rt}). \tag{4.98}$$

It is easy to show that in the uniform case this equation coincides with the BCS equation for the gap at $T = 0$.

At finite temperatures

we use the same methods as in Chap. 3. Due to the analyticity of the retarded Green's function in the upper half-plane, we just put $\omega \to \omega + i0$ and obtain

$$G^R(\mathbf{p}, \omega) = \frac{|u_p|^2}{\omega - E_p + i0} + \frac{|v_p|^2}{\omega + E_p + i0}. \tag{4.99}$$

The *spectral density* (Fig. 4.12) is given by

$$\Gamma(\mathbf{p}, \omega) \equiv -2\Im G^R(\mathbf{p}, \omega) = 2\pi \left(u_p^2 \delta(\omega - E_p) + v_p^2 \delta(\omega + E_p) \right). \tag{4.100}$$

The causal Green's function is obtained from the retarded one with the use of the relation between their real and imaginary parts in equilibrium:

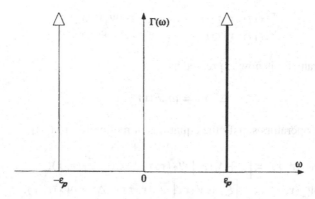

Fig. 4.12 Spectral density of the retarded Green's function in the superconductor. Note that the quasiparticles (bogolons) have infinite lifetime in spite of interactions

$$G(\mathbf{p}, \omega) = \mathcal{P}\left(\frac{|u_p|^2}{\omega - E_p} + \frac{|v_p|^2}{\omega + E_p}\right)$$

$$-i\pi \tanh \frac{E_p}{2T} \cdot \left(|u_p|^2 \delta(\omega - E_p) - |v_p|^2 \delta(\omega + E_p)\right). \quad (4.101)$$

Since $\tanh \frac{E_p}{2T} = 1 - 2n_F(E_p)$, this can be represented as

$$G(\mathbf{p}, \omega)|_{T \neq 0} = G(\mathbf{p}, \omega)|_{T=0} + 2\pi i n_F(E_p)\left(u_p^2 \delta(\omega - E_p) - v_p^2 \delta(\omega + E_p)\right),$$
$$(4.102)$$

while the anomalous Green's function is now

$$F^+(\mathbf{p}, \omega)|_{T \neq 0} = F^+(\mathbf{p}, \omega)|_{T=0}$$
$$- \frac{\Delta^*(T)}{\omega + \xi_p} 2\pi i n_F(E_p)\left(u_p^2 \delta(\omega - E_p) - v_p^2 \delta(\omega + E_p)\right). \quad (4.103)$$

Again, this equation leads to the BCS equation for $\Delta(T)$ at a finite temperature.

4.4.3.1 Matsubara Functions for the Superconductor

All the modifications of Green's functions techniques discussed earlier can be generalized to Nambu-Gor'kov Green's functions. For example, a Keldysh Green's function becomes a 4×4 matrix (because $G^{A,R,K}$ are 2×2 Nambu matrices) (see Rammer and Smith [7]). In equilibrium it is, though, more convenient to use a less cumbersome Matsubara formalism.

The temperature anomalous Green's functions are defined by

$$\mathcal{F}(\mathbf{r}_1\tau_1; \mathbf{r}_2\tau_2) = \langle T_\tau \Psi_\uparrow(\mathbf{r}_1\tau_1)\Psi_\downarrow(\mathbf{r}_2\tau_2)\rangle;$$
$$\overline{\mathcal{F}}(\mathbf{r}_1\tau_1; \mathbf{r}_2\tau_2) = -\langle T_\tau \overline{\Psi}_\downarrow(\mathbf{r}_1\tau_1)\overline{\Psi}_\uparrow(\mathbf{r}_2\tau_2)\rangle. \tag{4.104}$$

The order parameter is now expressed as

$$\Delta^*(\mathbf{r}) = |g|\overline{\mathcal{F}}(\mathbf{r}\tau^+; \mathbf{r}\tau). \tag{4.105}$$

Matsubara operators satisfy the equations of motion [cf. (4.64)]

$$-\tfrac{\partial}{\partial\tau}\Psi_\uparrow(\mathbf{r}, \tau) = \left[\hat{\xi} + V(\mathbf{r})\right]\Psi_\uparrow(\mathbf{r}, \tau) - \Delta(\mathbf{r})\overline{\Psi}_\downarrow(\mathbf{r}, \tau);$$
$$-\tfrac{\partial}{\partial\tau}\overline{\Psi}_\downarrow(\mathbf{r}, \tau) = -\left[\hat{\xi}_c + V(\mathbf{r})\right]\overline{\Psi}_\downarrow(\mathbf{r}, \tau) - \Delta^*(\mathbf{r})\Psi_\uparrow(\mathbf{r}, \tau). \tag{4.106}$$

The Gor'kov equations for the temperature Green's functions are thus

$$\left(-\tfrac{\partial}{\partial\tau} - \left(\hat{\xi} + V\right)_\mathbf{r}\right)\mathcal{G}(\mathbf{r}\tau, \mathbf{r}'\tau') + \Delta(\mathbf{r})\overline{\mathcal{F}}(\mathbf{r}\tau, \mathbf{r}'\tau') = \delta(\mathbf{r} - \mathbf{r}')\delta(\tau - \tau');$$
$$\left(-\tfrac{\partial}{\partial\tau} + \left(\hat{\xi}_c + V\right)_\mathbf{r}\right)\overline{\mathcal{F}}(\mathbf{r}\tau, \mathbf{r}'\tau') + \Delta^*(\mathbf{r})\mathcal{G}(\mathbf{r}\tau, \mathbf{r}'\tau') = 0. \tag{4.107}$$

In the uniform case we Fourier transform this to get

$$(i\omega_s - \xi_p)\mathcal{G}(\mathbf{p}\omega_s) + \Delta\overline{\mathcal{F}}(\mathbf{p}\omega_s) = 1;$$
$$(i\omega_s + \xi_p)\overline{\mathcal{F}}(\mathbf{p}, \omega_s) + \Delta^*\mathcal{G}(\mathbf{p}\omega_s) = 0. \tag{4.108}$$

The solution is

$$\mathcal{G}(\mathbf{p}, \omega_s) = -\frac{i\omega_s + \xi_p}{\omega_s^2 + E_p^2}; \tag{4.109}$$

$$\overline{\mathcal{F}}(\mathbf{p}, \omega_s) = \frac{\Delta^*}{\omega_s^2 + E_p^2}$$
$$= F^+(\mathbf{p}, i\omega_s). \tag{4.110}$$

The equation for the order parameter now follows from

$$\Delta^* = |g|\overline{\mathcal{F}}(\mathbf{r} = 0, \tau = 0^+) \equiv |g|T \sum_{s=-\infty}^{\infty} \int \frac{d^3p}{(2\pi)^3}\overline{\mathcal{F}}(\mathbf{p}, \omega_s).$$

Then

$$1 = \frac{|g|T}{(2\pi)^3} \sum_s \int \frac{d^3p}{\omega_s^2 + E_p^2}. \tag{4.111}$$

This equation immediately yields the BCS equation for $\Delta(T)$. The sum is evaluated with the help of the formula

$$\sum_s [(2s+1)^2\pi^2 + a^2]^{-1} = \frac{1}{2a} \tanh \frac{a}{2}.$$

4.4.4 Current-Carrying State of the Superconductor

So far we have not considered the most remarkable feature of the superconducting state: the supercurrent. We have suggested that it is carried by the superconducting condensate, i.e., by Cooper pairs, but all the same we have limited our analysis to the case of the Cooper pair at rest. Therefore, we would seemingly have to repeat everything from the beginning, this time allowing for the finite momentum of a Cooper pair.

Fortunately, instead of following this lengthy route, we can modify the many-body Gor'kov equations to describe a current-carrying state of the superconductor.

First, let us suppose the condensate moves as a whole with a uniform velocity (superfluid velocity) $\mathbf{v}_s$. Then the Galileo transformation yields the obvious changes:

$$a_\mathbf{p} \,\rightarrow\, a_{\mathbf{p}-m\mathbf{v}_s}; \quad a_\mathbf{p}^\dagger \rightarrow a_{\mathbf{p}-m\mathbf{v}_s}^\dagger;$$

$$\psi_\sigma(\mathbf{r}) \rightarrow e^{im\mathbf{v}_s\mathbf{r}}\psi_\sigma(\mathbf{r}); \tag{4.112}$$

$$\psi_\sigma^\dagger(\mathbf{r}) \rightarrow e^{-im\mathbf{v}_s\mathbf{r}}\psi_\sigma^\dagger(\mathbf{r}). \tag{4.113}$$

Thus the field operators gain the phase factor

$$\exp(im\chi(\mathbf{r})),$$

where (in the absence of the external field)

$$\mathbf{v}_s = \nabla\chi(\mathbf{r}).$$

The Gor'kov functions accordingly acquire the phase factors

$$G(X, X') \rightarrow \exp(im(\chi(\mathbf{r}) - \chi(\mathbf{r}')) \tilde{G}(X, X'); \tag{4.114}$$

$$F^+(X, X') \rightarrow \exp(-im(\chi(\mathbf{r}) + \chi(\mathbf{r}')) \tilde{F}^+(X, X'); \tag{4.115}$$

$$\Delta^*(\mathbf{r}) \rightarrow \exp(-2im\chi(\mathbf{r}))|\Delta(\mathbf{r})| \equiv \exp(-2im\chi(\mathbf{r}))\Delta(\mathbf{r}). \tag{4.116}$$

In the case of spatially uniform superflow, $\chi(\mathbf{r}) = \mathbf{v}_s \cdot \mathbf{r}(+\text{const})$, but once we have related the superfluid velocity to the phase of the order parameter, we no longer depend on this uniformity assumption and can consider arbitrary $\chi(\mathbf{r})$ and $\mathbf{v}_s(\mathbf{r})$.

Let us choose the transverse gauge $\nabla \cdot \mathbf{A} = 0$. It simplifies the calculations, since then the vector potential and the momentum operator commute (see, e.g.,

Landau and Lifshitz [4]). Indeed, for arbitrary coordinate functions, $f(\mathbf{r})$, $g(\mathbf{r})$, we find $[\hat{\mathbf{p}}, f]g = -i\hbar(\nabla f)g$. Therefore, $[\hat{\mathbf{p}}, \mathbf{A}] = -i\hbar\nabla \cdot \mathbf{A} = 0$ in this gauge.

Substituting (4.115) into Gor'kov's equations, we then find that they obey the equations

$$\left(i\frac{\partial}{\partial t} - \frac{1}{2m}(\hat{\mathbf{p}} + m\mathbf{v}_s)^2 - \mu\right)\tilde{G}(X, X') + \Delta(\mathbf{r})\tilde{F}^+(X, X') = \delta(X - X');$$
$$\left(i\frac{\partial}{\partial t} + \frac{1}{2m}(\hat{\mathbf{p}} - m\mathbf{v}_s)^2 + \mu\right)\tilde{F}^+(X, X') + \Delta(\mathbf{r})\tilde{G}(X, X') = 0, \tag{4.117}$$

where the order parameter phase and the vector potential of the electromagnetic field enter through the gauge-invariant combination, superfluid velocity

$$\mathbf{v}_s(\mathbf{r}) = \nabla\chi(\mathbf{r}) - \frac{e}{mc}\mathbf{A}(\mathbf{r}); \tag{4.118}$$

$$\nabla \times \mathbf{v}_s(\mathbf{r}) = -\frac{e}{mc}\mathbf{B}(\mathbf{r}). \tag{4.119}$$

The supercurrent is a "thermodynamic current": it flows in equilibrium, it is a property of the ground state of the system. Therefore, it can be calculated from thermodynamics only, where the current density is expressed as a variational derivative of any of thermodynamic potentials with respect to the vector potential:

$$-\frac{1}{c}\mathbf{j}(\mathbf{r}) = \left(\frac{\delta E}{\delta\mathbf{A}(\mathbf{r})}\right)_{S,V,N} = \left(\frac{\delta F}{\delta\mathbf{A}(\mathbf{r})}\right)_{T,V,N}$$

$$= \left(\frac{\delta W}{\delta\mathbf{A}(\mathbf{r})}\right)_{S,P,N} = \left(\frac{\delta\Phi}{\delta\mathbf{A}(\mathbf{r})}\right)_{T,P,N} \tag{4.120}$$

$$= \left(\frac{\delta\Omega}{\delta\mathbf{A}(\mathbf{r})}\right)_{T,V,\mu}.$$

Therefore, we find

$$\mathbf{j}(\mathbf{r}) = -c\frac{\delta\Omega}{\delta\mathbf{A}(\mathbf{r})}$$

$$= \frac{ie}{2m}\sum_\sigma \left\langle\left(\nabla\psi_\sigma^\dagger(\mathbf{r})\right)\psi_\sigma(\mathbf{r}) - \psi_\sigma^\dagger(\mathbf{r})(\nabla\psi_\sigma(\mathbf{r}))\right\rangle - \frac{e^2}{mc}\mathbf{A}(\mathbf{r})\sum_\sigma \left\langle\psi_\sigma^\dagger(\mathbf{r})\psi_\sigma(\mathbf{r})\right\rangle. \tag{4.121}$$

This definition satisfies the charge conservation law:

$$\nabla \cdot \mathbf{j}(\mathbf{r}) = 0. \tag{4.122}$$

The current can be expressed through the normal Green's function (compare to Sect. 2.1.4):

$$\mathbf{j}(\mathbf{r}) = \frac{i\,e}{m} \lim_{\mathbf{r}'\to\mathbf{r}} (\nabla_{\mathbf{r}'} - \nabla_{\mathbf{r}})\tilde{\mathcal{G}}(\mathbf{r}0, \mathbf{r}'0^+) + 2e\mathbf{v}_s(\mathbf{r})\tilde{\mathcal{G}}(\mathbf{r}0, \mathbf{r}'0^+)$$

$$= \frac{i\,e}{m} \lim_{\mathbf{r}'\to\mathbf{r}} (\nabla_{\mathbf{r}'} - \nabla_{\mathbf{r}})T \sum_{s=-\infty}^{\infty} \tilde{\mathcal{G}}(\mathbf{r}, \mathbf{r}', \omega_s)$$

$$+ 2e\mathbf{v}_s(\mathbf{r})T \sum_{s=-\infty}^{\infty} \tilde{\mathcal{G}}(\mathbf{r}, \mathbf{r}', \omega_s), \tag{4.123}$$

which follows directly from its definition.

4.4.4.1 Gradient Expansion: Local Approximation

Here we will do basically the same thing as in Chap. 3, when we derived the quantum kinetic equation: the gradient expansion. Again we assume that the field $\mathbf{A}(\mathbf{r})$, the superfluid velocity, and the order parameter are slow functions of the coordinates, and introduce the Wigner representation:

$$\mathbf{R} = (\mathbf{r} + \mathbf{r}')/2; \quad \rho = \mathbf{r} - \mathbf{r}'; \tag{4.124}$$

$$f(\mathbf{r}, \mathbf{r}') \to f(\mathbf{R}, \mathbf{q}) = \int d^3\rho\, e^{-i\mathbf{q}\rho} f(\mathbf{R} + \rho/2, \mathbf{R} - \rho/2); \tag{4.125}$$

$$\hat{\mathbf{p}} = -i\nabla \to \mathbf{q} - \frac{i}{2}\nabla_{\mathbf{R}}; \mathbf{r} \to \mathbf{R} + \frac{i}{2}\nabla_{\mathbf{q}}. \tag{4.126}$$

Then the Gor'kov's equations read

$$\left\{ i\omega_s - 1/2m \left(\mathbf{q} - \frac{i}{2}\nabla_{\mathbf{R}} + m\mathbf{v}_s \left(\mathbf{R} - \frac{i}{2}\nabla_{\mathbf{q}} \right) \right)^2 + \mu \right\} \mathcal{G}(\mathbf{R}, \mathbf{q}, \omega_s)$$

$$+ \Delta(\mathbf{R} - \frac{i}{2}\nabla_{\mathbf{q}})\overline{\mathcal{F}}(\mathbf{R}, \mathbf{q}, \omega_s) = 1;$$

$$\left\{ i\omega_s + 1/2m \left(\mathbf{q} - \frac{i}{2}\nabla_{\mathbf{R}} - m\mathbf{v}_s \left(\mathbf{R} - \frac{i}{2}\nabla_{\mathbf{q}} \right) \right)^2 - \mu \right\} \overline{\mathcal{F}}(\mathbf{R}, \mathbf{q}, \omega_s)$$

$$+ \Delta(\mathbf{R} - \frac{i}{2}\nabla_{\mathbf{q}})\mathcal{G}(\mathbf{R}, \mathbf{q}, \omega_s) = 0.$$

In zeroth order in gradients we have thus the set of algebraic equations

$$\left\{ i\omega_s - 1/2m(\mathbf{q} + m\mathbf{v}_s(\mathbf{R}))^2 + \mu \right\} \mathcal{G}(\mathbf{R}, \mathbf{q}, \omega_s) + \Delta(\mathbf{R})\overline{\mathcal{F}}(\mathbf{R}, \mathbf{q}, \omega_s) = 1;$$

$$\left\{ i\omega_s + 1/2m(\mathbf{q} - m\mathbf{v}_s(\mathbf{R}))^2 - \mu \right\} \overline{\mathcal{F}}(\mathbf{R}, \mathbf{q}, \omega_s) + \Delta(\mathbf{R})\mathcal{G}(\mathbf{R}, \mathbf{q}, \omega_s) = 0,$$

while for the current we have the expression

$$\mathbf{j}(\mathbf{R}) = \frac{2e}{m} T \sum_s \int \frac{d^3\mathbf{q}}{(2\pi)^3} (\mathbf{q} + m\mathbf{v}_s(\mathbf{R})) \mathcal{G}(\mathbf{R}, \mathbf{q}, \omega_s). \tag{4.127}$$

The solutions to the Gor'kov equations are thus

$$\mathcal{G}(\mathbf{R}, \mathbf{q}, \omega_s) = \frac{1}{2} \left\{ \frac{1 + \tilde{\xi}_q(\mathbf{R})/E_q(\mathbf{R})}{i\omega_s - \mathbf{q} \cdot \mathbf{v}_s(\mathbf{R}) - E_q(\mathbf{R})} \right.$$
$$\left. + \frac{1 - \tilde{\xi}_q(\mathbf{R})/E_q(\mathbf{R})}{i\omega_s - \mathbf{q} \cdot \mathbf{v}_s(\mathbf{R}) + E_q(\mathbf{R})} \right\}; \tag{4.128}$$

$$\overline{\mathcal{F}}(\mathbf{R}, \mathbf{q}, \omega_s) = \frac{\Delta(\mathbf{R})}{2E_q(\mathbf{R})} \left\{ \frac{1}{i\omega_s - \mathbf{q} \cdot \mathbf{v}_s(\mathbf{R}) + E_q(\mathbf{R})} \right.$$
$$\left. - \frac{1}{i\omega_s - \mathbf{q} \cdot \mathbf{v}_s(\mathbf{R}) - E_q(\mathbf{R})} \right\}. \tag{4.129}$$

The only change brought to these formulae by the supercurrent is that instead of $i\omega_s$ we have

$$i\omega_s - \mathbf{q}\mathbf{v}_s,$$

and the chemical potential is shifted as well:

$$\mu \to \mu(\mathbf{R}) = \mu - \frac{mv_s^2}{2}, \tag{4.130}$$

so that the kinetic energy term becomes

$$\tilde{\xi}_q(\mathbf{R}) = \frac{q^2}{2m} - \mu(\mathbf{R}) = \xi_q(\mathbf{R}) + \frac{mv_s^2}{2}.$$

Taking this into account, the relation between the kinetic energy and excitation energy is the same as before, locally:

$$E_q(\mathbf{R})^2 = \tilde{\xi}_q(\mathbf{R})^2 + \Delta(\mathbf{R})^2. \tag{4.131}$$

The shift of $i\omega_s$ (or ω, in the formalism of time-dependent Green's functions) means that the *energy gap* in the superconductor is diminished by the maximum value of $\mathbf{q}\mathbf{v}_s$, i.e., by qv_s: the elementary excitation can now be created with energy of only $\Delta - qv_s$. But the question of whether and how the *order parameter* is changed must be addressed separately.

4.4.4.2 Supercurrent: Explicit Expression

Using the formula for summation over odd (fermionic) Matsubara frequencies,

$$T \sum_{s=-\infty}^{\infty} \frac{1}{i\omega_s - \alpha} = n_F(\alpha),$$

we find that

$$T \sum_{s=-\infty}^{\infty} \tilde{\mathcal{G}}(\mathbf{q}, \omega_s)$$

$$= \frac{1}{2} \left\{ \left(1 + \frac{\tilde{\xi}_q}{E_q}\right) n_F(\mathbf{q}\mathbf{v}_s + E_q) + \left(1 - \frac{\tilde{\xi}_q}{E_q}\right) n_F(\mathbf{q}\mathbf{v}_s - E_q) \right\}, \qquad (4.132)$$

so that the supercurrent equals

$$\mathbf{j}(\mathbf{R}) = \frac{e}{m} \int \frac{d^3q}{(2\pi)^3} (\mathbf{q} + m\mathbf{v}_s(\mathbf{R}))$$

$$\times \left\{ \left(1 + \frac{\tilde{\xi}_q}{E_q}\right) n_F(\mathbf{q}\mathbf{v}_s + E_q) + \left(1 - \frac{\tilde{\xi}_q}{E_q}\right) n_F(\mathbf{q}\mathbf{v}_s - E_q) \right\} \qquad (4.133)$$

The total density of electrons is evidently given by

$$n(\mathbf{R}) \equiv 2 \int \frac{d^3q}{(2\pi)^3} T \sum_{s=-\infty}^{\infty} \tilde{\mathcal{G}}(\mathbf{q}, \omega_s)$$

$$= \int \frac{d^3q}{(2\pi)^3} \left\{ \left(1 + \frac{\tilde{\xi}_q}{E_q}\right) n_F(\mathbf{q}\mathbf{v}_s + E_q) \right.$$

$$\left. + \left(1 - \frac{\tilde{\xi}_q}{E_q}\right) n_F(\mathbf{q}\mathbf{v}_s - E_q) \right\}. \qquad (4.134)$$

Therefore, the term proportional to the superfluid velocity in (4.133) is simply

$$\mathbf{j}_2 = ne\mathbf{v}_s.$$

For the supercurrent, we should rather expect that $\mathbf{j} = n_s e \mathbf{v}_s$, where $n_s \leq n$, because not all electrons can enter the condensate at finite temperatures and thus participate in the supercurrent. This formula would serve as a definition of n_s, the density of "superconducting electrons", whatever this may mean, if we could show that the other term in (4.133) equals

$$\mathbf{j}_1 = -n_n e \mathbf{v}_s \equiv -\mathbf{j}_n,$$

in order to eliminate the contribution of the "normal" electrons (n_n is then the density of the normal component, $n_n + n_s = n$).

Noting that

$$n_F(-x) = 1 - n_F(x),$$

and

$$\int d^3q \, \mathbf{q} \left(1 \pm \frac{\tilde{\xi}_q}{E_q} \right) = 0$$

(the latter because the energies are direction independent), we can write this term as follows:

$$
\begin{aligned}
\mathbf{j}_1(\mathbf{R}) &= \frac{e}{m} \int \frac{d^3q}{(2\pi)^3} \mathbf{q} \left\{ \left(1 + \frac{\tilde{\xi}_q}{E_q} \right) n_F(E_q + \mathbf{q}\mathbf{v}_s) \right. \\
&\qquad\qquad \left. - \left(1 - \frac{\tilde{\xi}_q}{E_q} \right) n_F(E_q - \mathbf{q}\mathbf{v}_s) \right\} \\
&\approx -\frac{2e}{m} \int \frac{d^3q}{(2\pi)^3} \mathbf{q} n_F(E_q - \mathbf{q}\mathbf{v}_s).
\end{aligned}
\tag{4.135}
$$

But $n_F(E_q - \mathbf{q}\mathbf{v}_s)$ is simply the distribution function of Fermi particles (in the event, bogolons) moving with velocity $\mathbf{v}_s$. Thus the term $\mathbf{j}_1$ indeed is minus the current of the elementary excitations moving with the velocity of the condensate:

$$\mathbf{j}_1 = -\mathbf{j}_n \equiv -n_n e \mathbf{v}_s.$$

Since the current carried by elementary excitations is a dissipative (i.e., normal) current, this gives us the effective density of the "normal component" in the superconductor. Then we see that the net current in equilibrium really can be written as the current of the superconducting component,

$$\mathbf{j}(\mathbf{R}) = \mathbf{j}_s = n e \mathbf{v}_s - n_n e \mathbf{v}_s \equiv n_s e \mathbf{v}_s(\mathbf{R}), \tag{4.136}$$

where

$$n_n + n_s = n. \tag{4.137}$$

As we mentioned, this should be understood as a definition of the density of the superconducting component.

4.4.5 Destruction of Superconductivity by Current

We noted that while the energy gap in the current-carrying state linearly drops with the supercurrent, the behavior of the order parameter should be determined from the self-consistency relation

$$\Delta(\mathbf{R}) = |g|T \sum_s \int \frac{d^3q}{(2\pi)^3} \overline{\overline{\mathcal{F}}}(\mathbf{R}, \mathbf{q}, \omega_s).$$ (4.138)

Substituting there the expression for the anomalous Green's function (4.129), we obtain the integral equation

$$1 = |g| \int \frac{d^3q}{(2\pi)^3} \frac{1}{2E_q} (1 - n_F(E_q + \mathbf{q}\mathbf{v}_s) - n_F(E_q - \mathbf{q}\mathbf{v}_s)).$$ (4.139)

After integration over the angles we get

$$1 = |g|N(0) \int_0^{\omega_D} \frac{d\xi}{E(\xi)} \left(1 + \frac{T}{p_F v_s} \ln \frac{1 + \exp\left\{-\frac{E(\xi)+p_F v_s}{T}\right\}}{1 + \exp\left\{-\frac{E(\xi)-p_F v_s}{T}\right\}} \right).$$ (4.140)

At zero temperature, this unpleasant equation becomes tractable. First, we observe that the second term in the brackets is zero, provided

$$p_F v_s < \Delta_0,$$ (4.141)

where Δ_0 is the order parameter in the state with $\mathbf{j} = 0$. Indeed, then both exponents are zero. Therefore, the equation for the order parameter stays the same as in the absence of the supercurrent. Equation (4.141) is the celebrated *Landau criterion*, telling at what v_s the energy gap (not the order parameter!) first goes to zero. But, unlike the case of Bose superfluid, the superconducting state in three dimensions is not immediately destroyed when v_s reaches $v_{s,\text{Landau}} = \Delta_0/p_F$: there still exists *gapless* superconductivity up to somewhat higher $v_{s,c}$, which can be seen from (4.140) (see [8]).

Let the order parameter be $\Delta < \Delta_0 < p_F v_s$, and introduce $\omega_\Delta = \sqrt{(p_F v_s)^2 - \Delta^2}$. The r.h.s. of (4.140) is then

$$|g|N(0) \left(\int_0^{\omega_\Delta} \frac{d\xi}{p_F v_s} + \int_{\omega_\Delta}^{\omega_D} \frac{d\xi}{\sqrt{\xi^2 + \Delta^2}} \right)$$

$$= |g|N(0) \left(\sqrt{1 - \frac{\Delta^2}{(p_F v_s)^2}} + \int_{\omega_\Delta}^{\omega_D} \frac{d\xi}{\sqrt{\xi^2 + \Delta^2}} \right),$$

while the l.h.s. can be identically rewritten as

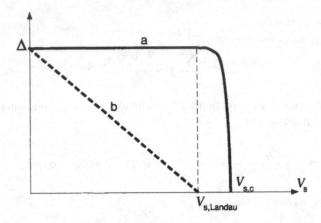

Fig. 4.13 Order parameter (**a**) and energy gap (**b**) dependence on the superfluid velocity

$$|g|N(0)\int_0^{\omega_D}\frac{d\xi}{\sqrt{\xi^2+\Delta_0^2}}.$$

The integrals can be taken explicitly, and we finally obtain

$$\sqrt{1-\frac{\Delta^2}{(p_Fv_s)^2}}+\ln\left[\frac{1+\sqrt{1-\frac{\Delta^2}{\omega_D^2}}}{1+\sqrt{1-\frac{\Delta^2}{(p_Fv_s)^2}}}\right]$$

$$=\ln\frac{p_Fv_s}{\Delta_0}+\ln\left[1+\sqrt{1-\frac{\Delta_0^2}{\omega_D^2}}\right].$$

The superconductivity is thus destroyed by the current when $\Delta=0$; that is,

$$1=\ln\frac{p_Fv_{s,c}}{\Delta_0}+\ln\left[1+\sqrt{1-\frac{\Delta_0^2}{\omega_D^2}}\right].$$

Therefore, the critical velocity is

$$v_{s,c}=\frac{\Delta_0}{p_F}e^{1-\ln\left[1+\sqrt{1-\frac{\Delta_0^2}{\omega_D^2}}\right]}\approx v_{s,\text{Landau}}e^{1-\ln 2},\tag{4.142}$$

since $(\Delta_0^2/\omega_D^2)\ll 1$.

We have established the existence of the region of v_s between $v_{s,\text{Landau}}$ and $v_{s,c}\approx 1.359v_{s,\text{Landau}}$, where the superconductivity exists in spite of the possibility of the creation of elementary excitations with arbitrarily small energies (see Fig. 4.13).

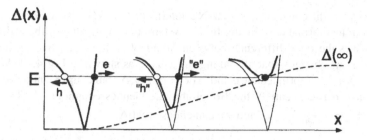

Fig. 4.14 Andreev reflection of the quasiparticle from the N-S interface. At the point where $E = \Delta(x)$ the quasiparticle changes the branch (particle to hole, e.g.), thus changing the velocity to its opposite with the minimal possible momentum change

(It is straightforward to check that the order parameter in the end point behaves as $\Delta(v_{s,c} - v_s) \propto \sqrt{v_{s,c} - v_s}$.) The reason why such a regime of gapless superconductivity can exist is that unlike the Bose superfluid, the elementary excitations in the superconductor are fermions. Since the destruction of the supercurrent can occur only after all of its momentum has been transferred to the quasiparticles, there must be a certain number of them generated. For fermions, this demands a finite phase space, with zero energy gap. But in three dimensions, after the Landau criterion is met, the gap is zero only on an infinitesimally small portion of the Fermi sphere. This portion, and the quasiparticle generation rate, grows as the superfluid velocity grows, while the order parameter continuously drops until it becomes zero at $v_{s,c}$.

The dimensionality is here crucial: if, e.g., formally consider the one-dimensional case, it is evident from (4.139), that the order parameter will drop to zero immediately at $v_{s,\text{Landau}}$ (Bagwell [14]). Small wonder: in this case there is no room for growth of the phase space volume available to quasiparticle generation.

4.5 Andreev Reflection

Andreev reflection is a remarkable phenomenon that takes place at the boundary between a superconductor and a normal conductor.

Let us ask a naïve question we have never asked before: How can the electric current flow from a normal-metal lead to a superconductor? In the normal state, at arbitrarily small applied voltage, current is carried by quasiparticles near the Fermi surface, but how can the quasiparticles cross an NS interface? There is a gap in the density of states of the superconductor. Electron-and hole-like bogolons can exist only above this gap: $E > \Delta$, so that subgap quasielectrons (and quasiholes) simply cannot penetrate the superconductor. Nevertheless, since current *can* flow through such an interface at subgap voltages, then real electrons can pass the boundary. Quasielectrons of the normal metal disappear in the process. What appears instead?

To begin with, consider a planar NS interface. Let $\Delta(x)$ be a slowly varying function of the normal coordinate. In our system, the local value of the order parameter makes the only difference between S-and N-regions, and the quasiparticle dispersion law changes spatially along the x-axis, as shown in Fig. 4.14. We have $\Delta(x = -\infty) = 0$ (normal state), and $\Delta(x = \infty) = \Delta$ (bulk superconductor).

The quasiparticles satisfy the Bogoliubov-de Gennes equations (4.143), where now $\hat{\xi}_c = \hat{\xi} = -\frac{\hbar^2}{2m}\nabla^2 - \mu$, and we can choose real Δ:

$$
\begin{bmatrix} \hat{\xi} & \Delta(\mathbf{r}) \\ \Delta(\mathbf{r}) & -\hat{\xi} \end{bmatrix} \begin{bmatrix} u(\mathbf{r}) \\ v(\mathbf{r}) \end{bmatrix} = E \begin{bmatrix} u(\mathbf{r}) \\ v(\mathbf{r}) \end{bmatrix}. \tag{4.143}
$$

We seek a solution in the form

$$
u_{\mathbf{q}}(\mathbf{r}) = e^{ik_F \mathbf{n}\cdot\mathbf{r}}\eta(\mathbf{r});
$$
$$
v_{\mathbf{q}}(\mathbf{r}) = e^{ik_F \mathbf{n}\cdot\mathbf{r}}\chi(\mathbf{r}),
$$

where $\eta(\mathbf{r})$ and $\chi(\mathbf{r})$ vary slowly in comparison to the exponential factors. Neglecting second derivatives, we thus obtain

$$
(i\hbar v_F(\mathbf{n}\cdot\nabla) + E)\eta(\mathbf{r}) + \Delta(\mathbf{r})\chi(\mathbf{r}) = 0;
$$
$$
(i\hbar v_F(\mathbf{n}\cdot\nabla) - E)\chi(\mathbf{r}) + \Delta(\mathbf{r})\eta(\mathbf{r}) = 0. \tag{4.144}
$$

When $x \to \infty$, propagating solutions exist only if $E > \Delta$, since they depend on coordinates as

$$
e^{q(E)\mathbf{n}\cdot\mathbf{r}}; \quad \hbar v_F q = \pm\sqrt{E^2 - \Delta^2}. \tag{4.145}
$$

So much we already guessed: the quasiparticle with subgap energy inside the gap cannot enter the superconductor, where there is no available place for it. Nor can it change its momentum significantly, since the effective potential varies too slowly. The only possible reflection process is thus to change the direction of its group velocity, changing the branch of the excitation spectrum (see Fig. 4.14).

When $x \to -\infty$, we find that

$$
\eta(\mathbf{r}) = Ae^{ik\mathbf{n}\cdot\mathbf{r}}; \tag{4.146}
$$
$$
\chi(\mathbf{r}) = Be^{-ik\mathbf{n}\cdot\mathbf{r}}; \tag{4.147}
$$
$$
k = \frac{E}{\hbar v_F}. \tag{4.148}
$$

A and B are integration constants. Equation (4.146) describes an electron-like, and (4.147) a hole-like, quasiparticle. If $n_x > 0$, a quasiparticle comes from $-\infty$ to the boundary, and a quasihole is reflected, and vice versa.

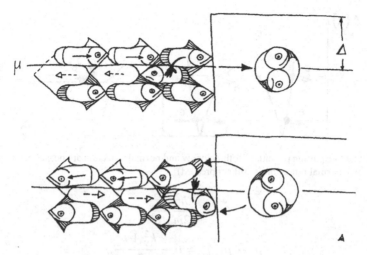

Fig. 4.15 Andreev reflection: the physical picture

Note that unlike the usual reflection, Andreev reflection changes *all* the velocity components of the incident quasiparticle (simultaneously transforming it into a quasihole), and vice versa.

Due to this feature, Andreev reflection provides a mechanism for *current flow through the NS interface*. A quasielectron with velocity **v** is transformed into a quasihole with velocity −**v** which carries the same current in the same direction. In terms of real electrons this means that an electron above the Fermi surface in the normal conductor forms a Cooper pair with another electron (below the Fermi surface) and leaves for the superconducting condensate (Fig. 4.15), thus transferring the dissipative quasiparticle current on the N-side into the condensate supercurrent on the S-side. The Andreev-reflected hole is the hole left by the second electron in the Fermi sea. In the conjugated process, a Cooper pair pounces at an unsuspecting hole wandering too close to the interface; one electron fills the hole, and the other moves away into the normal region.

To investigate this picture in some detail, let us now consider the case of steplike pairing potential (Fig. 4.16):

$$\Delta(x, y, z) = \Delta e^{i\phi}\theta(x). \tag{4.149}$$

This is, of course, an approximation (though a very often used one). Since pairing potential must be determined self-consistently from (4.71), in reality it sags a little near to the SN interface (at distances about ξ_0). This is sometimes called the *proximity effect* (later we will discuss another meaning of this term); anyway, it is irrelevant for our problem, since it can bring only small corrections.

Since the system is homogeneous in the y, z-directions, we can separate a single mode with dependence on y, z given by $e^{ik_y y + ik_z z}$. The Bogoliubov-de Gennes equations for such a mode will have the same form as (4.143), with

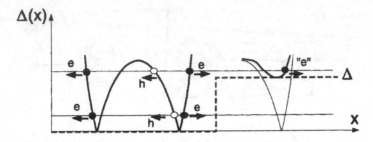

Fig. 4.16 Steplike pairing potential. Both Andreev and normal reflection are possible. (We will see that in the gap, normal reflection is still suppressed)

$$\hat{\xi} \to -\frac{\hbar^2}{2m}\frac{d^2}{dx^2},$$

$$\mu \to \mu_{k_y,k_z} = \mu - \frac{k_y^2 + k_z^2}{2m},$$

and

$$k_F \to k_{F,k_y,k_z} = \sqrt{k_F^2 - (k_y^2 + k_z^2)}.$$

(We will keep this in mind and omit trivial y, z-dependence and $k_{y,z}$ subscripts to simplify notation.)

In the normal part of the system there exist electrons and holes, described by the (non-normalized) vectors

$$\psi_e^\pm(x) = \begin{pmatrix} 1 \\ 0 \end{pmatrix} e^{\pm ik_+x}; \tag{4.150}$$

$$\psi_h^\pm(x) = \begin{pmatrix} 0 \\ 1 \end{pmatrix} e^{\pm ik_-X}, \tag{4.151}$$

where as is easy to see from the Bogoliubov-de Gennes equations, the wave vectors satisfy the dispersion law

$$k_\pm(E) = k_F\sqrt{1 \pm \frac{E}{\mu}}. \tag{4.152}$$

In the superconductor there exist electron-like and hole-like bogolons,

$$\Psi_e^\pm(x) = \begin{pmatrix} ue^{i\phi/2} \\ ve^{-i\phi/2} \end{pmatrix} e^{\pm iq_+x}; \tag{4.153}$$

$$\Psi_h^\pm(x) = \begin{pmatrix} ve^{i\phi/2} \\ ue^{-i\phi/2} \end{pmatrix} e^{\pm iq_-X}. \tag{4.154}$$

Here the dispersion law and expressions for u, v also follow from the Bogoliubov-de Gennes equations (plus the condition $u^2 + v^2 = 1$) after some exercise in elementary algebra, and look as follows:

$$u(E) = \sqrt{\frac{1 + \sqrt{1 - \Delta^2/E^2}}{2}}; \qquad (4.155)$$

$$v(E) = \sqrt{\frac{1 - \sqrt{1 - \Delta^2/E^2}}{2}}; \qquad (4.156)$$

$$q \pm (E) = k_F \sqrt{1 \pm \frac{\sqrt{E^2 - \Delta^2}}{\mu}}. \qquad (4.157)$$

For subgap excitations ($E < \Delta$), u, v, and q acquire imaginary parts (we will take $\sqrt{-1} = +i$). Physically possible subgap solutions must exponentially decay into the bulk of the superconductor (in our case, at $x \to \infty$, which allows only Ψ_e^+ and Ψ_h^-).

Now we can solve the stationary scattering problem for the Bogoliubov-de Gennes equations. Let an electron impinge on the boundary from the left. Then the wave function at $x < 0$ will be

$$\psi_e^+ + r_{ee}\psi_e^- + r_{eh}\psi_h^+,$$

while at $x > 0$ it is

$$t_{ee}\Psi_e^+ + t_{eh}\Psi_h^-.$$

(We keep physical subgap states.)

Of the transmission/reflection amplitudes, we are especially interested in r_{eh}, the amplitude of Andreev electron-hole reflection. Matching the wave functions and their derivatives at $x = 0$ will, after some tedious but straightforward calculations, give the following results:

$$t_{eh} = \frac{1}{u}\frac{q_+ - k_-}{q_+ + k_-}e^{i\phi/2}r_{eh}; \qquad (4.158)$$

$$t_{ee} = \frac{1}{v}\frac{q_- + k_-}{q_+ + k_-}e^{i\phi/2}r_{eh}; \qquad (4.159)$$

$$r_{ee} = \left(\frac{u}{v}\frac{q_+ + k_-}{q_+ + q_-} + \frac{v}{u}\frac{q_+ - k_-}{q_+ + q_-}\right)e^{i\phi}r_{eh} - 1, \qquad (4.160)$$

while

$$r_{eh} = \frac{2e^{-i\phi}}{\frac{u}{v}\frac{q_- - k_-}{q_+ + q_-}\left(1 + \frac{q_+}{k_+}\right) + \frac{v}{u}\frac{q_+ - k_-}{q_+ + q_-}\left(1 - \frac{q_-}{k_+}\right)}. \qquad (4.161)$$

This somewhat cumbersome expression greatly simplifies in the so-called Andreev approximation, that is, in the lowest nonvanishing order in $\frac{\max(\Delta.E)}{\mu}$: quite reasonable if we recall the orders of magnitude of these energies. Thus, we take $k_\pm \approx q_\pm \approx k_F$, while

$$k_+ - q_- \approx q_+ - k_- \approx k_F \frac{u}{v} \frac{\Delta}{2\mu}.$$

Therefore, the Andreev and normal reflection amplitudes are respectively

$$r_{\text{eh}} \approx e^{-i\phi} \frac{E - \sqrt{E^2 - \Delta^2}}{\Delta} \tag{4.162}$$

and

$$r_{\text{ee}} \approx \frac{E - \sqrt{E^2 - \Delta^2}}{4\mu}. \tag{4.163}$$

Note that the Andreev electron-hole amplitude acquires the phase $-\phi$. We will not repeat the calculations, but the hole-electron amplitude acquires phase $+\phi$:

$$r_{\text{he}} = \frac{2e^{i\phi}}{\frac{u}{v}\frac{q_- + k_-}{q_+ + q_-}\left(1 + \frac{q_+}{k_+}\right) + \frac{v}{u}\frac{q_+ - k_-}{q_+ + q_-}\left(1\frac{q_-}{k_+}\right)}$$

$$\approx e^{i\phi} \frac{E - \sqrt{E^2 - \Delta^2}}{\Delta}. \tag{4.164}$$

We will see the importance of this in the next section.

In the Andreev approximation the Andreev reflection amplitude (4.162, 4.164) can be written as

$$r_{\text{eh(he)}} = e^{\mp i\phi} \times \begin{cases} e^{-i\arccos \frac{E}{\Delta}}, & E \leq \Delta; \\ e^{-\operatorname{arccosh} \frac{E}{\Delta}}, & E > \Delta. \end{cases} \tag{4.165}$$

For subgap particles, we thus have total Andreev reflection $|r_{\text{eh(he)}}(E)|^2 = 1$, in complete agreement with our qualitative reasoning, and the sharp change in pairing potential notwithstanding. The latter leads to small corrections $o(\Delta/\mu)$, leaving place for finite, but small, normal reflection (evidently, for $E < \Delta$, $|r_{\text{eh(he)}}(E)|^2 + |r_{\text{ee(hh)}}(E)|^2 = 1$), which can be usually neglected. It becomes significant if there is a (normal) potential barrier at the NS interface, or if the Fermi vectors in normal and superconducting regions differ (see Blonder et al. [16] for a detailed discussion). Instead, we will consider the Andreev levels and Josephson effect in SNS junctions, which is far more exciting.

4.5.1 The Proximity Effect in a Normal Metal in Contact with a Superconductor

We have mentioned the proximity effect: a suppression of order parameter in the superconductor close to a boundary with normal conductor. Probably more often *proximity effect* is called the effect of a superconductor inducing superconducting correlations between electrons and holes in the normal conductor. More formally, the anomalous Green's function $F(x, x')$ is nonzero in the normal region, though it decays as we move away from the boundary (in our case, as $x, x' \to -\infty$). This does not mean that there always appears finite pairing potential, alias order parameter, in the normal part of the system: since $\Delta \propto gF$, it can be identically zero for a normal conductor with $g = 0$ (the case we are considering). Such a conductor never becomes superconducting by itself, but externally induced superconducting correlations in it may survive.

To see this, we will not solve the Gorkov equations explicitly; it is enough to look at the solutions of our scattering problem.

We have seen that inside the gap, the wave function of an Andreev reflected electron (hole) is (in Andreev approximation)

$$\Psi_{\text{eh}}(x; E) = \psi_{\text{e}}^+ + r_{\text{eh}}\psi_{\text{h}}^+ \approx \begin{pmatrix} e^{iEx/\hbar v_F x} \\ e^{-i\phi - i\arccos\frac{E}{\Delta} - \frac{iEx}{\hbar v_F}} \end{pmatrix} e^{ik_F X}; \qquad (4.166)$$

$$\Psi_{\text{he}}(x; E) = \psi_{\text{e}}^- + r_{\text{hc}}\psi_{\text{h}}^- \approx \begin{pmatrix} e^{i\phi - i\arccos\frac{E}{\Delta} - \frac{iEx}{\hbar v_F}} \\ e^{iEx/\hbar v_F} \end{pmatrix} e^{-ik_F x}. \qquad (4.167)$$

Here we have expanded $k_\pm(E) \approx k_F \pm E/\hbar v_F$.

The anomalous Green's function at the point x can be expressed through products like, e.g.,

$$[\Psi_{\text{eh}}(x; E)]_2([\Psi_{\text{eh}}(x; E)]_1)^* = e^{iEx/\hbar v_F} \cdot e^{i\phi + i\arccos\frac{E}{\Delta} + tEx/\hbar v_F} \propto e^{2iEx/\hbar v_F}$$

[cf. Eq. (3.163)]. The only coordinate dependence enters this expression via the phase factor, $2Ex/(\hbar v_F)$, which has the sense of the relative phase shift of electron and hole components of the wave function $\Psi_{\text{eh}}(x; E)$. If $E = 0$, then these components keep constant relative phase $\phi + \arccos\frac{E}{\Delta}$ (and thus superconducting coherence) all the way to $x = -\infty$, where no pairing interactions exist ($g = 0$)! At finite energy, this coherence decays when the relative phase becomes of order unity, that is, at a distance from the boundary $\approx l_E = \frac{\hbar v_F}{2E}$. At temperature T, for a thermal electron-hole pair correlations decay at a distance $\approx l_T = \frac{\hbar v_F}{k_B T}$. The latter length is usually called the *normal metal coherence length* (in the clean case: we completely neglected impurity scattering).

The strange coherence of electrons and holes in the absence of any pairing interaction can be understood if we return to the picture of Andreev reflection. A hole into

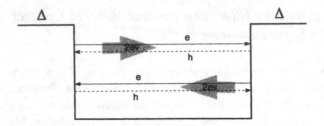

Fig. 4.17 Andreev levels in an SNS junction

which the incident electron has been transformed has exactly the opposite velocity (if $E = 0$), and will thus retrace *exactly the same path* all the way to minus infinity! Little wonder that the correlations are conserved. (They will be conserved even in the presence of nonmagnetic scatterers, but this is not a book on superconductivity theory.) At $E \neq 0$ the paths of electron and hole will diverge, and l_E is exactly the measure of when they diverge irreparably: the proximity effect is essentially a kinematic phenomenon.

4.5.2 Andreev Levels and Josephson Effect in a Clean SNS Junction

Since the subgap particle in the normal region is reflected by the pairing potential as a hole, and vice versa, in an SNS junction, when a normal region, of width L, is sandwiched between two superconductors, there should appear quantized *Andreev levels* [22]. This is illustrated in Fig. 4.17. In the normal pail of the system the solution of the Bogoliubov-de Gennes equations will be

$$\psi(x; E) = a\psi_e^+(x, E) + b\psi_h^+(x, E) + c\psi_e^-(x, E) + d\psi_h^-(x, E).$$

The coefficients a, b, c, d are found easily, since

$$c\psi_e^-(0, E) = r_{ee}(E)a\psi_e^+(0, E) + r_{he}(E)d\psi_h^-(0, E);$$
$$b\psi_h^+(0, E) = r_{eh}(E)a\psi_e^+(0, E) + r_{hh}(E)d\psi_h^-(0, E);$$
$$a\psi_e^+(-L, E) = r_{ee}(E)c\psi_e^-(-L, E) + r_{he}(E)b\psi_h^+(-L, E);$$
$$d\psi_h^-(-L, E) = r_{eh}(E)c\psi_e^-(-L, E) + r_{hh}(E)b\psi_h^+(-L, E).$$

In the gap, the above system yields the discrete set of allowed energy levels of the bound states, the Andreev levels, which must satisfy

$$-2\arccos\frac{E_n^{\pm}}{\Delta} \pm (\phi_1 - \phi_2) + (k_+(E_n^{\pm}) - k_-(E_n^{\pm}))L = 2\pi n; \quad n = 0, \pm 1, \ldots. \quad (4.168)$$

Here there are two sets of levels, with $\pm(\phi_1 - \phi_2)$, depending on the direction of motion of the electron. Both depend explicitly on the superconducting phase difference between the superconductors, due to the phase acquired by the wave function during Andreev reflection. Note that the phase itself is irrelevant, as it should be: it is the difference that counts. You may have noticed that (4.168) could be written immediately from the quasiclassical quantization condition in the pairing potential well,

$$\oint p(E)dq = 2\pi n,$$

if we take into account scattering phases ($\pm\phi_j + \arccos\frac{E}{\Delta}$) at the interfaces.

Unlike bound states in a usual rectangular well, Andreev levels carry electric current. This fact, together with phase sensitivity of their positions, is responsible for the *Josephson effect* in SNS junctions, which is a very remarkable phenomenon.

The transcendental equation (4.168) can be easily solved in two limiting cases: $L = 0$ and $L \to \infty$. (Actually, in the second case we must have $L \gg \xi_0$, but still $L \ll l_T$).

In the first case,

$$\arccos\frac{E_n^\pm}{\Delta} = \frac{1}{2}(\perp\Delta\psi - 2\pi n), \tag{4.169}$$

where $\Delta\phi \equiv \phi_1 - \phi_2$; therefore,

$$E(\Delta\phi) = \Delta\cos\frac{\phi_1 - \phi_2}{2} \equiv \Delta\cos\frac{\Delta\phi}{2}. \tag{4.170}$$

The contact contains a single, twice-degenerate level. (If we now recall that we considered a single k_y, k_z mode, the degeneracy is $2 \times N$ $\perp$-fold, where $N_\perp \approx A/\lambda_F^2$ is the number of transverse modes in the contact of area A, in the spirit of the Landauer formula.)

In the second case, for $E \ll \Delta$, we can expand $k_+(E) - k_-(E) \approx k_F E/\mu$ and set $\arccos(E/\Delta) = \pi/2$, to yield

$$E_n^\pm = \frac{\hbar v_F}{2L}[\pi(2n + 1) \pm \Delta\phi]. \tag{4.171}$$

In this case there are many Andreev levels in each mode. (Exactly how many we cannot tell, because the above formula doesn't work when $E^\pm$ is close to the top of the well.)

The knowledge of Andreev levels will allow us to calculate the *Josephson current* in the SNS junction. We will do this, using different approaches in the cases $L = 0$ and $L \to \infty$.

The most popular version of the *Josephson effect* is the one in SIS (superconductor-insulator-superconductor) tunneling junctions. Josephson's great achievement was the discovery of *coherent tunneling* of Cooper pairs-and thus supercurrent flow - through the insulating barrier between superconductors, and its dependence on their superconducting phase difference. As you know, the Josephson current is then given by

$$I_J(\phi) = I_c \sin \phi. \qquad (4.172)$$

(We will not discuss it here: the topic was covered in great detail, e.g., in Barone and Paterno [2]). But the Josephson effect is not limited to SIS junctions and $\sin \phi$ dependence. It appears wherever supercurrent can flow due to coherent transport of Cooper pairs through a *weak link-a* layer of normal conductor, for example, where superconducting correlations are not supported dynamically. The specific mechanism of the effect is, though, different, which leads-as we shall see shortly -to drastic deviations from (4.172).

4.5.3 Josephson Current in a Short Ballistic Junction: Quantization of Critical Current in Quantum Point Contact

First we must decide whether the limit $L = 0$ for an SNS contact corresponds to a physically sensible situation. It would seem that this limit corresponds simply to a bulk superconductor, and a stationary jump of the superconducting phase is as impossible to realize as a finite voltage drop between two banks of such a contact in the normal state. But we have already encountered the latter situation. in the point contact, which *is* a weak link of exactly the sort we need for the Josephson effect (it is often called an ScS junction, c for constriction). Therefore, the limit $L = 0$ can be considered as an approximation to the case of a superconducting point contact.

The Josephson current can be calculated if we know a thermodynamic potential of the system, Π (for example, $G, F, \Omega, \ldots$), as a function of the phase difference across the junction, $\Delta\phi$ (this is a general formula, valid for any sort of Josephson junction):

$$I = \frac{2e}{\hbar} \frac{d\Pi}{d(\Delta\phi)}. \qquad (4.173)$$

We have already noted that the supercurrent is a "thermodynamic current," and its density can be obtained by taking a variational derivative of any of the thermodynamic potentials over the vector potential, A (4.120). Since the latter always enters through the gauge invariant supercurrent velocity,

$$m v_s(\mathbf{r}) = m \nabla \chi(\mathbf{r}) - \frac{e}{c} \mathbf{A}(\mathbf{r}),$$

$2m\chi \equiv \phi$ being the phase of the order parameter, then

$$\frac{\delta}{\delta \mathbf{A}(\mathbf{r})} = -\frac{2e}{c}\frac{\delta}{\delta \nabla \phi(\mathbf{r})}.$$

Therefore, the supercurrent density is given by

$$\mathbf{j}(\mathbf{r}) = 2e\frac{\delta \Pi}{\delta \nabla \phi(\mathbf{r})}.$$

The variational derivative in this expression is the coefficient that appears when we write Π as

$$\Pi = \int d^3 r \frac{\delta \Pi}{\delta \nabla \phi(\mathbf{r})} \cdot \nabla \phi(\mathbf{r}) + \cdots.$$

(We have dropped the term independent on $\nabla \phi(\mathbf{r})$.)

On the other hand, a thermodynamic potential of the Josephson junction can be expanded in powers of $\Delta \phi$:

$$\Pi = \frac{\partial \Pi}{\partial \Delta \phi}\Delta \phi + \cdots.$$

In this system the phase changes abruptly by $\Delta \phi$, so that

$$\nabla \phi(\mathbf{r}) = \Delta \phi \delta(x)\mathbf{e}_x.$$

Writing down the identity (A being the area of the contact)

$$\Pi = \frac{1}{A}\int dA \Pi = \frac{1}{A}\int dy dz \int dx \delta(x) \Pi$$
$$= \frac{1}{A}\int dx dy dz \frac{\partial \Pi}{\partial \Delta \phi}\mathbf{e}_x \cdot \Delta \phi \delta(x)\mathbf{e}_x + \cdots,$$

we notice that

$$\frac{\delta \Pi}{\delta \nabla \phi(\mathbf{r})} = \frac{1}{A}\frac{\partial \Pi}{\partial \Delta \phi}\mathbf{e}_x. \tag{4.174}$$

Then the total Josephson current is indeed given by (4.173).

We use here the grand potential Ω, because then we can immediately use the results of Sect. 4.3.4, namely (4.77)

$$\Omega[\Delta, \Delta^*] = \frac{1}{|g|} \int d^3\mathbf{r}|\Delta(\mathbf{r})|^2 + \sum_q \int d^3\mathbf{r}\{v_q(\mathbf{r})(\hat{\xi} + V(\mathbf{r}))v_q^*(\mathbf{r})$$

$$+ u_q^*(\mathbf{r})(\hat{\xi} + V(\mathbf{r}))u_q(\mathbf{r})\}$$

$$- \frac{2}{\beta} \sum_q \ln\left(2\cosh\frac{\beta E_q}{2}\right).$$

In this formula, the first two lines are independent of $\Delta\phi$. Therefore, the Josephson current can be written as follows:

$$I = -\frac{2e}{\hbar} \sum_p \tanh\frac{E_p}{2k_BT} \cdot \frac{dE_p}{d\Delta\phi}$$

$$- \frac{2e}{\hbar} \cdot 2k_BT \int_{\Delta_0}^{\infty} dE \ln(2\cosh\frac{E}{2k_BT}) \cdot \frac{dN_c(E)}{d\Delta\phi}.$$

The first term contains the contributions from the discrete Andreev levels in the gap, while the second term accounts for the excited states in the continuum, with energies exceeding $|\Delta|$, $N_c(E)$ being the density of states in the continuum. The latter—since we consider the case $L = 0$—is evidently the same as in a bulk superconductor, and thus is $\Delta\phi$ independent. Therefore *only* the discrete Andreev levels contribute to the Josephson current:

$$I = -\frac{2e}{\hbar} \sum_p \tanh\frac{E_p}{2k_BT}\frac{dE_p}{d\Delta\phi}. \qquad (4.175)$$

Substituting here (4.170), we see that [1]

$$I = \frac{2N_\perp e}{\hbar} \cdot \frac{\Delta_0}{2} \sin\frac{\Delta\phi}{2} \cdot \tanh\frac{\Delta_0\cos\Delta\phi/2}{2k_BT}. \qquad (4.176)$$

At zero temperature this reduces to

$$I = \frac{\pi\Delta_0 G_N}{e} \sin\frac{\Delta\phi}{2}, \qquad (4.177)$$

where $G_N = N_\perp e^2/(\pi\hbar)$ is the normal (Sharvin) conductance of the contact. The latter formula was derived by Kulik and Omelyanchouk [23], for a classical superconducting point contact. Note the nonsinusoidal and moreover, discontinuous phase dependence of the Josephson current (Fig. 4.18), as well as its proportionality to the normal conductance of the system. We already know that in the case of normal *quantum* point contact G_N is quantized in the units of $e^2/(\pi\hbar)$. The result (4.176)

Fig. 4.18 Phase dependence
of the Josephson current in a
superconducting point contact

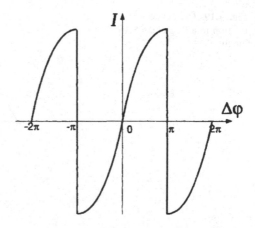

tells us that in the superconducting quantum point contact the critical current is also
quantized, in the (nonuniversal) units of $e\Delta_0\beta$.[5]

An important property of the short ScS contact, which made quantization possible,
was the $N_\perp$-fold degeneracy of Andreev levels (4.170), which ensured that each
transverse mode in the contact gives the same contribution to the current-as it did in
the normal case. Unfortunately, for an ScS junction this holds only in the zero length
limit, as we will see in the next subsection.

4.5.4 Josephson Current in a Long SNS Junction

Let us now consider along ballistic SNS junction. Here the normal layer width is still
much smaller than both normal metal coherence length and elastic scattering length,
but exceeds the superconducting coherence length: $l_T, l_e \gg L \gg \xi_0$.

We begin with an insightful picture due to Bardeen and Johnson [15]. Suppose the
temperature is zero and there is a supercurrent through the junction, with superfluid
velocity v_s in the x-direction. In a long channel, we can neglect the boundary effects
and simply relate v_s to the phase difference:

$$mv_s = \frac{\hbar\Delta\phi}{L}. \tag{4.178}$$

We have seen earlier, Eq. (4.136), that the supercurrent can be presented as the current
due to macroscopic flow of the condensate, minus the current of the elementary
excitations, carried with velocity v_s:

---

[5] The effect was recently observed on experiment [25].

Fig. 4.19 Current-phase
relation in a long clean SNS
junction

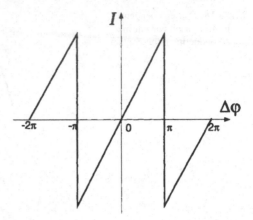

$$I = nev_s - I_{n,x} = nev_s - \sum_{k_q} \frac{ek_{q,x}}{mL} n_F(E_q^{\pm} - k_{q,x}v_s). \qquad (4.179)$$

The factor of $1/L$ appears because the u, v-components of the wave function must be normalized to the width of the normal layer. The quasiparticle energies in this expression are measured in the reference frame of the moving condensate. Therefore, they correspond to $v_s = 0$ and are simply the energies of the Andreev levels at $\Delta\phi = 0$.

At zero temperature, the Fermi distribution function is identically zero for all "–" levels (with $k_{q,x} < 0$, when the electron moves to the left). Let us now increase $\Delta\phi$ (and v_s). At first, the contribution from the "+"-levels will be zero as well, and the current grows linearly with the phase difference. When for the lowest Andreev level

$$E_q^+ - k_{q,x}v_s = 0,$$

then $n_F = \frac{1}{2}$. It can be shown that this happens at $\Delta\phi = \pi$, and the contribution of this level exactly cancels the first term in (4.179), zeroing the supercurrent. For an infinitesimally larger v_s, $n_F = 1$, and the current changes sign. Then the current will grow linearly again, until at $\Delta\phi = 3\pi$ the same happens for the second lowest level, ad infinitum. Of course, the picture holds for negative $\Delta\phi$, v_s, only this time "–"-levels are involved.

Thus the characteristic sawtooth dependence arises at $T = 0$ (Fig. 4.19). It was first obtained by Ishii [19] with the use of a more sophisticated method.

The remarkable feature of this result is that the 2π-periodic $I(\Delta\phi)$ dependence is again nonsinusoidal and discontinuous at odd multiples of π, as in the case of a short ScS junction.

We will neither reproduce here the details of the Bardeen-Johnson or Ishii calculations, nor apply Eq. (4.173). Instead we derive $I(\Delta\phi)$ at arbitrary temperature by direct calculation of the density of Andreev states and the supercurrent. We will, though, use the insight that the low-lying Andreev states dominate the supercurrent in a long SNS junction.

We use the basic formula (4.73) to calculate the current in the normal part of the junction, and for the time being consider a single transverse mode, with fixed k_y, k_z. Taking normalization into account, we see that the current is given by

$$I = \frac{e}{Lm} \int_0^\infty dE (\nu_+(E) - \nu_-(E))[n_F(E)k_+(E) - (1 - n_F(E))k_-(E)]. \quad (4.180)$$

Here in the brackets $k_\pm(E)$ is the momentum of the electron (hole) excitation. The parentheses contain the densities of "$\pm$"-states, which take into account both bound Andreev levels and the continuum. At $E < |\Delta|$ they we simply sets of δ-functions at energies $E_q^\pm$.

Since we expect that the main contribution in the current is from the lowest Andreev levels, we substitute k_F (that is, $k_{F,x}$) for of $k_\pm(E)$. Then a simple manipulation with Fermi distribution functions allows us to extend the integration to $(-\infty, \infty)$:

$$I = \frac{ev_{F,x}}{L} \int_{-\infty}^\infty dE (\nu_+(E) - \nu_-(E))[2n_F(E) - 1]$$

$$= -\frac{ev_{F,x}}{L} \int_{-\infty}^\infty dE (\nu_+(E) - \nu_-(E)) \tanh \frac{\beta E}{2}, \quad (4.181)$$

$\beta = 1/k_B T$.

Now we find the density of the excited states. Using the Weierstrass formula, we write

$$\nu_\pm(E) = \sum_{-\infty}^\infty \delta(E - E_k^\pm)$$

$$\equiv -\frac{1}{\pi} \Im \sum_{-\infty}^\infty \frac{1}{E - E_k^\pm + i0} \rightarrow -\frac{1}{\pi} \Im \sum_{-\infty}^\infty \frac{1}{E - E_k^\pm + i\frac{\hbar}{\tau}}, \quad (4.182)$$

where on the right-hand side we immediately recognize the retarded Green's function in Källén-Lehmann representation.

Here we introduced a finite lifetime, τ, due, e.g., to weak nonmagnetic impurity scattering. In the limit $\tau \rightarrow \infty$ it reduces to the infinitesimal $i0$ term. It is convenient to parameterize by

$$\tau = \frac{L}{\epsilon v_{F,x}}; \quad \epsilon = \frac{L}{v_{F,x}\tau} \equiv \frac{L}{l_e} \ll 1,$$

where $l_e = v_{F,x}\tau$ is the transport scattering length in a given mode.

After substituting the low-energy approximation for Andreev levels, the densities of states become

$$\nu \pm (E) = \frac{\epsilon L}{\pi \hbar v_{F,x}} \sum_{q=-\infty}^{\infty} \left| \frac{LE}{\hbar v_{F,x}} - \left((q + \frac{1}{2})\pi \pm \frac{1}{2}\Delta\phi \right) + i\epsilon \right|^{-2}.$$

The sum is easily taken using the *Poisson summation* formula, which transforms the functional series to the series of Fourier transforms:

$$\sum_{n=-\infty}^{\infty} f(n) = \int_{-\infty}^{\infty} dx f(x) \sum_{n=-\infty}^{\infty} \delta(x-n)$$

$$= \int_{-\infty}^{\infty} dx f(x) \sum_{p=-\infty}^{\infty} e^{2\pi i p x} \equiv \sum_{p=-\infty}^{\infty} \tilde{f}(p). \tag{4.183}$$

It is very convenient, e.g. if the latter series converges faster.

The integration over x is executed using the methods of complex analysis, and we obtain

$$\nu_{\pm}(E) = \frac{L}{\pi \hbar v_{F,x}} \sum_{p=-\infty}^{\infty} e^{-2|p|\epsilon} e^{2ip(\frac{LE}{\hbar v_{F_x}} - \frac{\pi \pm \Delta\phi}{2})};$$

$$\nu_+(E) - \nu_-(E) = \frac{-2iL}{\pi \hbar v_{F,x}} \sum_{p=-\infty, p\neq 0}^{\infty} (-1)^p e^{-2|p|\epsilon + 2ip\frac{LE}{\hbar v_{F,x}}} \sin p\Delta\phi. \tag{4.184}$$

Substituting the last expression in (4.181) and taking the tabular integral

$$\int_{-\infty}^{\infty} dE e^{\frac{2ipLE}{\hbar v_{F,x}}} \tanh \frac{\beta E}{2} = \frac{2\pi i}{\beta \sinh \frac{2\pi L p}{\hbar \beta v_{F,x}}}$$

and summing up the contributions from all transverse modes, we finally obtain the expression for the Josephson current in a long clean SNS junction:

$$I(\Delta\phi) = \sum_{k_F} \frac{e v_{F,x}}{L} \frac{2}{\pi} \sum_{p=1}^{\infty} (-1)^{p+1} e^{-2p\frac{L}{l_C(k_F)}} \frac{L}{l_T(k_F)} \frac{\sin p\Delta\phi}{\sinh \frac{pL}{l_T(k_F)}}. \tag{4.185}$$

Here $l_T(k_F) = \frac{\hbar v_{F,x}}{2\pi k_B T}$, and $l_e(k_F) = v_{F,x}\tau$.

This is a remarkable expression. At zero temperature, and in the ballistic limit, all L/l_e, L/l_T are zero. You can check that the series

$$\frac{2}{\pi} \sum_{p=1}^{\infty} (-1)^{p+1} \frac{\sin p\Delta\phi}{p}$$

converges to the 2π-periodic sawtooth of unit amplitude. Thus we indeed recover the result of Ishii. Finite temperature and elastic scattering smoothes it over, and eventually, as is clear from (4.185), only the lowest harmonic survives, leaving us the "standard" $I \propto \sin \Delta\phi$-dependence.

The total Josephson current is given by the sum of contributions of different modes, $\mathbf{k}_F$. To calculate it in a classical SNS junction we should integrate over all $\mathbf{k}_F$ (with positive projection on the x-axis), which will give the critical current as proportional to R_N. In the case of a quantum junction, when a few allowed modes are present, we see that the critical current is not quantized even at zero temperature: the amplitudes $ev_{F,x}/L$ depend on the direction of $\mathbf{k}_F$; that is, opening an extra mode will bring a mode-dependent increase to the critical Josephson current.

4.5.4.1 SND Junction: Case of d-Wave Pairing Symmetry

A curious situation arises if the pairing symmetry in one bank of a long SNS junction is different from that in the other. We have already discussed superconducting transition in case of arbitrary orbital symmetry of pairing. The only complication was that the order parameter $\langle a_{\mathbf{k}} a_{-\mathbf{k}} \rangle = \Delta_{\hat{\mathbf{k}}}$ becomes dependent on the relative momentum direction on the Fermi surface, $\hat{\mathbf{k}}$.

It is now firmly established that in superconducting cuprates the order parameter has d-wave symmetry [3, 10, 11, 26]. This means that under an appropriate choice of coordinate axes, $\Delta_{\hat{\mathbf{k}}} \propto k_x^2 - k_y^2$. Therefore, it is negative for some directions. If we insist on writing the order parameter as $|\Delta| e^{i\phi}$, we are now compelled to add π to the superconducting phase. As a result, the Andreev reflected electron may now (depending on its direction of propagation) acquire an extra phase of π.

What happens in a long SNS junction between conventional (s-wave) and d-wave superconductors (a so-called SND junction)? There will now be two kinds of Andreev levels (Fig. 4.20): zero- and π-levels, so called because of additional phase acquired when reflected from the d-wave bank. The Josephson current is thus a sum of two contributions, each given by (4.185), with momentum summation limited to appropriate states [28]. The magnitudes of these terms of course depend on the orientation of the d-wave crystal. In the most symmetric case, when each zero-level has its π-counterpart, we would find that

$$I(\Delta\phi) = \sum_{\mathbf{k}_F}^{\text{zerolevels}} \frac{ev_{F,x}}{L} \frac{2}{\pi} \sum_{p=1}^{\infty} (-1)^{p+1} e^{-2p\frac{L}{l_e(\mathbf{k}_F)}}$$

$$\times \frac{L}{l_T(\mathbf{k}_F)} \frac{\sin p\Delta\phi + \sin p(\Delta\phi + \pi)}{\sinh \frac{pL}{l_T(\mathbf{k}_F)}}. \tag{4.186}$$

This is a π-periodic sawtooth of $\Delta\phi$ (Fig. 4.21): the Josephson effect period is thus halved.[6]

---

[6] This phenomenon can occur in different types of Josephson junctions between superconductors with different pairing symmetry; see Zagoskin [28] and references therein.

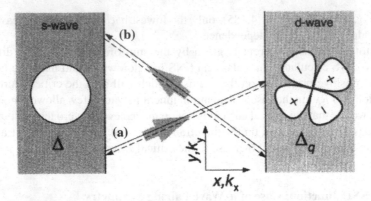

Fig. 4.20 SND junction: **a** zero- and **b** π-Andreev levels

Fig. 4.21 Current-phase dependence in a symmetric SND junction

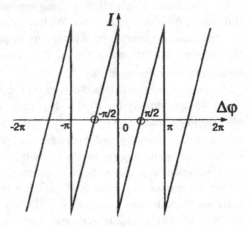

From the relation $I(\Delta\phi) = (2e/\hbar)dE/d\Delta\phi$, we see that the equilibrium is achieved not at $\Delta\phi = 0$, but at $\Delta\phi = \pm\pi/2$.[7] The Josephson current $I(\pm\pi/2)$ is indeed zero, because contributions from zero-and π-levels cancel. But the current component *parallel* to the SN boundary does not vanish. On the contrary, the contributions to this component from zero- and π-levels add up, thus creating *spontaneous currents* in the normal layer [18].

---

[7] Generally, depending on relative contributions of zero- and π-levels to the current, the equilibrium phase can take any value from $-\pi$ to π.

4.5.5 * Transport in Superconducting Quantum Point Contact: The Keldysh Formalism Approach

We have already mentioned the possibility of generalizing the Keldysh formalism to Nambu matrices. Since this makes Green's functions 4×4 matrices (each Keldysh component being a Nambu matrix), this is warranted only if other approaches fail (in an essentially nonequilibrium situation) or if the problem allows some significant simplifications. Examples can be found in Rammer and Smith [7].

Here we present some recent results by Cuevas et al. [17] for transport in superconducting point contact, obtained using a combination of Keldysh techniques and the method of tunneling Hamiltonian. We have discussed this approach in application to normal contact in Sect. 3.7.

The Hamiltonian is again presented as a sum (3.200)

$$\mathcal{H} = \mathcal{H}_L + \mathcal{H}_R + \mathcal{H}_T, \tag{4.187}$$

where the tunneling term is convenient to write as [cf. (3.203), (3.225)]

$$\mathcal{H}_T = \sum_\sigma (\text{T} e^{i\phi(t)/2} c_\sigma^\dagger d_\sigma + \text{T}^* e^{-i\phi(t)/2} d_\sigma^\dagger c_\sigma). \tag{4.188}$$

Here the phase

$$\phi(t) = \phi_0 + \frac{2eV}{\hbar} t \tag{4.189}$$

describes the time dependence of field operators due to voltage applied to one of the banks [as in (3.207)]. We write it explicitly because in the superconducting junction, (4.189) gives the observable *nonstationary (ac) Josephson effect:*. oscillations of superconducting phase difference across a junction (and therefore Josephson current) with *Josephson frequency*

$$\omega_J = \frac{2eV}{\hbar}, \tag{4.190}$$

which is determined by the applied voltage. (Physically, this is simply the quantum frequency associated with a Cooper pair gaining or losing energy $2eV$ in transfer across a finite voltage drop V.[8])

The current through the contact is derived in the same way as (3.208), and the analogue to (3.210) is given by the (11)-component of a Nambu matrix:

$$I(t) = \frac{2e}{\hbar} [\text{T} \mathbf{F}_1^{+-}(t, t) - \text{T}^* \mathbf{F}_2^{+-}(t, t)]_{11}. \tag{4.191}$$

---

[8] In an SND junction of the previous section (4.186) the frequency of the nonstationary Josephson effect doubles to $2\omega_J = 4eV/\hbar$, because the period of $I(\Delta\phi)$ was halved.

Here Nambu matrices

$$T = \begin{pmatrix} T & 0 \\ 0 & -T^* \end{pmatrix}, \tag{4.192}$$

$$F_1^{+-}(t', t) = \begin{pmatrix} \langle d_\uparrow^\dagger(t') c \uparrow (t) \rangle & \langle d_\downarrow(t') c \uparrow (t) \rangle \\ \langle d_\uparrow^\dagger(t') c_\downarrow^\dagger(t) \rangle & \langle d_\downarrow(t') c_\downarrow^\dagger(t) \rangle \end{pmatrix}, \tag{4.193}$$

and in $F_2^{+-}(t', t)$ the c's and d's are transposed. Expression (4.191) contains both superconducting (Josephson) current and normal (quasiparticle) current; the latter can flow if there is a finite voltage drop across the contact.

The "hybrid" Green's functions F^{+-} are then calculated as a $(+-)$-component of a matrix series over the Keldysh-Nambu matrix $\hat{T}$ and unperturbed Green's functions in the banks (assuming them identical):

$$g^{R,A}(\omega) = \frac{\pi N(0)}{\sqrt{\Delta^2 - (\omega \pm i\zeta)^2}} \begin{pmatrix} -\omega \pm i\zeta & \Delta \\ \Delta & -\omega \pm i\zeta \end{pmatrix}; \tag{4.194}$$

$$g^{+-}(\omega) = 2\pi i n_F(\omega) \left(-\frac{1}{\pi} \Im g^R(\omega) \right). \tag{4.195}$$

Here ζ is the small energy dissipation rate in the banks due to inelastic scattering.

The calculations are of necessity much more involved than those of Sect. 3.7. Therefore, we present here only the results.

In the linear response regime, $V \to 0$, the quasiparticle current through the quantum contact is characterized by phase-dependent conductance ($\beta = 1/k_B T$)

$$G(\Delta\phi) = \frac{2e^2}{h} \frac{\pi\beta}{16\zeta} \left[\frac{\Delta\alpha \sin \Delta\phi}{\sqrt{1 - \alpha \sin^2(\Delta\phi/2)}} \operatorname{sech} \frac{\beta E_a(\Delta\phi)}{2} \right]^2. \tag{4.196}$$

The Josephson current, in its turn, is given by the well-known expression

$$I(\Delta\phi) = \frac{e\Delta^2\alpha}{2\hbar} \frac{\sin \Delta\phi}{E_a(\Delta\phi)} \tanh \frac{\beta |E_a(\Delta\phi)|}{2}. \tag{4.197}$$

In the above formulae, $\alpha = \frac{(2\pi N(0)T)^2}{1+(\pi N(0)T)^2}$ is nothing but the effective Landauer transparency of the barrier, T_{Landauer}, that we calculated in (3.222). Energies

$$E_a(\Delta\phi) = \pm \Delta \sqrt{1 - \alpha \sin^2(\Delta\phi/2)} \tag{4.198}$$

are energies of Andreev levels in the contact. If $\alpha = 1$, we are back to our previous results (4.170) and (4.176) for an ideal short Josephson junction in a one-mode limit.

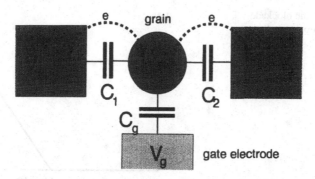

Fig. 4.22 Single-grain tunnel junction

4.6 Tunneling of Single Electrons and Cooper Pairs Through a Small Metallic Dot: Charge Quantization Effects

In this, the final section of the book, we are going to make good on our promise to discuss the case when a single electron can make a difference in a many-body system.[9]

This can occur if the picture of independent quasiparticles is no longer valid, and we need to take into account their correlations. Roughly, this happens when the criterion of the mean-field approximation applicability, Eq. (1.10), fails. This criterion,

$$\frac{\hbar v_F}{e^2} \gg 1, \tag{4.199}$$

shows explicitly that the faster the particles move, the better is the MFA description -something we have discussed already. On the other hand, for correlations to become important, the particles must be slowed down. One way to do this is to create "bumps" in their way: tunneling barriers, for example, or point contacts, or whatever weak links we can invent. Consider, e.g., the system shown in Fig. (4.22), a *single-grain tunnel junction*. Here electrons can travel between two massive banks through a small grain, separated from it by tunneling barriers. (The role of gate electrode will become clear in a moment.) Though v_F is not affected by the presence of the barriers, the characteristic time electrons spend on the grain (and effective travel velocity) is, which allows correlations to develop. Alternatively, we could rewrite the left-hand side of (4.199) as $\frac{\hbar v_F/l}{e^2/l}$, where l is some characteristic length. This is the ratio of interlevel spacing in the grain due to spatial quantization (remember the similar result for Andreev levels?) to a characteristic Coulomb energy. If the ratio is large, then correlations that are due to Coulomb interactions are irrelevant, and MFA will work. In the opposite limit we need a more refined approach.

---

[9] In this section I follow an unpublished lecture by R. Shekhter.

Fig. 4.23 Coulomb blockade of a single-electron tunneling

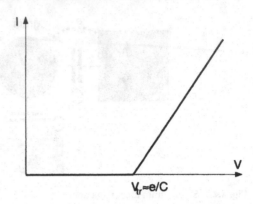

4.6.1 Coulomb Blockade of Single-Electron Tunneling

First let us consider a normal system. Denote the capacitance of our grain by C (for an isolated spherical grain of radius ρ in the medium with dielectric constant κ, $C = \kappa\rho$). Then the electrostatic energy due to the extra electron on the grain is

$$E = \frac{e^2}{2C}.$$ (4.200)

If $\rho \sim 10^3\,°\text{A}$, then $E \approx 10\,\text{K}$. Therefore, at a low enough temperature the Coulomb energy quantization due to charge discreteness is observable and leads to a *Coulomb blockade of single-electron tunneling*[10]: in the situation of Fig. 4.22, we should expect that no current can flow between the banks until the voltage is high enough to offset the electrostatic energy due to charging of the grain (Fig. 4.23).

But we did not yet take into account quantum mechanical properties of the electrons. The energy of the system with n extra electrons on the grain is expressed through C and mutual capacitances C_1, C_2, C_g as

$$E(n) = \frac{n^2 e^2}{2C} + en(\frac{C_1}{C}V_1 + \frac{C_2}{C}V_2 + \frac{C_g}{C}V_g) \approx \text{const} + \frac{e^2}{2C}[n - n^*(V_1, V_2, V_g)]^2.$$ (4.201)

Here the parameter n^* is given by

$$n^* = -\frac{1}{e}[C_1 V_1 + C_2 V_2 + C_g V_g].$$ (4.202)

---

[10] See, e.g., Tinkham [9], Chap. 7, Zagoskin [13], 2.4 and references therein.

Note that though there is no charge transfer between the gate electrode and the rest of the system, the gate voltage V_g critically affects the electrostatic energy and the transport through the grain.

Indeed, at a certain combination of parameters two the lowest-lying states with n and $n \pm 1$ electrons on the dot become degenerate (Fig. 4.24). This degeneracy is equivalent to lifting the Coulomb blockade, deblocking the current. It appears as periodic conductance dependence on V_g (*single-electron oscillations*) with period

$$\delta V_g \approx \frac{e}{C}. \tag{4.203}$$

Note that in a normal system, degeneracy between states with n and $n \pm 2$ extra electrons on the grain is impossible: the system will always drop to a lower-lying state with $n \pm 1$ electrons (Fig. 4.24c).

Observation of the Coulomb blockade and single-electron oscillations is possible, if we can resolve the levels differing by the charging energy $E(n + 1) - E(n) \approx \frac{e^2}{2C} = U_c$. On the other hand, the level width is

$$\delta E \approx \frac{\hbar}{\tau}, \tag{4.204}$$

where τ is the characteristic lifetime of the extra electron on the grain. If the conductance of the system is G, this time will be of order the discharge time of an RC-contour,

$$\tau = C/G. \tag{4.205}$$

Therefore, we have the observability criterion

$$\frac{e^2}{2C} > \frac{\hbar G}{C}, \tag{4.206}$$

leading to the condition on the conductance of our system

$$G < \frac{e^2}{2\hbar} \approx \frac{2e^2}{h}. \tag{4.207}$$

The latter is the same quantum conductance unit ($\approx (13 k\Omega)^{-1}$ in more conventional units) that we have met before. As we remarked, the transport through the grain must be hindered enough by the barriers in order to make correlation effects important. The quantum resistance unit provides a quantitative measure of this hindrance.

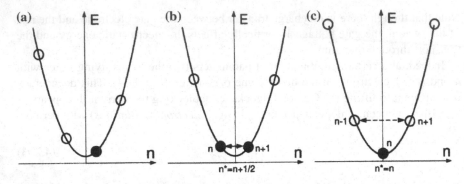

Fig. 4.24 a Electrostatic energy quantization. **b** Single-electron degeneracy of the ground state: deblocking of the single-electron tunneling. **c** No double-electron degeneracy of the ground state

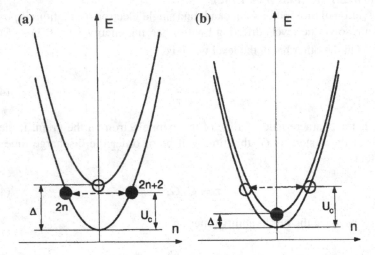

Fig. 4.25 Splitting of the Coulomb parabola due to parity effect: **a** 2e-degeneracy is possible if $\Delta > U_c = e^2/2C$. **b** No 2e-degeneracy otherwise

4.6.2 Superconducting Grain: When One Electron Is Too Many

If the grain becomes superconducting, there appear interesting new possibilities. As we know, in the ground state of a superconductor all electrons are bound in Cooper pairs (and therefore the ground state can contain only an even number of electrons). Any odd electron will thus occupy an excited state, as a bogolon, and its minimum energy, measured from the ground state energy, will be Δ.

This is the *parity effect* in superconductivity. Of course, in a bulk superconductor it is of no importance, but not so in our small system, where charging effects enter the game.

Let us write the number of electrons on the grain as

$$n = 2n_C + n_q, \tag{4.208}$$

where n_C is the number of Cooper pairs, and $n_q = 0$ or 1 is the number of unpaired elecrons. The energy of the system is thus (Fig. 4.25)

$$E = U_c(2n_C + n_q - n^*)^2 + n_q \Delta. \tag{4.209}$$

The situation depends significantly on whether $\Delta > U_c$ or $\Delta < U_c$. In the former case the 2e-degeneracy of the ground state becomes possible, and it occurs at odd values of $n^*(V_g)$. The period of corresponding oscillations is now

$$\delta V_g^{(SC)} \approx \frac{2e}{C}. \tag{4.210}$$

Let us now make the entire system superconducting. Then the grain will play the role of a weak link between the banks 1 and 2, allowing for a superconducting phase difference and Josephson current to appear. The latter is carried by condensate, i.e., Cooper pairs. Therefore, one should expect that 2e-degeneracy of the ground state will lead to enhancement of the critical current, which thus oscillates as a function of V_g. Moreover, since at $\Delta < U_c$ 2e-degeneracy becomes impossible, the effect must disappear, e.g., in a magnetic field strong enough to suppress Δ below the charging energy.

A quantitative consideration must take into account that we are dealing with a system of bosons—Cooper pairs—which can be described by a couple of complementary operators, number, and phase (like in Sect. 1.4.3). It is convenient to work in the basis of phase eigenstates, where [see (1.132)]

$$\hat{n}_C = \frac{1}{i} \frac{\partial}{\partial \phi}.$$

Here ϕ is the superconducting phase of the grain. The eigenstates of the operator $\hat{n}_C$ are evidently

$$\langle \phi | n \rangle \propto e^{in\phi}, \tag{4.211}$$

since $(1/i)\partial e^{in\phi}/\partial \phi = ne^{in\phi}$. The phases of the massive superconducting banks, $\phi_{1,2}$, are external parameters, and we can choose them to be

$$\phi_{1,2} = \pm \frac{\phi_0}{2}.$$

The Josephson current through the grain is thus

$$I = \frac{2e}{\hbar} \frac{\partial E_0(\phi_0)}{\partial \phi_0}, \tag{4.212}$$

where E_0 is the lowest eigenstate of the Hamiltonian

$$\mathcal{H} = 4U_c(-i\frac{\partial}{\partial\phi} + \frac{n_q}{2} - \frac{n^*}{2})^2 + n_q\Delta - E_J(\cos(\phi_1 - \phi) + \cos(\phi_2 - \phi)). \quad (4.213)$$

In the above formula, the first term is the charging energy expressed through the Cooper pair number operator; the second term is the odd electron contribution. The third term is the so-called *Josephson coupling energy* between the banks and the grain (we assume that the system is symmetric). By itself, a term like $-E_J\cos(\Delta\phi)$ would yield a usual Josephson current $(2eE_J/\hbar)\sin(\Delta\phi)$ between the bank and the grain [see Eq. (4.173)].

The Hamiltonian (4.213) is convenient to rewrite as

$$\mathcal{H} = U_c(-i\frac{\partial}{\partial\phi} + n_q - n^*)^2 + n_q\Delta - 2\tilde{E}_J(\phi_0)\cos(\phi), \quad (4.214)$$

where $\tilde{E}_J(\phi_0) = E_J\cos(\phi_0/2)$ is the only parameter dependent on the superconducting phase difference ϕ_0 between the banks.

The coupling term mixes states with n and $n \pm 1$ Cooper pairs on the grain:

$$\langle n|\cos\phi|n'\rangle \propto \int d\phi e^{-in\phi}\cos\phi e^{in'\phi} \propto \delta_{n,n'\pm 1}. \quad (4.215)$$

We can thus write the Hamiltonian in the basis of two states, e.g., $|n\rangle$ and $|n + 1\rangle$:

$$\mathcal{H} = \begin{pmatrix} E(2n) & -\tilde{E}_J(\phi_0) \\ -\tilde{E}_J(\phi_0) & E(2n + 2) \end{pmatrix}. \quad (4.216)$$

Here $E(2n)$, $E(2n + 2)$ are corresponding eigenvalues of the charging part of the Hamiltonian. The eigenvalues of matrix (4.216) are easily found, with the ground state energy

$$E_0(\phi_0) = \frac{1}{2}\left\{(E(2n) + E(2n + 2)) - \sqrt{(E(2n) - E(2n + 2))^2 + 4\tilde{E}_J^2(\phi_0)}\right\}$$

$$= \frac{1}{2}\left\{(E(2n) + E(2n + 2))\right. \quad (4.217)$$

$$\left. - \sqrt{(E(2n) - E(2n + 2))^2 + 4E_J^2\cos^2(\frac{\phi_0}{2})}\right\}.$$

The Josephson current is thus

$$I(\phi_0) = \frac{2e}{h}\frac{(1/4)E_J\sin\phi_0}{\sqrt{(\frac{E(2n)-E(2n+2)}{2E_J})^2 + \cos^2(\frac{\phi_0}{2})}}. \quad (4.218)$$

It depends on the gate voltage V_g through $E(2n) - E(2n+2)$, and it is clear that near the values $V_g = V_{g,n}$ where these two energies are degenerate, it peaks. Near these values of V_g,

$$I(V_g, \phi_0) \approx \frac{2e}{\hbar} \frac{(1/4)E_J \sin \phi_0}{\sqrt{[\frac{e(V_g - V_{g,n})}{E_J}]^2 + \cos^2(\frac{\phi_0}{2})}}. \tag{4.219}$$

We obtain a periodic in V_g enhancement of the Josephson current through a super-conducting grain, with period $2e/C$.

The effect exists, of course, only while 2e-degeneracy is possible. Therefore when Δ becomes lower than the charging energy, the critical current sharply drops. It retains dependence on V_g, but its period becomes e/C, which corresponds to e-degeneracy, as in the normal case. This behavior was predicted by Matveev et al. [24], and it was confirmed experimentally a year later (Joyez et al. [20]).

4.7 Problems

- *Problem 1*
  Show that the following equation for the anomalous Green's function at finite temperature leads to the BCS equation for $\Delta(T)$:

  $$F^+(\mathbf{p}, \omega)|_{T \neq 0}$$
  $$= F^+(\mathbf{p}, \omega)|_{T=0} - \frac{\Delta^*(T)}{\omega + \xi_p} \cdot 2\pi i n_F(E_p)(|u_p|^2 \delta(\omega - E_p) - |v_p|^2 \delta(\omega + E_p)).$$

  Use the self-consistency relation between Δ and F^+ and the expressions for F^+ at zero from Sect. 4.4.3.
- *Problem 2*
  Write the analytical expressions for the following Nambu diagram both in the matrix form and in components.

(a)   (b)   (c)

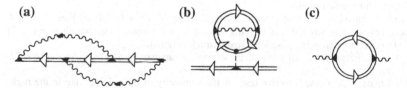

- *Problem 3*
  Prove that the expression for the superconducting current

$$\mathbf{j}(\mathbf{r}) = \frac{ie}{2m} \sum_{\sigma} \langle (\nabla \psi_{\sigma}^{\dagger}(\mathbf{r})) \psi_{\sigma}(\mathbf{r}) - \psi_{\sigma}^{\dagger}(\mathbf{r})(\nabla \psi_{\sigma}(\mathbf{r})) \rangle$$

$$- \frac{e^2}{mc} \mathbf{A}(\mathbf{r}) \sum_{\sigma} \langle \psi_{\sigma}^{\dagger}(\mathbf{r}) \psi_{\sigma}(\mathbf{r}) \rangle,$$

satisfies the charge conservation law,

$$\nabla \cdot \mathbf{j}(\mathbf{r}) = 0.$$

Use the Bogoliubov transformation, then Bogoliubov-de Gennes equations, and finally, the equation for the order parameter (self-consistency relation).

References

Books and Reviews

1. Beenakker, C.W.J., van Houten, H.: Phys. Rev. Lett. **66**, 3056 (1991)
2. Barone, A., Paternó, G.: Physics and Applications of the Josephson Effect. Wiley, New York (1982)
3. Basov, D.N., Tinusk, T.: Electrodynamics of high-T_c superconductors. Rev. Mod. Phys. **77**, 721 (2005) (Section VIII)
4. Landau, L.D., Lifshitz, E.M.: Quantum Mechanics, Non-relativistic Theory (Landau and Lifshitz Course of Theoretical Physics, v. III.). Pergamon Press, Oxford (1989)
5. Lifshitz, E.M., Pitaevskii, L.P.: Statistical Physics pt.II (Landau and Lifshitz Course of theoretical physics, v. IX.). Pergamon Press, Oxford (1980) (Ch. 5. Theory of Superconducting Fermi Gas)
6. Maiti, S., Chubukov, A.V.: Superconductivity from repulsive interaction, . In: Proceedings of the XVII Training Course in the Physics of Strongly Correlated Systems, Vietri sul Mare (Salerno), Italy (2013)
7. Rammer, J., Smith, H.: Quantum field-theoretical methods in transport theory of metals. Rev. Mod. Phys. **58**, 323 (1986)
8. Svidzinskii A.V.: Spatially Inhomogeneous Problems in the Theory of Superconductivity. Nauka, Moscow (1982) (in Russian) (An excellent book, using both functional integration and Green's functions formalism)
9. Tinkham M.: Introduction to Superconductivity, 2nd edn. McGraw Hill, New York (1996) (A classical treatise on superconductivity. This second edition contains a special chapter (Ch.7) devoted to effects in small (mesoscopic) Josephson junctions)
10. Tsuei, C.C., Kirtley, J.R.: Pairing symmetry in cuprate superconductors. Rev. Mod. Phys. **72**, 969 (2000)
11. Van Harlingen, D.J.: Phase-sensitive tests of the symmetry of the pairing state in the high-temperature superconductors-evidence for $d_{x^2-y^2}$ symmetry. Rev. Mod. Phys. **67**, 515 (1995)
12. Vonsovsky, S.V., Izyumov Yu A., Kurmaev, E.Z.: Superconductivity of Transition Metals, Their Alloys and Compounds. Springer, Berlin (1982) (Chapter 2 of this comprehensive book provides both formulation of Nambu-Gor'kov formalism and detailed derivation and analysis of Eliashberg equations. In Chapter 3 the formalism is generalized on case of magnetic metal)
13. Zagoskin, A.M.: Quantum Engineering: Theory and Design of Quantum Coherent Structures. Cambridge University Press, Cambridge (2011). (Chapters 2 and 4 contain a review of theory and experiment on superconducting quantum bits (qubits) and qubit arrays)

Articles

14. Bagwell, P.F.: Phys. Rev. B **49**, 6481 (1994)
15. Bardeen, J., Johnson, J.L.: Phys. Rev. B **5**, 72 (1972)
16. Blonder, G.E., Tinkham, M., Klapwijk, T.M.: Phys. Rev. B **25**, 4515 (1982)
17. Cuevas, J.C., Martín-Rodero, A., Levy Yeyati A.: Phys. Rev. B **54**, 7366 (1996)
18. Huck, A., van Otterlo, A., Sigrist, M.: Phys. Rev. B **56**, 14163 (1997)
19. Ishii, G.: Prog. Theor. Phys. **44**, 1525 (1970)
20. Joyez, P., Lafarge, P., Filipe, A., Esteve, D., Devoret, M.H.: Phys. Rev. Lett. **72**, 2458 (1994)
21. Kohn, W., Luttinger, J.M.: Phys. Rev. Lett. **15**, 524 (1965)
22. Kulik, I.O.: Sov. Phys. JETP **30**, 944 (1970)
23. Kulik, I.O., Omelyanchuk, A.N.: Fiz. Nizk. Temp. **3**, 945; **4**, 296 (Sov. J. Low Temp. Phys. **3**, 459; **4**, 142) (1977)
24. Matveev, K.A., Gisselfält, M.. Glazman, L.I., Jonson, M., Shekhter, R.I.: Phys. Rev. Lett. **70**, 2940 (1993)
25. Takayanagi, H., Akazaki, T., Nitta, J.: Surf. Sci. **361–362**, 298 (1996)
26. Tsuei, C.C., et al.: Science **271**, 329 (1996)
27. Yang, C.N.: Rev. Mod. Phys. **34**, 694 (1962)
28. Zagoskin, A.M.: J. Phys.: Condens. Matter **9**, L419 (1997)

Chapter 5
Many-Body Theory in One Dimension

Stay on the Path. Never step off!
Ray Bradbury. "A Sound of Thunder"

Abstract Orthogonality catastrophe and its treatment in a one-dimensional model. Anderson orthogonality exponent. Tomonaga-Luttinger model for fermions in one dimension. Bosonization: describing one-dimensional fermions in terms of bosons and vice versa. Interacting one-dimensional fermions: Tomonaga-Luttinger liquid. Spin-charge separation. Elements of conformal field theory. Conformal field theory approach to the orthogonality catastrophe in one dimension. General relation between the orthogonality exponent and the scattering phase.

5.1 Orthogonality Catastrophe and Related Effects

5.1.1 Dynamical Charge Screening by a System of Fermions

We have seen, that the existence of the Fermi surface gives rise to subtle and important effects. One example is the Cooper pairing, where the instability of the normal ground state in the presence of an arbitrarily small electron–electron attraction near the Fermi surface was due to the effectively two-dimensional character of the problem. Another is provided by the charge screening, where instead of an exponential Thomas–Fermi screened potential one obtains Friedel oscillations with a much slower, power-law potential drop. Mathematically, the latter followed from branch cuts—as opposed to simple poles—in the complex frequency plane of the response function of the electron systems (in this case, the polarization operator). It is natural to expect that such non-analyticity may also produce a non-exponential time dependence in the response of a system of fermions to a perturbation. This is actually the case, e.g., if a point charge is suddenly introduced into a sea of electrons.

A. Zagoskin, *Quantum Theory of Many-Body Systems,*
Graduate Texts in Physics, DOI: 10.1007/978-3-319-07049-0_5,
© Springer International Publishing Switzerland 2014

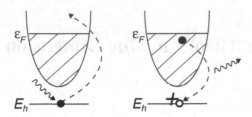

Fig. 5.1 Schematics of X-ray spectroscopy of metals. A core hole potential is instantaneously created or eliminated with, respectively, the absorption or emission of an X-ray photon

Such a situation arises in the X-ray spectroscopy of metals (see Fig. 5.1). An electron in a core level of an ion in the metal absorbs an X-ray photon and is (practically instantaneously) ejected on top of the conduction band, leaving behind produces a positively charged core hole in the middle of a sea of conduction electrons, which rush to screen it. The filling in of the hole by a conduction electron, with the emission of a photon, will later eliminate this charge. This process can be described (see, e.g., [3], Sect. 8.3.B) by the Hamiltonian

$$\mathcal{H} = E_h b^\dagger b + \sum_{\mathbf{k}} \epsilon_{\mathbf{k}} c_{\mathbf{k}}^\dagger c_{\mathbf{k}} + b^\dagger b \sum_{\mathbf{kq}} V_{\mathbf{kq}} c_{\mathbf{k}}^\dagger c_{\mathbf{q}}. \tag{5.1}$$

Here $b^\dagger$, b are the core hole creation/annihilation operators, $c_{\mathbf{k}}^\dagger$, $c_{\mathbf{k}}$ ditto for the band electrons, E_h is the core hole energy, $\epsilon_{\mathbf{k}}$ the band dispersion law, and $V_{\mathbf{kq}}$ the matrix elements of the core hole scattering potential. We consider here spinless fermions, for the sake of simplicity.

At zero temperature the X-ray emission and absorption spectra obviously have the coinciding threshold, ω_0: $\omega_{ab} \geq \omega_0 \geq \omega_{em}$, where $\omega_0 = \epsilon_F - E_h$ (with the energies measured from the bottom of the conduction band). The absorption and emission spectra near the threshold have a power-law shape. For example, the absorption intensity

$$A(\omega) \propto (\omega - \omega_0)^\alpha. \tag{5.2}$$

The *Fermi-edge singularity (FES) exponent*, α, was calculated in several different ways ([3], Sect. 8.3.C) and turned out to be

$$\alpha = 1 - \left(\frac{\delta(\epsilon_F)}{\pi}\right)^2, \tag{5.3}$$

where $\delta(\epsilon_f)$ is the phase shift of an electron at the Fermi surface, scattered by the core-hole potential.

In order to understand this, we start from the result of [11], who related the X-ray absorption and emission rates to the Green's function of the core hole,

$$G_h(t) = -i\langle Tb(t)b^\dagger(0)\rangle = -i\theta(t)\langle\Phi_0|e^{\frac{i}{\hbar}\mathcal{H}_0 t}be^{-\frac{i}{\hbar}(\mathcal{H}_1+\hbar\omega_0)t}b^\dagger|\Phi_0\rangle$$

$$= -i\theta(t)e^{-i\omega_0 t}\langle\Phi_0|e^{\frac{i}{\hbar}\mathcal{H}_0 t}e^{-\frac{i}{\hbar}\mathcal{H}_1 t}|\Phi_0\rangle. \tag{5.4}$$

Here $|\Phi_0\rangle$ is the ground state of the system with no core hole; obviously, $\langle\Phi_0|bb^\dagger|\Phi_0\rangle = 1$; and we made use of the hole being created and annihilated instantaneously, so that the Hamiltonian (5.1) is either

$$\mathcal{H}_0 = \sum_{\mathbf{k}} \epsilon_{\mathbf{k}}c_{\mathbf{k}}^\dagger c_{\mathbf{k}}, \tag{5.5}$$

or

$$\hbar\omega_0 + \mathcal{H}_1 = \hbar\omega_0 + \sum_{\mathbf{k}} \epsilon_{\mathbf{k}}c_{\mathbf{k}}^\dagger c_{\mathbf{k}} + \sum_{\mathbf{kq}} V_{\mathbf{kq}}c_{\mathbf{k}}^\dagger c_{\mathbf{q}}. \tag{5.6}$$

Now, as usual, we insert in (5.4) the closure relation,

$$\sum_m |\tilde{\Phi}_m\rangle\langle\tilde{\Phi}_m| = \mathcal{I}, \tag{5.7}$$

where $\left\{|\tilde{\Phi}_m\rangle\right\}_{m=0,1,\dots}$ is the full set of states of the electron system *with* the core hole potential. The result is

$$G_h(t) = -i\theta(t)e^{-i\omega_0 t}\sum_m \langle\Phi_0|e^{\frac{i}{\hbar}\mathcal{H}_0 t}|\tilde{\Phi}_m\rangle\langle\tilde{\Phi}_m|e^{-\frac{i}{\hbar}\mathcal{H}_1 t}|\Phi_0\rangle$$

$$= -i\theta(t)e^{-i\omega_0 t}\sum_m e^{-\frac{i}{\hbar}(\tilde{E}_m - E_0)t}\left|\langle\tilde{\Phi}_m|\Phi_0\rangle\right|^2. \tag{5.8}$$

If the core hole potential produces a bound state, there are two possibilities: it will be filled by one of the conduction electrons, or will remain empty (Fig. 5.2). The above expression is easily modified to take both cases into account:

$$G_h(t) = -i\theta(t)e^{-i\omega_0 t}\sum_m \sum_{s=e,f} e^{-\frac{i}{\hbar}(\tilde{E}_m^s - E_0)t}\left|\langle\tilde{\Phi}_m^s|\Phi_0\rangle\right|^2. \tag{5.9}$$

Here the index s labels empty or filled bound state of the core hole potential. We will see what difference it makes.

The overlap $\langle\tilde{\Phi}_0^s|\Phi_0\rangle$ between the ground state wave functions of the N-electron systems with and without the core hole potential tends to zero as the number of electrons grows. We have quoted [14] on a similar situation in Sect. 2.2 when discussing an approximate calculation of an N-body wave function in the limit $N \to \infty$: even a small error in a one-particle wave function leads to the approximate N-particle function being orthogonal to the exact one. This situation is called the *orthogonality catastrophe*.

Fig. 5.2 X-ray spectroscopy, when the core hole potential has a bound state

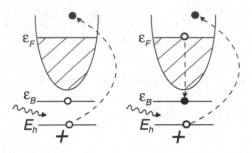

The wave function of N spinless fermions can be written as a Slater determinant (1.105), $|\Phi\rangle = \frac{1}{\sqrt{N!}} \det [(\phi_k(\mathbf{r}_m)]$, where $\phi_k(\mathbf{r}_m)$ is the kth one-particle wave function taken at the position of the mth particle. The overlap $\langle \tilde{\Phi}_0^s | \Phi_0 \rangle$ is thus

$$\langle \tilde{\Phi}_0^s | \Phi_0 \rangle = \frac{1}{N!} \int d\mathbf{r}_1 \int d\mathbf{r}_2 \cdots \int d\mathbf{r}_N \det \left[(\tilde{\phi}_k^{s*}(\mathbf{r}_m) \right] \det \left[(\phi_q(\mathbf{r}_m) \right]$$

$$\equiv \det \left[\langle \tilde{\phi}_k^{s*} | \phi_q \rangle \right], \tag{5.10}$$

that is, a determinant composed of the single-particle overlaps. Anderson [13] demonstrated that actually

$$\langle \tilde{\Phi}_0^s | \Phi_0 \rangle \propto N^{-x}, \tag{5.11}$$

where the *Anderson orthogonality exponent*

$$x = \frac{1}{2} \left(\frac{\delta(\epsilon_F)}{\pi} \right)^2 \tag{5.12}$$

also appears in the expression (5.3) for the FES exponent.

5.1.2 One-Dimensional Tight-Binding Model

Calculating overlaps of N-particle wave functions in (5.9) and (5.11) requires certain approximations. For example, one can assume the band and the core-hole potential to be spherically symmetric and consider only s-wave scattering. This effectively reduces the problem to one dimension, with spinless (for simplicity) fermions confined to a ray, $r > 0$, and a scattering potential placed at the origin. Let us go one step farther and consider a 1D chain of finite length $(l - 1)$, with nearest-neighbour hopping, free boundary conditions and a potential at the first site, which represents the core hole charge and can be switched on and off (Fig. 5.3); [12]:

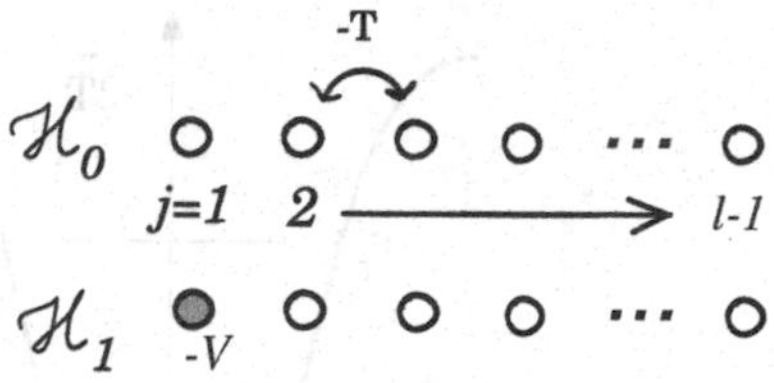

Fig. 5.3 1D tight-binding
chain

$$\mathcal{H}_0 = -T \sum_{i=1}^{l-2} \left(\psi_i^\dagger \psi_{i+1} + \psi_{i+1}^\dagger \psi_i \right);$$ (5.13)

$$\mathcal{H}_1 = \mathcal{H}_0 - V \psi_1^\dagger \psi_1.$$ (5.14)

Here $\psi_i^\dagger$, ψ_i creates/annihilates an electron on the ith site, and the negative signs at the tunneling amplitude T and the "hole potential" V are chosen for convenience. The one-particle eigenstates of $\mathcal{H}_0$ and $\mathcal{H}_1$ are produced by acting with the creation operators on the vacuum state, $|0\rangle$, and can be written as

$$|k\rangle = \sum_{j=1}^{l-1} \Psi_{kj} \psi_j^\dagger |0\rangle \equiv C \sum_{j=1}^{l-1} \sin\left[k(j-l)\right] \psi_j^\dagger |0\rangle.$$ (5.15)

Here C is the normalization constant. The free boundary condition at the last (($l-1$)th) site is satisfied: the wave function is zero at the $j = l$. Since $\sin[k(j+1-l)] + \sin[k(j-1-l)] = 2\sin[k(j-l)]\cos k$, the Schrödinger equation $\mathcal{H}|k\rangle = \epsilon(k)|k\rangle$ is also automatically satisfied everywhere, except the first site, for any k, if only

$$\epsilon(k) = -2T \cos k$$ (5.16)

(the standard tight-binding dispersion law). The allowed values of k are determined from the contribution to the Schrödinger equation from site $j = 1$:

$$-T\Psi_{k2} - V\Psi_{k1} = -2T \cos k \; \Psi_{k1}.$$

Substituting here $\Psi_{kj} \propto \sin k(j-l)$, after an exercise in trigonometry we find the implicit expression for the spectrum:

$$\frac{\sin k(l-1)}{\sin kl} = \frac{T}{V}.$$ (5.17)

Without the core hole, $V = 0$, this yields $\sin kl = 0$ and the obvious set of ($l-1$) band states with $k_n = \pi n/l$; $n = 1, 2, \ldots (l-1)$. Small enough attractive potential ($V > 0$) does not change this situation qualitatively. Nevertheless at some V a bound state can be formed (Fig. 5.4). This happens when Eq. (5.17) acquires an imaginary root, $k = i\kappa$, i.e.,

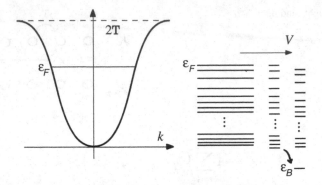

Fig. 5.4 Band spectrum and the bound state in the tight-binding model

$$\frac{\sinh \kappa(l - 1)}{\sinh \kappa l} = \frac{T}{V}. \tag{5.18}$$

Then the one-particle wave function amplitude is $\Psi_B \propto \sinh \kappa(l - j)$. The wave function should decay away from the origin. Therefore $\kappa > 0$, and there is only one bound state in our model. Its energy

$$\epsilon_B = -2T \cosh \kappa. \tag{5.19}$$

In the limit of an infinitely long chain we can introduce the scattering phase shift for the band states via

$$\Psi_{kj} \sim \sin [kj + \delta(k)], \quad j \gg 1. \tag{5.20}$$

Comparing this to (5.15), we see, that

$$\delta(k) + \pi n = -kl, \quad l \to \infty. \tag{5.21}$$

Therefore

$$\frac{\sin[\delta(k) + k]}{\sin \delta(k)} = \frac{T}{V}, \tag{5.22}$$

and

$$\delta(k) = \arctan \left[\frac{\sin k}{(T/V) - \cos k} \right]. \tag{5.23}$$

In the same limit for the bound state we find

$$e^{-\kappa} = \frac{T}{V}; \quad \epsilon_B = -\frac{T^2 + V^2}{V}. \tag{5.24}$$

The advantage of our tight-binding model is that it allows a simple expression for the overlap (5.10). We will use the trigonometric identities

$$\sum_{j=1}^{l-1} \sin^2 k(j-l) = \frac{1}{2}\left[(l-1) - \frac{\sin k(l-1)\cos kl}{\sin k}\right],$$

$$\sum_{j=1}^{l-1} \sinh^2 \kappa(j-l) = -\frac{1}{2}\left[(l-1) - \frac{\sinh \kappa(l-1)\cosh \kappa l}{\sinh \kappa}\right],$$

$$\sum_{j=1}^{l-1} \sin k(j-l) \sin q(j-1) = \frac{1}{2}\frac{\sin k(l-1)\sin ql - \sin q(l-1)\sin kl}{\cos q - \cos k}.$$

From the two first equations the normalizations of one-particle states are easily found. If substitute in the last one the wave vectors k and $\tilde{k}$, which satisfy Eq. (5.17) with the potential U and $\tilde{U}$, respectively, we obtain

$$\sum_{j=1}^{l-1} \sin k(j-l) \sin \tilde{k}(j-1) = \frac{(\tilde{U} - U)\sin k(l-1)\sin \tilde{k}(l-1)}{2T(\cos \tilde{k} - \cos k)}$$

$$= \frac{(\tilde{U} - U)\sin k(l-1)\sin \tilde{k}(l-1)}{\epsilon(\tilde{k}) - \epsilon(k)}. \tag{5.25}$$

The one-particle overlaps are thus given by

$$\langle \tilde{\Psi}_{\tilde{k}} | \Psi_k \rangle = \frac{(\tilde{U} - U)C(\tilde{k})C(k)}{\epsilon(k) - \epsilon(\tilde{k})}, \tag{5.26}$$

where $C(k) = \sqrt{2}\sin k(l-1)\left[(l-1) - (\sin k(l-1)\cos kl)/\sin k\right]^{-1/2}$. For the overlap with a bound state (if it exists) we just need to replace $\epsilon(\tilde{k})$ with ϵ_B Eq. (5.19) and $C(\tilde{k})$ with

$$C_B = \sqrt{2}\sinh \kappa(l-l)\left[(\sinh \kappa(l-1)\cosh \kappa l)/\sinh \kappa - (l-1)\right]^{-1/2}.$$

In our problem, $U = 0$ (no core hole), $\tilde{U} = V$, $\sin k(l-1) = \pm \sin k$ and, in the limit $l \to \infty$, $\sin \tilde{k}(l-1) = (T/V)\sin \tilde{k}l = \pm(T/V)\sin \delta(k)$.

Equation (5.26) is exact. We now linearize the dispersion relation near the Fermi surface:

$$\epsilon(k) = \epsilon(k_F + \Delta k) \approx \epsilon(k_F) + \hbar v_F \Delta k = -2T\cos k_F + 2T\sin k_F \cdot \Delta k. \tag{5.27}$$

and find

$$\langle \tilde{\Psi}_{\tilde{k}} | \Psi_k \rangle \approx \pm \frac{2V \sin k_F (\mathrm{T}/V) \sin(\delta(\tilde{k}))}{2l\mathrm{T} \sin k_F ([k - \tilde{k} - \delta(\tilde{k})/l]} = \pm \frac{\sin(\pi \delta_{\tilde{n}})}{\pi} \frac{1}{n - \tilde{n} + \delta_{\tilde{n}}}, \quad (5.28)$$

where $k = \pi n/l$, and $\delta_n = \delta(\pi n/l)/\pi$. Substituting this in (5.10), we obtain the scalar product of N-particle wave functions with and without core hole potential

$$\langle \tilde{\Phi}_0 | \Phi_0 \rangle = \pm \left(\prod_{m=1}^{N} \frac{\sin(\delta_m)}{\pi} \right) \det \left[\frac{1}{n - \tilde{n} + \delta_{\tilde{n}}}, \right]. \quad (5.29)$$

This is the same formula, which was obtained by Anderson [13] in the first-order perturbation theory for s-wave scattering. Following his approach towards deriving the orthogonality exponent (5.12), we transform it to a more convenient form using the Cauchy formula,

$$\det \left[\frac{1}{a_m - b_n} \right] = \frac{\prod_{m>n}(a_m - a_n) \prod_{m>n}(b_m - b_n)}{\prod_{m,n}(a_m - b_n)}. \quad (5.30)$$

This allows to calculate the logarithm of $|\langle \tilde{\Phi}_0 | \Phi_0 \rangle|$:

$$\ln |\langle \tilde{\Phi}_0 | \Phi_0 \rangle| \approx - \sum_n \ln \left(\frac{\sin \pi \delta_n}{\pi \delta_n} \right) - \sum_{m<n} \left[\ln \left(1 + \frac{\delta_n - \delta_m}{n - m} \right) \right.$$
$$\left. - \ln \left(1 + \frac{\delta_n}{n - m} \right) - \ln \left(1 - \frac{\delta_m}{n - m} \right) \right]. \quad (5.31)$$

Using also the Euler's product formula

$$\frac{\sin x}{x} = \prod_{j=1}^{\infty} \left(1 - \frac{x^2}{\pi^2 j^2} \right), \quad (5.32)$$

we can expand (5.31) in powers of δ_n (assuming $\delta_n/\pi \ll 1$) and check that the linear terms cancel, while the quadratic terms yield the Anderson's result (5.11, 5.12): up to a factor,

$$|\langle \tilde{\Phi}_0 | \Phi_0 \rangle| \approx N^{-\frac{1}{2} \left(\frac{\delta(k_F)}{\pi} \right)^2}.$$

This formula holds also if the core potential has a bound state, *if* this bound state is filled. Otherwise the Anderson exponent is given by

$$x_e = \frac{1}{2} \left(1 - \frac{\delta(k_F)}{\pi} \right)^2. \quad (5.33)$$

Instead of following the original derivation of the latter [9, 10] we will use some generalizations of our one-dimensional approach, which do not rely on such assumptions as smallness of the phase shift, provide us with some powerful theoretical tools and, in the end, better illuminate the physical meaning of these results.

5.2 Tomonaga-Luttinger Model

5.2.1 Spinless Fermions in One Dimension

Let us return to the tight-binding Hamiltonian of Eq. (5.13), adding to the model the particle–particle interaction on the neighbouring sites:

$$\mathcal{H} = -T \sum_{i=1}^{l-2} \left(\psi_i^\dagger \psi_{i+1} + \psi_{i+1}^\dagger \psi_i \right) + g \sum_{i=1}^{l-2} \left(\psi_i^\dagger \psi_i - \overline{n_i} \right) \left(\psi_{i+1}^\dagger \psi_{i+1} - \overline{n_{i+1}} \right).$$

(5.34)

Here $\overline{n_j} \equiv \langle 0|\psi_j^\dagger \psi_j|0\rangle$ is the ground state expectation value of the site occupation number. Besides modeling various one-dimensional conductors with interactions (Eggert 2009) and impurity scattering (with a given orbital momentum) in a 3D electron gas, this Hamiltonian also describes the XXZ spin-1/2 chain. It turns out that the latter's Hamiltonian,

$$\mathcal{H}_S = \frac{1}{4} J_\perp \sum_i \left(\sigma_i^x \sigma_{i+1}^x + \sigma_i^y \sigma_{i+1}^y \right) + \frac{1}{4} J_z \sum_i \sigma_i^z \sigma_{i+1}^z,$$

(5.35)

where $\sigma^{x,y,z}$ are the Pauli matrices, can be transformed to the form (5.34) with $T = J_\perp, g = J_z$ and $\overline{n} = 1/2$, i.e., the Fermi level in the middle of the band (see, e.g., [4], 1.2). The fermionic operators create and remove kinks in the spin configuration. Thus the model we are going to investigate has quite versatile applications.

Substituting the Fourier-transforms of the site operators ($\psi_j = C \sum_k e^{ikx_j} c_k$), we find for the non-interacting part of the Hamiltonian:

$$\mathcal{H}_0 = \sum_{k=-k_{\max}}^{k_{\max}} \epsilon(k) c_k^\dagger c_k = \sum_q \epsilon(-k_F + q) c_{-k_F+q}^\dagger c_{-k_F+q}$$
$$+ \sum_{q'} \epsilon(k_F + q') c_{k_F+q'}^\dagger c_{k_F+q'}.$$

(5.36)

Since linearizing the dispersion law near the Fermi surface (5.27) brought useful simplifications earlier, we will do the same here:

$$\mathcal{H}_0 \approx \sum_q (-\hbar v_F) q c^\dagger_{-k_F+q} c_{-k_F+q} + \sum_{q'} \hbar v_F q' c^\dagger_{k_F+q'} c_{k_F+q'} \qquad (5.37)$$

(measuring energy from the Fermi level). We can denote

$$d_q = c_{k_F+q}, \; d^\dagger_q = c^\dagger_{k_F+q}; \; s_q = c_{-k_F+q}, \; s^\dagger_q = c^\dagger_{-k_F+q} \qquad (5.38)$$

and rewrite (5.37) as

$$\mathcal{H}_0 = \hbar v_F \sum_q q \left(d^\dagger_q d_q - s^\dagger_q s_q \right). \qquad (5.39)$$

The anticommutation relations between the new operators are, obviously, as follows:

$$\{d_k, d^\dagger_q\} = \delta_{kq}; \; \{s_k, s^\dagger_q\} = \delta_{kq}; \; \{d_k, s^\dagger_q\} = \delta_{k,q-2k_F}. \qquad (5.40)$$

This is just a different notation, as long as we keep $k_F + q > 0$ for the "right-movers", and $-k_F + q < 0$ for the "left-movers". The crucial step is now to consider them as *two different* kinds of fermions, and let q run from $-\infty$ to ∞ both for s_q and d_q (Fig. 5.5). This is, of course, an approximation. As long as we are interested in low-energy excitations, it is reasonable enough: situations, when the anticommutators of right- and left-movers would be nonzero, require the momentum transfer of $2k_F$. The advantage of this approximation is that the Hamiltonian (5.34) becomes exactly solvable [3].

The electron annihilation operator at the position x_j is now written as

$$\psi_j = \frac{1}{\sqrt{L}} \sum_q e^{ik_F x_j} e^{iq x_j} d_q + \frac{1}{\sqrt{L}} \sum_q e^{-ik_F x_j} e^{iq x_j} s_q$$

$$= e^{ik_F x_j} \frac{1}{\sqrt{L}} \sum_q e^{iq x_j} d_q + e^{-ik_F x_j} \frac{1}{\sqrt{L}} \sum_q e^{iq x_j} s_q$$

$$\equiv e^{ik_F x_j} d(x_j) + e^{-ik_F x_j} s(x_j), \qquad (5.41)$$

where L is the length of our chain. The anticommutation relations between the right- or left-movers are

$$\{d(x_j), d^\dagger(x_m)\} = L^{-1} \sum_{q,q'} e^{iq x_j - iq' x_m} \{d_q, d^\dagger_{q'}\} = L^{-1} \sum_q e^{iq(x_j-x_m)}$$

$$= L^{-1} \sum_{n=-\infty}^{\infty} e^{2\pi i n(x_j - x_m)/L} = \sum_{n'=-\infty}^{\infty} \delta(x_j - x_m - n'L). \quad (5.42)$$

Assuming periodic boundary conditions, the allowed wave numbers are

$$k_n = \frac{2\pi n}{L}, n = 0, \pm 1, \pm 2, \ldots$$

and we have used the identity

$$\sum_{n=-\infty}^{\infty} e^{i\frac{2\pi n}{L}y} = L \sum_{n'=-\infty}^{\infty} \delta(y - n'L).$$

(5.43)

In the limit $L \to \infty$ the summation over the wave numbers is replaced by integration,

$$\sum_q \to \frac{L}{2\pi} \int dq.$$

On the right-hand side of Eq. (5.42) will remain only the term with $n' = 0$:

$$\{d(x), d^\dagger(y)\} = \{s(x), s^\dagger(y)\} = \delta(x-y); \{d(x), s^\dagger(y)\} = \{d(x), d(y)\} = \cdots = 0.$$

(5.44)

Then the fermion field operators $\psi(x)$ will satisfy the standard anticommutation relations,

$$\{\psi(x), \psi^\dagger(y)\} = \delta(x - y), \quad \{\psi(x), \psi(y)\} = \{\psi^\dagger(x), \psi^\dagger(y)\} = 0,$$

if

$$\psi(x) = \frac{1}{\sqrt{2}} \left(e^{ik_Fx}d(x) + e^{-ik_Fx}s(x) \right).$$

(5.45)

Note that the interaction term of Hamiltonian (5.34) enters the expression

$$\left(\psi_i^\dagger \psi_i - \overline{n_i} \right) \equiv \, : \psi_i^\dagger \psi_i :,$$

(5.46)

the normally ordered product of Fermi operators, and its ground state expectation value is zero. In the continuous limit then

$$: \psi^\dagger(x)\psi(x) : = \frac{1}{2} : \left(d^\dagger(x)d(x) + s^\dagger(x)s(x) \right) :$$
$$+ \frac{1}{2} : \left(d^\dagger(x)s(x)e^{-2ik_Fx} + s^\dagger(x)d(x)e^{2ik_Fx} \right) :$$

(5.47)

We would like to get rid of electron–electron interactions (quartic terms in (5.34)) and deal with free particles. In the BCS theory this was made possible by the presence of superconducting condensate, but at a price of having to solve a self-consistent equation for the order parameter. Here a different approach is possible, due to the strictly one-dimensional character of the problem (and the approximations we made up to this point). The goal of the following excercise is to express the Hamiltonian of 1D fermions in terms of bosonic *fermion density operators*,

$$\rho^d(x) = d^\dagger(x)d(x), \quad \rho^s(x) = s^\dagger(x)s(x).$$

(5.48)

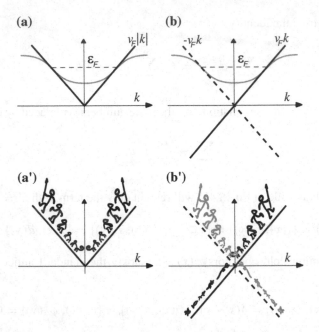

Fig. 5.5 Tomonaga (**a**) and Luttinger (**b**) models of a one-dimensional system of fermions. Note that in the Tomonaga model the spectrum is bounded from below, and the total number of fermions is finite. In the Luttinger model the spectrum is not bounded, the number of particles is infinite, and we have two distinct "species" of fermions (*right*- and *left*-movers)

The interaction terms, which are quartic in terms of fermions, will become quadratic in terms of these new operators, i.e., yielding the Hamiltonian for *non-interacting* bosons. This is the idea of *bosonization*.

5.2.2 Bosonization

For the right-moving fermion density we find[1]

$$\rho^d(x) = L^{-1} \sum_k \sum_q e^{-i(k_F+k)x+i(k_F+q)x} d_k^\dagger d_q = L^{-1} \sum_Q e^{iQx} \rho_Q^d, \qquad (5.49)$$

where

$$\rho_Q^d = \sum_k d_k^\dagger d_{k+Q} \equiv \sum_{k'} d_{k'-Q}^\dagger d_{k'}; \quad \left(\rho_Q^d\right)^\dagger = \rho_{-Q}^d, \qquad (5.50)$$

---

[1] See [3], Sect. 4.4; [4], Sect. 2.1; [2, 6] for more details and original references. Be aware of significant differences in notation and conventions.

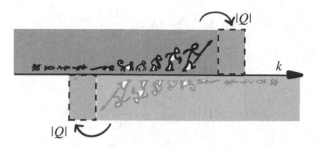

Fig. 5.6 Action of the fermion density operators on the ground state of the Luttinger model

and therefore $\rho^d(x)$ is Hermitian.

In the ground state of our model all states of right-movers with $k < 0$ (i.e., below the Fermi surface) are occupied, and all those with $k > 0$ are empty. Therefore the action of ρ_Q^d on the ground state will give zero for all $Q > 0$, while for $Q < 0$ this operator shifts the ground state by $|Q|$ to the right (Fig. 5.6). Of course, the same relations hold, mutatis mutandis, for ρ^s.

The operators ρ_Q^d and $\rho_{Q'}^d$, Q, $Q' \neq 0$, do not commute. Indeed,

$$[\rho_Q^d, \rho_{Q'}^d] = \sum_{k,k'} [d_k^\dagger d_{k+Q}, d_{k'}^\dagger d_{k'+Q'}] = \sum_k \left(d_k^\dagger d_{k+Q+Q'} - d_{k-Q'}^\dagger d_{k+Q} \right).$$

Substituting here

$$d_k^\dagger d_{k+Q+Q'} = :d_k^\dagger d_{k+Q+Q'}: + \langle 0|d_k^\dagger d_{k+Q+Q'}|0\rangle;$$
$$d_{k-Q'}^\dagger d_{k+Q} = :d_{k-Q'}^\dagger d_{k+Q}: + \langle 0|d_{k-Q'}^\dagger d_{k+Q}|0\rangle,$$

we see that, since the contributions from normal products cancel,

$$[\rho_Q^d, \rho_{Q'}^d] = \delta_{Q,-Q'} \sum_k \left(\langle 0|d_k^\dagger d_k|0\rangle - \langle 0|d_{k+Q}^\dagger d_{k+Q}|0\rangle \right) = \delta_{Q,-Q'} \frac{QL}{2\pi}. \quad (5.51)$$

In the same way, we obtain the relations for the left movers. In particular, for Q, $Q' \neq 0$

$$[\rho_Q^s, \rho_{Q'}^s] = -\delta_{Q,-Q'} \frac{Q}{2\pi}. \quad (5.52)$$

For the position-dependent densities this yields

$$[\rho^d(x), \rho^d(x')] = L^{-2} \sum_{Q,Q'} \delta_{Q,-Q'} \frac{QL}{2\pi} e^{iQx+iQ'x'} \xrightarrow[L\to\infty]{}$$

$$-\frac{i}{2\pi} \frac{\partial}{\partial x} \delta(x - x'); \tag{5.53}$$

$$[\rho^s(x), \rho^s(x')] \xrightarrow[L\to\infty]{} \frac{i}{2\pi} \frac{\partial}{\partial x} \delta(x - x'). \tag{5.54}$$

We should include in the expansions (5.49) for $\rho^{d,s}(x)$ also the contributions from $\rho^{d,s}_{Q=0}$. The case $Q = 0$ (so called *zero mode*) requires a special consideration. Indeed, the expectation value of ρ^d_0 is infinite. But the normally ordered operator $: \rho^d_0 :$

$$: \rho^d_0 := \sum_k : d^\dagger_k d_k := \sum_k \left(c^\dagger_{k_F+k} c_{k_F+k} - \overline{n_{k_F+k}} \right) = \mathcal{N}_d, \tag{5.55}$$

is the number operator for the right-moving particles in the given state relative to the ground state. Same reasoning applies to the left-moving zero mode. Note, that $: \rho^{d,s}_0 :$ commute with each other and with all $\rho^{d,s}_k$ with $k \neq 0$.

It is clear from (5.51, 5.52), that the Fourier components $\rho^{d,s}_k$ can be expressed through standard Bose operators $b^{d,s}_k$ with $[b_k, b^\dagger_{k'}] = \delta_{kk'}$: for $k > 0$

$$\rho^d_k = -i\sqrt{\frac{kL}{2\pi}} b^d_k; \quad \rho^d_{-k} = i\sqrt{\frac{kL}{2\pi}} (b^d_k)^\dagger; \tag{5.56}$$

$$\rho^s_k = i\sqrt{\frac{kL}{2\pi}} (b^s_k)^\dagger; \quad \rho^s_{-k} = -i\sqrt{\frac{kL}{2\pi}} b^s_k. \tag{5.57}$$

Again, it is straightforward to check that left and right bosons commute with each other and with both right- and left-moving zero modes.

5.2.2.1 "Bosonic" Hamiltonian of Non-Interacting Fermions in One Dimension

The non-interacting part of the Hamiltonian, (5.39) can be expressed in terms of the Bose operators $b, b^\dagger$. The least cumbersome way of doing this is to find first its commutator with the densities of right- and left movers. For example,

$$[\rho^d_Q, \mathcal{H}_0] = \hbar v_F \sum_{k,q} [d^\dagger_k d_{k+Q}, q d^\dagger_q d_q]$$

$$= \hbar v_F \sum_k ((k + Q) - k) d^\dagger_k d_{k+Q} = \hbar v_F Q \rho^d_Q,$$

which for positive Q's means

$$[b_Q^d, \mathcal{H}_0] = (\hbar v_F Q) b_Q^d, \tag{5.58}$$

and for negative Q's (since $\rho_{-Q} = (\rho_Q)^\dagger$)

$$[(b_{|Q|}^d)^\dagger, \mathcal{H}_0] = -(\hbar v_F |Q|)(b_{|Q|}^d)^\dagger. \tag{5.59}$$

(You see that the linear dispersion law was here important). In the same way we can show that ($Q > 0$)

$$[b_Q^s, \mathcal{H}_0] = (\hbar v_F Q) b_Q^s \tag{5.60}$$

and

$$[(b_Q^s)^\dagger, \mathcal{H}_0] = -(\hbar v_F Q)(b_Q^s)^\dagger. \tag{5.61}$$

Recalling, that for Bose operators $[b, b^\dagger b] = b$ and $[b^\dagger, b^\dagger b] = -b^\dagger$, we see that the Hamiltonian $\mathcal{H}_0$ can be written as

$$\mathcal{H}_0 = \hbar v_F \sum_{Q>0} Q \left\{ (b_Q^d)^\dagger b_Q^d + (b_Q^s)^\dagger b_Q^s \right\} + \mathcal{H}_0'. \tag{5.62}$$

The term $\mathcal{H}_0'$ is the part of the expression (5.39), which commutes with all Bose operators. Actually, it equals

$$\mathcal{H}_0' = \frac{\pi \hbar v_F}{L} \left(N_d(N_d + 1) + N_s(N_s + 1) \right) \tag{5.63}$$

and is simply the energy of zero modes. Indeed, in the right-moving zero mode all the states up to the state with momentum $k_F + k_{max}$ are occupied. The additional number of right-moving particles is $N_d = \langle \mathcal{N}_d \rangle$ (which can have either sign, see Eq. (5.55)), and the additional energy respective to the (infinite) energy of states filled up to k_F is

$$E_{0,d} = \hbar v_F \sum_Q Q = \hbar v_F \frac{2\pi}{L} \sum_{n=1}^{N_d} n = \hbar v_F \frac{2\pi}{L} \frac{N_d(N_d + 1)}{2}$$

for positive N_d's, and

$$E_{0,d} = -\hbar v_F \frac{2\pi}{L} \sum_{n=N_d+1}^{0} n = \hbar v_F \frac{2\pi}{L} \frac{N_d(N_d + 1)}{2}$$

for negative N_d's (see Fig. 5.6). In the same way, for the left-moving zero mode

$$E_{0,s} = \hbar v_F \frac{2\pi}{L} \frac{N_s(N_s + 1)}{2},$$

which is in full agreement with Eq. (5.63). The rest of the Hamiltonian (5.62) describes the electron-hole excitations against the background of zero modes.

5.2.2.2 Bosonization Formulas

Not only the Hamiltonian of a system of free fermions in one dimension can be expressed in terms of Bose operators. *Single* Fermi operators can be expressed in terms of Bose operators as well. In order to do so, let us first introduce the "phase" operators $\phi^{d,s}(x)$, such that

$$\rho^{d,s}(x) = \frac{\mathcal{N}_{d,s}}{L} + \frac{1}{2\pi}\frac{\partial}{\partial x}\phi^{d,s}(x). \tag{5.64}$$

Then we can write

$$\frac{\phi^{d,s}(x)}{2\pi} = \frac{1}{L}\sum_{k>0}\frac{1}{ik}\left\{e^{ikx}\rho_k^{d,s} - e^{-ikx}\rho_{-k}^{d,s}\right\}e^{-k\alpha/2}. \tag{5.65}$$

This expression reminds the formula (2.17) for the phonon field, especially if substitute here the Bose operators from (5.56, 5.57):

$$\phi^d(x) = -\sqrt{\frac{2\pi}{L}}\sum_{k>0}\frac{1}{\sqrt{k}}\left\{e^{ikx}b_k^d + e^{-ikx}(b_k^d)^\dagger\right\}e^{-k\alpha/2} \equiv \varphi^d(x) + \varphi^{d\dagger}(x);$$
$$\tag{5.66}$$

$$\phi^s(x) = \sqrt{\frac{2\pi}{L}}\sum_{k>0}\frac{1}{\sqrt{k}}\left\{e^{ikx}(b_k^s)^\dagger + e^{-ikx}b_k^s\right\}e^{-k\alpha/2} \equiv \varphi^{s\dagger}(x) + \varphi^s(x). \tag{5.67}$$

The factor $\exp[-k\alpha/2]$, where we will eventually put $\alpha \to 0$, enables the Abel regularization (Sect. 2.2.1) in case the resulting expressions diverge. It is straightforward to check that the only nonzero commutators of φ-operators are

$$[\varphi^d(x), \varphi^{d\dagger}(y)] = \frac{2\pi}{L}\sum_{k>0}\frac{1}{k}e^{ik(x-y)-k\alpha} = \sum_{n=1}^{\infty}\frac{1}{n}\left[e^{\frac{2\pi}{L}(i(x-y)-\alpha)}\right]^n$$
$$= -\ln\left[1 - e^{\frac{2\pi}{L}(i(x-y)-\alpha)}\right] \xrightarrow{L\to\infty} -\ln\left[\frac{2\pi i}{L}(y - x - i\alpha)\right],$$
$$\tag{5.68}$$

and

$$[\varphi^s(x), \varphi^{s\dagger}(y)] = \frac{2\pi}{L} \sum_{k>0} \frac{1}{k} e^{-ik(x-y)-k\alpha}$$

$$= -\ln\left[1 - e^{\frac{2\pi}{L}(-i(x-y)-\alpha)}\right] \xrightarrow{L\to\infty} -\ln\left[\frac{2\pi i}{L}(x - y - i\alpha)\right],$$

$$(5.69)$$

Here we have used the expansion

$$\ln(1 - y) = -\sum_{n=1}^{\infty} \frac{y^n}{n}. \tag{5.70}$$

For such operators $\mathcal{A}$, $\mathcal{B}$, that $[\mathcal{A}, \mathcal{B}]$ commutes with either of them, the *Baker-Hausdorff formula* holds:

$$e^{\mathcal{A}} e^{\mathcal{B}} = e^{\mathcal{A}+\mathcal{B}} e^{\frac{1}{2}[\mathcal{A},\mathcal{B}]}. \tag{5.71}$$

Then in the limit $L \to \infty$ for either right- or left-moving fields

$$e^{i\varphi^\dagger(x)} e^{i\varphi(x)} = e^{i[\varphi^\dagger(x)+\varphi(x)]} e^{-\frac{1}{2}\ln\left[\frac{2\pi i}{L}(-i\alpha)\right]} = (L/2\pi\alpha)^{1/2} e^{i\phi(x)}; \tag{5.72}$$

$$e^{i\varphi^\dagger(x)} e^{i\varphi(x)} = e^{i[\varphi(x)+\varphi^\dagger(x)]} e^{+\frac{1}{2}\ln\left[\frac{2\pi i}{L}(-i\alpha)\right]} = (L/2\pi\alpha)^{-1/2} e^{i\phi(x)} \tag{5.73}$$

We can also calculate the commutator of the fields $\phi(x)$. For example,

$$[\phi^d(x), \phi^d(y)] = [\varphi^d(x), \varphi^{d\dagger}(y)] - [\varphi^d(y), \varphi^{d\dagger}(x)]$$

$$= -\ln(1 - e^{\frac{2\pi i}{L}(x-y+i\alpha)})$$

$$+ \ln(1 - e^{\frac{2\pi i}{L}(y-x+i\alpha)}) \xrightarrow{\alpha\to 0} \ln \frac{1 - e^{-\frac{2\pi i}{L}(x-y)}}{1 - e^{\frac{2\pi i}{L}(x-y)}} \xrightarrow{L\to\infty}$$

$$i\pi \text{sgn}(x - y), \tag{5.74}$$

where $\text{sgn}(x) = \pm 1$ if x is positive (negative), and $\text{sgn}(0) = 0$. We have used here the formula

$$\ln \frac{1 - e^{-ix}}{1 - e^{ix}} = \begin{cases} i(\pi - x), & x > 0; \\ 0, & x = 0; \\ -i(\pi + x), & x < 0. \end{cases} \tag{5.75}$$

Indeed, $\ln(1 - e^{-z}) - \ln(1 - e^z) = \ln(-e^{-z})$ in the complex plane, but $z = 0$ is a branch point. A branch cut along the negative real semi-axis imposes the choice of $-1 = \exp[i\pi]$ for the upper and $-1 = \exp[-i\pi]$ for the lower complex half-plane.

It follows from Eqs. (5.44, 5.48), that the field operators and fermion density operators satisfy the conditions

$$[d(x), \rho^d(y)] = d(x)\delta(x - y); \quad [s(x), \rho^s(y)] = s(x)\delta(x - y). \quad (5.76)$$

Using the *Baker-Hausdorff theorem* (of which the formula (5.71) is a corollary),

$$e^{-B}Ae^{B} = \sum_{n=0}^{\infty} \frac{1}{n!}[A, B]_n$$

$$\equiv A + [A, B] + \frac{1}{2!}[[A, B], B] + \cdots + \frac{1}{n!}[\cdots[[A, \underbrace{B], B]\cdots B]}_{n \text{ times}} + \cdots, \quad (5.77)$$

we see that, since $[\phi, \phi^\dagger]$ is not an operator, the exponent of the "phase operator" satisfies the conditions

$$[e^{i\phi^d(x)}, \rho^d(y)] = \left(e^{i\phi^d(x)}\rho^d(y)e^{-i\phi^d(x)} - \rho^d(y)\right)e^{i\phi^d(x)}$$

$$= -i[\rho^d(y), \phi^d(x)]e^{i\phi^d(x)}$$

$$= \frac{-i}{2\pi}\frac{\partial}{\partial y}[\phi^d(y), \phi^d(x)]e^{i\phi^d(x)} = \frac{e^{i\phi^d(x)}}{2}\frac{\partial}{\partial y}\text{sgn}(y - x)$$

$$= e^{i\phi^d(x)}\delta(x - y), \quad (5.78)$$

(see Eqs. (5.64, 5.74)), and the same holds for $\phi^s(x)$, $\rho^s(y)$. Therefore $e^{i\phi(x)}$ could be a candidate for representing the right- or left-moving field operator.

Now we arrive at the central point of the bosonization technique. With the help of Baker-Hausdorff formula (5.71) written as

$$e^A e^B = e^B e^A e^{[A,B]}, \quad (5.79)$$

and the commutation relation (5.74) we see, that the exponentials of *bosonic* "phase" operators, $e^{\pm i\phi(x)}$, $e^{\pm i\phi(y)}$, *anti*commute:

$$e^{i\phi^d(x)}e^{i\phi^d(y)} = e^{i\phi^d(y)}e^{i\phi^d(x)}e^{[i\phi^d(x),i\phi^d(y)]} \xrightarrow[L\to\infty]{} e^{i\phi^d(y)}e^{i\phi^d(x)}e^{\pm i\pi}$$

$$= -e^{i\phi^d(y)}e^{i\phi^d(x)}; \quad (5.80)$$

$$e^{-i\phi^d(x)}e^{-i\phi^d(y)} \xrightarrow[L\to\infty]{} e^{-i\phi^d(y)}e^{-i\phi^d(x)}e^{\pm i\pi}$$

$$= -e^{-i\phi^d(y)}e^{-i\phi^d(x)};$$

$$\{e^{i\phi^d(x)}, e^{-i\phi^d(y)}\} = e^{i(\phi^d(x)-\phi^d(y))} \left(e^{-\frac{1}{2}[\phi^d(x),\phi^d(y)]} + e^{\frac{1}{2}[\phi^d(x),\phi^d(y)]} \right) \xrightarrow{L\to\infty}$$

$$e^{i(\phi^d(x)-\phi^d(y))} \left(e^{\frac{-i\pi}{2}\text{sgn}(x-y)} + e^{\frac{i\pi}{2}\text{sgn}(x-y)} \right) = \begin{cases} 0, & x \neq y; \\ 2, & x = y, \end{cases}$$

$$(5.81)$$

and the same for the left-movers. Therefore it is indeed possible to represent fermions in one dimension in terms of bosonic operators: $d(x) \sim e^{i\phi^d(x)}$, $s(x) \sim e^{i\phi^s(x)}$, with the appropriate normalization factors.

A more painstaking approach [6], reminiscent of what we did when introducing second quantization in §1.4, yields the *bosonization formulas*

$$d(x) = \frac{1}{\sqrt{2\pi\alpha}} \, \mathcal{F}_d \, e^{i\frac{2\pi}{L}\mathcal{N}_d x} e^{i\phi^d(x)} = \frac{1}{\sqrt{L}} \, \mathcal{F}_d \, e^{i\frac{2\pi}{L}\mathcal{N}_d x} e^{i(\varphi^d)^\dagger(x)} e^{i\varphi^d(x)}; \quad (5.82)$$

$$s(x) = \frac{1}{\sqrt{2\pi\alpha}} \, \mathcal{F}_s \, e^{i\frac{2\pi}{L}\mathcal{N}_s x} e^{i\phi^s(x)} = \frac{1}{\sqrt{L}} \, \mathcal{F}_s \, e^{i\frac{2\pi}{L}\mathcal{N}_s x} e^{i(\varphi^s)^\dagger(x)} e^{i\varphi^s(x)}. \quad (5.83)$$

The appearance of zero mode number operators $\mathcal{N}_{d,s}$ in the exponents does not contradict (5.78), since they commute with the "phase" operators. The operators $\mathcal{F}_\beta$ (so called *Klein factors*) satisfy the following (anti)commutation relations

$$\{\mathcal{F}_\beta, \mathcal{F}_\gamma^\dagger\} = 2\delta_{\beta\gamma}; \quad \mathcal{F}_\gamma \mathcal{F}_\gamma^\dagger = \mathcal{F}_\gamma^\dagger \mathcal{F}_\gamma = 1; \quad (5.84)$$

$$\{\mathcal{F}_\beta, \mathcal{F}_\gamma\} = \{\mathcal{F}_\beta^\dagger, \mathcal{F}_\gamma^\dagger\} = 0, \quad \beta \neq \gamma;$$

$$\left[\mathcal{N}_\beta, \mathcal{F}_\gamma^\dagger\right] = \delta_{\beta\gamma}\mathcal{F}_\gamma^\dagger; \quad [\mathcal{N}_\beta, \mathcal{F}_\gamma] = -\delta_{\beta\gamma}\mathcal{F}_\gamma. \quad (5.85)$$

They play a double role. First, they ensure that right- and left-moving fermions of (5.82, 5.83) anticommute (the relations (5.80, 5.81) only provide for the anticommutation inside each of the right- or left-moving groups). They also enforce the anticommutation relations between the fermions belonging to some other distinct species (e.g., with opposite spins). Second, the Klein factors change the number of fermions in the system by one. This is important: as one can show [6], the reason why bosonization works is that the Fock space of a one-dimensional Fermi system can be split into subspaces, each with a fixed particle number; the excitations in each of these subspaces are creating particle-hole pairs, and are thus bosonic. The Klein factors serve as *ladder operators*, i.e., they allow the transitions *between* such subspaces. (Note that the commutaton relations (5.85) between $\mathcal{F}$, $\mathcal{F}^\dagger$ and the number operators are the same as between creation/annihilation operators a, $a^\dagger$ and the number operator $a^\dagger a$.) The bosonic and fermionic descriptions of 1D systems are thus exactly equivalent[2] *if* the energy spectrum is not bound from below (as in Luttinger, but not Tomonaga, model—see Fig. 5.5). This is not going to create any

---

[2] One way of wrapping one's head around this counterintuitive fact is to recall that the difference between bosons and fermions comes from the wave function of the latter changing sign, when two fermions change places. But in one dimension there is no way to make two particles change places

problems, as long as we are only interested in low energy excitations (compared to the Fermi energy). Remarkably, this equivalence and the bosonization rules hold for an arbitrary dispersion relation (as long as the spectrum is not limited from below), and not necessarily a linear one. The linearity, though, is essential for most of the applications of bosonization techniques.

5.2.3 Tomonaga-Luttinger Liquid: Interacting Fermions in One Dimension

The non-interacting Hamiltonian (5.62, 5.63) can be written as

$$\mathcal{H}_0 = \pi \hbar v_F \int_0^L dx \left[\rho^d(x)^2 + \rho^s(x)^2 \right] \tag{5.86}$$

$$= \pi \hbar v_F \left\{ \int_0^L dx \left[\left(\frac{1}{2\pi} \frac{\partial \phi^d(x)}{\partial x} \right)^2 + \left(\frac{1}{2\pi} \frac{\partial \phi^s(x)}{\partial x} \right)^2 \right] + \frac{(\mathcal{N}_d)^2 + (\mathcal{N}_s)^2}{L} \right\}. \tag{5.87}$$

(We have used the bosonization formulas (5.64, 5.66, 5.67). The difference between $\mathcal{N}^2/L$ and $\mathcal{N}(\mathcal{N}+1)/L$ is negligible in the limit $L \to \infty$.) The fermion-fermion interaction term, following from the tight-binding Hamiltonian (5.34), with help of (5.47) and (5.48) becomes

$$\mathcal{H}_1 = \frac{g}{4} \int dx : \left\{ \rho^d(x)\rho^d(x+a) + \rho^s(x)\rho^s(x+a) + 2\rho^d(x)\rho^s(x)(1 - \cos 2k_F a) \right.$$

$$\left. + (\cdots)e^{-2ik_F x} + (\cdots)e^{2ik_F x} + (\cdots)e^{-4ik_F x} + (\cdots)e^{4ik_F x} \right\} :, \tag{5.88}$$

where $a \to 0$ is the lattice constant. The second line describes the backward scattering processes with momentum transfer $2k_F$, which contain products like $d^\dagger d d^\dagger s$, and the *Umklapp processes* (with momentum transfer $4k_F$ and terms like $d^\dagger d^\dagger s s$). These processes convert left-moving fermions into right-moving ones, and vice versa. They can be produced by sharp enough interaction potentials. But such potentials could also couple, within the right-moving (or left-moving) sector, the states just above the Fermi surface with the unphysical states below the bottom of the band (Fig. 5.5), which we have added in order to make bosonization possible, and with the understanding that they will not be excited at low energies we are interested in. It is possible to deal with such processes, but they requre special care. We will therefore drop the second line in (5.88), and understand the "pointlike" interactions in the first line as short-range, but smooth enough to justify such treatment (which is usually the case). Since the factor $(1 - \cos 2k_F a)$ is not quite controllable, and the next-neighbour coupling in (5.34) is just an approximation anyway, we will simply

---

without passing through each other—i.e., occupying the same state simultaneously, which fermions simply cannot do.

introduce two different coupling constants g_1, g_2, and finally write the Hamiltonian of the *Tomonaga-Luttinger liquid* as

$$
\mathcal{H}_{TLL} = \pi \hbar v_F \int_0^L dx \left[\rho^d(x)^2 + \rho^s(x)^2 + g_4(\rho^d(x)^2 + \rho^s(x)^2) + 2g_2 \rho^d(x) \rho^s(x) \right]
$$

$$
= \frac{\pi \hbar v_F}{2} \sqrt{(1+g_4)^2 - g_2^2} \int_0^L dx \left(\frac{1}{\tilde{g}} \rho_+(x)^2 + \tilde{g} \rho_-(x)^2 \right).
\tag{5.89}
$$

Here $\rho_{\pm}(x) = \rho_d(x) \pm \rho_s(x)$, and

$$
\tilde{g} = \sqrt{\frac{1 + g_4 - g_2}{1 + g_4 + g_2}}
\tag{5.90}
$$

is the *Luttinger liquid parameter*. Now using an appropriate Bogoliubov transformation we will diagonalize this Hamiltonian and reduce the system of *interacting fermions* to the one of noninteracting *bosons*. In the same way, as we obtained Eq. (5.87) from Eq. (5.86), we get

$$
\mathcal{H}_{TLL} = A \int_0^L dx \left\{ \left(\frac{1}{\tilde{g}} + \tilde{g} \right) \left[\frac{\mathcal{N}_d^2 + \mathcal{N}_s^2}{L^2} + \left(\frac{1}{2\pi} \frac{\partial \phi^d}{\partial x} \right)^2 + \left(\frac{1}{2\pi} \frac{\partial \phi^s}{\partial x} \right)^2 \right] + \right.
$$
$$
\left. \left(\frac{1}{\tilde{g}} - \tilde{g} \right) \left[\frac{2\mathcal{N}_d \mathcal{N}_s}{L^2} + 2 \left(\frac{1}{2\pi} \frac{\partial \phi^d}{\partial x} \right) \left(\frac{1}{2\pi} \frac{\partial \phi^s}{\partial x} \right) \right] \right\}.
$$

Here $A = \frac{\pi \hbar v_F}{2} \sqrt{(1+g_4)^2 - g_2^2}$. Substituting the expansions (5.66, 5.67), we find (always dropping the unphysical contribution of zero point oscillations)

$$
\mathcal{H}_{TLL} = \frac{A}{L} \left[\left(\frac{1}{\tilde{g}} + \tilde{g} \right) (\mathcal{N}_d^2 + \mathcal{N}_s^2) + \left(\frac{1}{\tilde{g}} - \tilde{g} \right) \cdot 2\mathcal{N}_d \mathcal{N}_s \right] +
$$
$$
\frac{A}{\pi} \sum_{k>0} k \left[\left(\frac{1}{\tilde{g}} + \tilde{g} \right) \left((b_k^d)^{\dagger} b_k^d + (b_k^s)^{\dagger} b_k^s \right) - \left(\frac{1}{\tilde{g}} - \tilde{g} \right) \left((b_k^d)^{\dagger} (b_k^s)^{\dagger} + b_k^d b_k^s \right) \right]
\tag{5.91}
$$

In the second line of (5.91) we can get rid of the off-diagonal terms by introducing new Bose operators, related to the old ones via a Bogoliubov transformation (cf. (4.54–4.57)):

$$
B_k^+ = b_k^d \cosh \theta - (b_k^s)^{\dagger} \sinh \theta; \quad B_k^- = b_k^s \cosh \theta - (b_k^d)^{\dagger} \sinh \theta;
$$
$$
b_k^d = B_k^+ \cosh \theta + (B_k^-)^{\dagger} \sinh \theta; \quad b_k^s = B_k^- \cosh \theta + (B_k^+)^{\dagger} \sinh \theta.
\tag{5.92}
$$

The chosen parametrization ensures the Bose commutation relations for B, $B^{\dagger}$-operators. Substituting (5.92) in (5.91), we see that the off-diagonal terms disappear, if

$$\tanh 2\theta = \frac{\frac{1}{\tilde{g}} - \tilde{g}}{\frac{1}{\tilde{g}} + \tilde{g}}, \tag{5.93}$$

i.e., $\exp[-2\theta] = \tilde{g}$. The remaining term is simply $\frac{2A}{\pi} \sum_{k>0} k[(B_k^+)^\dagger B_k^+ + (B_k^-)^\dagger B_k^-]$.

The first line in (5.91) is diagonalized by introducing $\mathcal{N}_\pm = (\mathcal{N}_d \pm \mathcal{N}_s)/2$. Finally we find, that

$$\mathcal{H}_{TLL} = \sum_{\mu=\pm} \mathcal{H}_\mu, \tag{5.94}$$

where

$$\mathcal{H}_\mu = \frac{2\pi\hbar\tilde{v}_F}{L} \left(\frac{1}{\tilde{g}} + \tilde{g}\right) (\mathcal{N}_\mu)^2 + \hbar\tilde{v}_F \sum_{k>0} k(B_k^\mu)^\dagger B_k^\mu. \tag{5.95}$$

The Hamiltonian of interacting 1D fermions with linear dispersion law is indeed reduced to the one of non-interacting 1D bosons, which allows an exact solution. Of course, the physical variables are defined in terms of the initial Fermi field $\psi(x)$, expressed through the operators $d(x)$, $s(x)$, the "old" Bose-operators b, $b^\dagger$, and eventually through $B^\pm$, $(B^\pm)^\dagger$, with the coefficients dependent on the Luttinger liquid parameter $\tilde{g}$. Besides that, and the change of the factors at new zero modes, the only effect of interactions is the renormalization of the Fermi velocity:

$$\tilde{v}_F = v_F\sqrt{(1 + g_4)^2 - g_2^2}. \tag{5.96}$$

This is the "speed of sound" of the "acoustic phonons" (the second term in (5.95)), which appear on top of zero modes.

As the final touch, we can introduce the new "phase" operators via (cf. (5.66, 5.67))

$$\Phi^+(x) = -\sqrt{\frac{2\pi}{L}} \sum_{k>0} \frac{1}{\sqrt{k}} \left\{ e^{ikx} B_k^+ + e^{-ikx} (B_k^+)^\dagger \right\} e^{-k\alpha/2}; \tag{5.97}$$

$$\Phi^-(x) = \sqrt{\frac{2\pi}{L}} \sum_{k>0} \frac{1}{\sqrt{k}} \left\{ e^{ikx} (B_k^-)^\dagger + e^{-ikx} B_k^- \right\} e^{-k\alpha/2}. \tag{5.98}$$

The Hamiltonian (5.95) can now be written as[3]

---

[3] Starting from $\Phi^\pm$ and $\mathcal{N}_\pm$, one can go backwards and introduce *new* Fermi operators. Such *refermionization* is sometimes useful ([6], §10.C).

$$\mathcal{H}_\mu = \pi\hbar\tilde{v}_F \int_0^L dx \left\{ 2\left(\frac{1}{\tilde{g}} + \tilde{g}\right)\left(\frac{\mathcal{N}_\mu}{L}\right)^2 + \left(\frac{1}{2\pi}\frac{\partial\Phi^\mu}{\partial x}\right)^2 \right\}. \tag{5.99}$$

5.2.4 Spin-Charge Separation

One more counterintuitive property of the Tomonaga-Luttinger model is the splitting of spin and charge degrees of freedom. Including spin in the model is trivial: we just add a spin index to the index distinguishing left- and right-movers. As mentioned before, the Klein factors will ensure that all different species of fermions anticommute, and the anticommutation relations within the same species are guaranteed by Eqs. (5.80, 5.81).

The total right(left)-moving charge density is then given by

$$\rho_C^{d(s)}(x) = \rho^{d(s)\uparrow}(x) + \rho^{d(s)\downarrow}(x), \tag{5.100}$$

and the net spin density by

$$\rho_S^{d(s)}(x) = \rho^{d(s)\uparrow}(x) - \rho^{d(s)\downarrow}(x). \tag{5.101}$$

We can now introduce new Bose operators, related to the ones of Eqs. (5.56, 5.57) for each spin species via

$$b_{C,k}^{d(s)} = \frac{b_k^{d(s)\uparrow} + b_k^{d(s)\downarrow}}{\sqrt{2}}; \quad b_{S,k}^{d(s)} = \frac{b_k^{d(s)\uparrow} - b_k^{d(s)\downarrow}}{\sqrt{2}}, \tag{5.102}$$

and new "phases" and number operators

$$\phi_C^{d(s)}(x) = \frac{\phi^{d(s)\uparrow}(x) + \phi^{d(s)\downarrow}(x)}{\sqrt{2}}; \quad \phi_S^{d(s)}(x) = \frac{\phi^{d(s)\uparrow}(x) - \phi^{d(s)\downarrow}(x)}{\sqrt{2}}; \tag{5.103}$$

$$\mathcal{N}_C^{d(s)}(x) = \frac{\mathcal{N}_{d(s)\uparrow}(x) + \mathcal{N}_{d(s)\downarrow}(x)}{\sqrt{2}}; \quad \mathcal{N}_S^{d(s)}(x) = \frac{\mathcal{N}_{d(s)\uparrow}(x) - \mathcal{N}_{d(s)\downarrow}(x)}{\sqrt{2}}. \tag{5.104}$$

Now the Hamiltonian (5.62, 5.63), trivially generalized to include contributions from both spin projections, can be written similarly to (5.87), and it splits into charge and spin parts, which commute with each other:

$$\mathcal{H}_0 = \pi\hbar v_F \int_0^L dx \left[\rho^{d\uparrow}(x)^2 + \rho^{d\downarrow}(x)^2 + \rho^{s\uparrow}(x)^2 + \rho^{s\downarrow}(x)^2 \right]$$

$$= \pi\hbar v_F \int_0^L dx \left[\rho_C^d(x)^2 + \rho_C^s(x)^2 + \rho_S^d(x)^2 + \rho_S^s(x)^2 \right] \qquad (5.105)$$

$$= \pi\hbar v_F \sum_{\mu=C,S} \left\{ \int_0^L dx \left[\left(\frac{1}{2\pi} \frac{\partial \phi_\mu^d(x)}{\partial x} \right)^2 + \left(\frac{1}{2\pi} \frac{\partial \phi_\mu^s(x)}{\partial x} \right)^2 \right] \right.$$

$$\left. + \frac{(\mathcal{N}_\mu^d)^2 + (\mathcal{N}_\mu^s)^2}{L} \right\}.$$

The physical right- or left-moving spinful fermion fields and densities are expressed in terms of new charge/spin operators through the generalizations of the bosonization formulas (5.64, 5.82, 5.83):

$$d(s)_{\uparrow/\downarrow}(x) \propto e^{i\frac{\phi_C^{d(s)}(x) \pm \phi_S^{d(s)}(x)}{\sqrt{2}}}; \quad \rho_{\uparrow/\downarrow}^{d(s)}(x) = \frac{1}{2\pi} \frac{\partial}{\partial x} \frac{\phi_C^{d(s)}(x) \pm \phi_S^{d(s)}(x)}{\sqrt{2}}. \quad (5.106)$$

Therefore the spin and charge degrees of freedom can be treated separately. This becomes interesting in the presence of interactions, which would couple to either spin or charge density and make their dynamics different (see, e.g., [5], Chap. 28; Nagaosa 1998, Sect. 3.2).

5.2.5 Green's Functions in Tomonaga-Luttinger Model

Let us return to free 1D fermions with a linear dispersion law. Their creation/ annihilation operators depend on position and time through $\exp[\pm(ix - iv_F t)]$ (right-movers) or $\exp[\pm(ix + iv_F t)]$ (left-movers). We will consider the Green's functions in imaginary time, like in Sect. 3.2, only here we denote

$$\tau = iv_F t. \qquad (5.107)$$

Then the free left-moving (right-moving) Fermi operators depend, respectively, on the complex variable $z = \tau + ix$ or its complex conjugate $z^* = \tau - ix$.

As in Sect. 3.2.1, here we replace the free Heisenberg field operators with Matsubara operators, and omit the superscript "M"

$$d_k(t) = e^{-iv_F k t} d_k(0) \rightarrow d_k(\tau) = e^{-k\tau} d_k(0);$$
$$d_k^\dagger(t) = e^{iv_F k t} d_k^\dagger(0) \rightarrow \bar{d}_k(\tau) = e^{k\tau} \bar{d}_k(0); \qquad (5.108)$$
$$s_k(t) = e^{iv_F k t} s_k(0) \rightarrow s_k(\tau) = e^{k\tau} s_k(0);$$
$$s_k^\dagger(t) = e^{-iv_F k t} s_k^\dagger(0) \rightarrow \bar{s}_k(\tau) = e^{-k\tau} \bar{s}_k(0).$$

Same relations hold for the right- and left-moving Matsubara Bose operators:

$$b_k^{d(s)}(\tau) = e^{\mp k\tau} b^{d(s)}(0); \quad \bar{b}_k^{d(s)}(\tau) = e^{\pm k\tau} \bar{b}^{d(s)}(0). \qquad (5.109)$$

Introducing the τ-ordered Green's function for the right-movers,

$$\mathcal{G}_d(\tau - ix) = -\langle T_\tau d(\tau - ix)\bar{d}(0)\rangle, \qquad (5.110)$$

we can immediately calculate it in equilibrium:

$$\mathcal{G}_d(\tau - ix) = -\theta(\tau)\frac{1}{L}\sum_k (1 - n_k)e^{-kz^*} + \theta(-\tau)\frac{1}{L}\sum_k n_k e^{-kz^*}, \qquad (5.111)$$

where $n_k = n_F(\hbar v_F k)$ (for right-movers; for left-movers, of course, $n_k = n_F(-\hbar v_F k)$). At zero temperature this reduces to

$$\mathcal{G}_d^0(\tau - ix) \equiv \mathcal{G}_d^0(z^*) = -\theta(\tau)\frac{1}{L}\sum_{k>0} e^{-kz^* - k\alpha} + \theta(-\tau)\frac{1}{L}\sum_{k<0} e^{-kz^* + k\alpha}$$

$$= -\theta(\tau)\frac{1}{L}\sum_{n=1}^\infty \left(e^{-\frac{2\pi}{L}(z^*+\alpha)}\right)^n + \theta(-\tau)\frac{1}{L}\sum_{n=1}^\infty \left(e^{\frac{2\pi}{L}(z^*-k\alpha)}\right)^n \qquad (5.112)$$

$$= -\frac{1}{L}\frac{e^{\frac{\pi}{L}\operatorname{sgn}(\tau)(\tau - ix + \alpha\operatorname{sgn}(\tau))}}{2\sinh\frac{\pi}{L}(\tau - ix + \alpha\operatorname{sgn}(\tau))} \xrightarrow{L\to\infty} -\frac{1}{2\pi}\cdot\frac{1}{z^* + \alpha\operatorname{sgn}(\tau)}.$$

Here $\alpha \to 0$ is the regularization parameter. In the same way, for the left-movers

$$\mathcal{G}_s^0(\tau + ix) \equiv \mathcal{G}_s^0(z) \xrightarrow{L\to\infty} -\frac{1}{2\pi}\cdot\frac{1}{z + \alpha\operatorname{sgn}(\tau)}. \qquad (5.113)$$

At a finite temperature $1/\beta$ we use the relation

$$1 - n_F(E) = 1 - \frac{1}{e^{\beta E} + 1} = n_F(-E)$$

and write

$$\mathcal{G}_d^{1/\beta}(\tau - ix) = -\text{sgn}(\tau)\frac{1}{L}\sum_k \frac{e^{-k(\tau-ix)-k\text{sgn}(\tau)\alpha}}{e^{-\beta\hbar k v_F \text{sgn}(\tau)} + 1} \xrightarrow{L\to\infty}$$

$$-\text{sgn}(\tau)\int_{-\infty}^{\infty}\frac{dk}{2\pi}\frac{e^{-k(\tau+\text{sgn}(\tau)\alpha)}e^{ikx}}{e^{-\beta\hbar k v_F \text{sgn}(\tau)} + 1} = -\frac{1}{2\pi}\cdot\frac{\pi/(\beta\hbar v_F)}{\sin\left(\frac{\pi(\tau-ix+\alpha\text{sgn}(\tau))}{\beta\hbar v_F}\right)}, \quad (5.114)$$

which in the limit $\beta \to \infty$ coincides with (5.112) (see Problem 2). For the left-movers, of course,

$$\mathcal{G}_s^{1/\beta}(\tau + ix) \xrightarrow{L\to\infty} -\frac{1}{2\pi}\cdot\frac{\pi/(\beta\hbar v_F)}{\sin\left(\frac{\pi(\tau+ix+\alpha\text{sgn}(\tau))}{\beta\hbar v_F}\right)}. \quad (5.115)$$

Now let us compute the equilibrium boson Green's functions. For the right-movers

$$\mathcal{D}_d(\tau - ix) = -\langle T_\tau \phi^d(\tau - ix)\phi^d(0)\rangle \quad (5.116)$$

$$= -\theta(\tau)\frac{2\pi}{L}\sum_{k>0}\frac{1}{k}\left[e^{-k(\tau-ix+\alpha\text{sgn}(\tau))}(n_B(\hbar v_F k) + 1)\right.$$

$$\left. +e^{k(\tau-ix+\alpha\text{sgn}(\tau))}n_B(\hbar v_F k)\right]$$

$$-\theta(-\tau)\frac{2\pi}{L}\sum_{k>0}\frac{1}{k}\left[e^{-k(\tau-ix+\alpha\text{sgn}(\tau))}n_B(\hbar v_F k)\right.$$

$$\left. +e^{k(\tau-ix+\alpha\text{sgn}(\tau))}(n_B(\hbar v_F k) + 1)\right].$$

At zero temperature $n_B = 0$, and (see (5.70))

$$\mathcal{D}_d^0(\tau - ix) = -\frac{2\pi}{L}\sum_{k>0}\frac{1}{k}e^{-k\,\text{sgn}(\tau)(\tau-ix+\alpha\text{sgn}(\tau))}$$

$$= -\sum_{n=1}^{\infty}\frac{1}{n}\left(e^{-(2\pi/L)\,\text{sgn}(\tau)(\tau-ix+\alpha\text{sgn}(\tau))}\right)^n \xrightarrow{L\to\infty}$$

$$\ln\left[\frac{2\pi}{L}(\text{sgn}(\tau)(\tau - ix) + \alpha)\right]. \quad (5.117)$$

At a finite temperature in the limit $L \to \infty$ the evaluation of (5.116) yields after rather more work, than for fermions see [6], H.2.b, for details

$$\mathcal{D}_d^{1/\beta}(\tau - ix) = \ln\left[\frac{2\beta\hbar v_F}{L}\,\sin\left(\frac{\pi}{\beta\hbar v_F}\left[\text{sgn}(\tau)(\tau - ix) + \alpha\right]\right)\right]. \quad (5.118)$$

The Green's functions for the left-movers are obtained by replacing $\tau - ix$ by $\tau + ix$.

Comparing (5.112) with (5.117) and (5.114) with (5.118) we see, that in the limit $L \to \infty$

$$\mathcal{G}_d(z^*) \to -\frac{1}{L}\,\mathrm{sgn}(\tau)\,e^{-\mathcal{D}_d(z^*)}. \tag{5.119}$$

This intuitively agrees with the bosonization formulas (5.82, 5.83). This intuition is fully justified. In order to demonstrate this we will need the following relation for the averages of exponents of a Bose operator $\mathcal{B}$:

$$\langle e^{\lambda \mathcal{B}} \rangle = e^{\frac{1}{2}\lambda^2 \langle \mathcal{B}^2 \rangle}. \tag{5.120}$$

It holds identically for any linear combination of bosons, $\mathcal{B} = \sum_{q>0}(\mu_q b_q^\dagger + \tilde{\mu}_q b_q)$, if the average is taken over the thermal state of the free boson Hamiltonian (5.62) (see [6], C.10, Theorem 4). In the thermodynamic limit it holds for an arbitrary state and Hamiltonian, as long as $\langle \mathcal{B}^{2n+1} \rangle = 0$. Indeed, then

$$\langle e^{\lambda \mathcal{B}} \rangle = \sum_{n=0}^{\infty} \frac{\lambda^{2n}}{(2n)!} \langle \mathcal{B}^{2n} \rangle = \sum_{n=0}^{\infty} \frac{\lambda^{2n}}{(2n)!} \frac{(2n)!}{2^n\, n!} \langle \mathcal{B}^2 \rangle^n = e^{\frac{1}{2}\lambda^2 \langle \mathcal{B}^2 \rangle}.$$

Here we have used the weak version of the Wick's theorem (Sect. 2.2.1) to reduce each macroscopic average $\langle \mathcal{B}^{2n} \rangle$ to the sum of $(2n)!/(2^n\, n!)$ identical fully contracted terms $\langle \mathcal{B}^2 \rangle^n$. Next, from (5.120) and the Baker-Hausdorff formula we find

$$\langle e^{\lambda_1 \mathcal{B}_1} e^{\lambda_2 \mathcal{B}_2} \rangle = e^{\frac{1}{2}(\lambda_1^2 \langle \mathcal{B}_1^2 \rangle + \lambda_2^2 \langle \mathcal{B}_2^2 \rangle) + \lambda_1 \lambda_2 \langle \mathcal{B}_1 \mathcal{B}_2 \rangle}. \tag{5.121}$$

Then, substituting (5.82), we find

$$\mathcal{G}_d(\tau - ix) = -\frac{1}{2\pi\alpha}\theta(\tau)\langle \mathcal{F}_d e^{-\frac{2\pi}{L}\mathcal{N}_d(\tau-ix)} e^{i\phi^d(x,\tau)} e^{-i\phi^d(0,0)} \mathcal{F}_d^\dagger \rangle$$
$$+ \frac{1}{2\pi\alpha}\theta(-\tau)\langle e^{-i\phi^d(0,0)} \mathcal{F}_d^\dagger \mathcal{F}_d e^{-\frac{2\pi}{L}\mathcal{N}_d(\tau-ix)} e^{i\phi^d(x,\tau)} \rangle$$

(the factor $\exp[-\frac{2\pi}{L}\mathcal{N}_d(\tau-ix)]$ reflects the dependence of the Klein factor on (imaginary) time, following from the commutation relations (5.85) and the Hamiltonian (5.62), (5.63)). Finally, using (5.84), dropping the $\mathcal{N}_d/L \to 0$ terms in the exponents, applying Eq. (5.121) and using $\mathcal{D}(0) = \ln(2\pi\alpha/L)$, we see, that indeed

$$\mathcal{G}_d(\tau - ix) \to -\frac{1}{2\pi\alpha}\mathrm{sgn}(\tau) e^{\langle \mathcal{T}_\tau \phi(x,\tau)\phi(0,0) \rangle - \frac{1}{2}(\langle \phi(0,0)\phi(0,0) \rangle + \langle \phi(x,\tau)\phi(x,\tau) \rangle)}$$
$$= -\frac{1}{2\pi\alpha}\mathrm{sgn}(\tau) e^{-\mathcal{D}_d(\tau-ix)+\mathcal{D}_d(0)} = -\frac{1}{L}\mathrm{sgn}(\tau) e^{-\mathcal{D}_d(\tau-ix)}.$$

For the left-movers, of course,

$$\mathcal{G}_s(z) \to -\frac{1}{L}\,\mathrm{sgn}(\tau)\,e^{-\mathcal{D}_s(z)}. \tag{5.122}$$

5.2.5.1 Tomonaga-Luttinger Versus Fermi Liquid

Taking the Fourier transform of the fermion Green's function with respect to the position, we obtain the single particle momentum distribution function (see (5.45)):

$$
n(k) = \int dx e^{-ikx} \langle \psi^\dagger(x, 0)\psi(0, 0)\rangle
$$

$$
= -\frac{i}{2} \lim_{t \to -0} \int dx \left(e^{-i(k-k_F)x} G_d(x, t) + e^{-i(k+k_F)x} G_s(x, t)\right), \quad (5.123)
$$

where $G_{d,s}$ are real time causal Green's functions (cf. Eq. (2.48)). Using the analytic continuations of zero-temperature thermal Green's functions (5.112, 5.113) (which in this case amounts to the replacement of τ with $i v_F t$) and taking the contour integrals, we find that the contributions of right- and left-movers are $n_d(k) = (1/2)\theta(k_F - k)$ and $n_s(k) = (1/2)\theta(k_F + k)$, as one would expect. In the presence of interactions the situation drastically changes. Expressing the physical Fermi operators in terms of the Bogoliubov-transformed Bose operators (5.92), and $n(k)$ in terms of their Green's functions, one discovers that in the presence of an *infinitesimally small* interaction $n(k) = 1/2$: not only the step at $|k| = k_F$ disappears, but the distribution becomes altogether momentum-independent ([15]; [3], Sect. 4.4.E). This is in a sharp contrast to the behaviour of the Fermi liquid, where the step in $n(k)$ at $|k| = k_F$ becomes less than unity, but still survives in the presence of interactions (Lifshits and Pitaevskii 1980, Sect. 10). Like in the case of the instability of the normal state of a superconductor considered in Chap. 4, the dependence of the Green's functions on the interaction strength is non-analytic and could not be reproduced by the perturbation theory.

5.3 Conformal Field Theory and the Orthogonality Catastrophe

5.3.1 Conformal Symmetry

A remarkable property of the Tomonaga-Luttinger model is that the field operators and Green's functions of left-movers depend only on the complex variable $z = \tau + ix$, and those of right-movers - only on its complex conjugate z^*, while the Hamiltonian splits into a sum of z- and z^*-dependent parts.[4] This provides significantly more than just the opportunity to treat the two sectors separately. Analytic functions of complex variable realize a special kind of symmetry of the complex plane - *local conformal invariance*.[5] Specifically, a conformal mapping is a one-to-one correspondence between the domains D and D' in the complex plain such, that in the vicinity

---

[4] z- (resp. z^*-) dependent quantities are called holomorphic (antiholomorphic), or analytic (antianalytic).

[5] *Global* conformal invariance exists in higher dimensions as well.

of any point in D it is an orthogonal, orientation-preserving transformation. In other words, it preserves the angles between intersecting curves, transforms infinitesimal circles into infinitesimal circles, and maintains the clockwise direction on them. Only the overall scale may be locally changed.

One of the basic theorems of complex analysis states that a function $f(z)$ realizes a conformal mapping of the domain D if and only if it is single-valued and analytic in D, and its derivative $df(z)/dz \neq 0$ everywhere in D. Thus *any* analytic function $f(z)$ (or $f(z^*)$) can be used for a mapping between *some* D and D', determined by the properties of the specific function. (In this context z and z^* should be treated as independent complex variables.) This is what makes the 2-dimensional case (one spatial dimension, plus (imaginary) time) so special and rich in possibilities. The *conformal field theory*, which investigates these possibilities and applies them to a broad range of physical problems, is a very large subject and well beyond the scope of this book.[6] We limit our acquaintance with it to what is necessary to derive the formulas (5.12, 5.33) for the Anderson orthogonality exponent.

In a conformally invariant theory there exist so called *primary fields*, that is, such operators $\mathcal{O}_{h,h^*}(z, z^*)$, that under the conformal mapping $z \to w(z)$, $z^* \to w^*(z^*)$ transform as

$$\mathcal{O}_{h,h^*}(z, z^*) \to \mathcal{O}_{h,h^*}(w, w^*) = \left(\frac{dw}{dz}\right)^{-h} \left(\frac{dw^*}{dz^*}\right)^{-h^*} \mathcal{O}_{h,h^*}(z, z^*) \quad (5.124)$$

In particular, their correlation functions in the infinite complex plane will have the form

$$\langle \mathcal{O}_{h,h^*}(z_1, z_1^*) \mathcal{O}_{h,h^*}^{\dagger}(z_2, z_2^*) \rangle = \left(\frac{1}{z_1 - z_2}\right)^{2h} \left(\frac{1}{z_1^* - z_2^*}\right)^{2h^*}. \quad (5.125)$$

The real integers h and h^* are called *conformal dimensions* of the field. As is clear from Eqs. (5.112), (5.113) and the bosonization formulas, for the system of free bosons (or fermions) in 1+1 dimensions these operators are left- and right-movers, $s(z) \propto e^{i\phi^s(z)}$ and $d(z^*) \propto e^{i\phi^d(z^*)}$, with the conformal dimensions $(1/2,0)$ and $(0,1/2)$ respectively. Using (eq:Last-100) and an appropriate analytic function, which maps the complex plane to a domain D, one can obtain from (5.125) the correlation function in this domain.[7]

---

[6] An introduction into it can be found in [5], Chap. 24, 25; [4], Sect. 2.2, and a massive exposition in [1].

[7] The conformal field theory with boundaries was developed by Cardy [8].

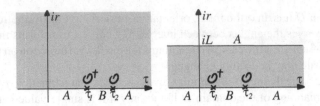

Fig. 5.7 Conformal field theory for a 1D system of infinite (*left*) and finite (*right*) length. The time axis is chosen to coincide with the real axis. Boundary condition changing operators act at τ_1 and τ_2

5.3.2 Conformal Dimensions, the Energy Spectrum and the Anderson Exponent

We have previously reduced the problem of the orthogonality catastrophe to the behaviour of a system of free one-dimensional fermions on a ray, $r \geq 0$, in the presence of a scattering potential at the boundary. In order to apply the methods of conformal field theory to this problem, it is necessary to consider such *boundary condition changing operators* [7, 8]. In the following we put $\hbar = 1$, $v_F = 1$.

Let us choose the coordinates in the complex plane $z = \tau + ir$, and consider the upper half-plane, $r \geq 0$ (Fig. 5.7). The real axis then represents the position of the boundary, where the scattering potential is located. It produces the scattering phase, relating the in- and outgoing waves via

$$\psi_{out}(0) = e^{2i\delta(k_F)}\psi_{in}(0). \tag{5.126}$$

Change of the scattering phase, which reflects the creation and filling in of the core hole, is produced by the operator $\mathcal{O}$, which changes the boundary condition from "A" (no core hole, zero phase shift) to "B" (core hole, phase shift δ). Assuming that $\mathcal{O}$ is a primary field with the conformal dimension x, its zero-temperature Green's function will be

$$\langle A|\mathcal{O}(\tau_1)\mathcal{O}^\dagger(\tau_2)|A\rangle = \frac{1}{(\tau_1 - \tau_2)^{2x}}. \tag{5.127}$$

Here $|A\rangle$ is the ground state of the system with the boundary condition "A" at the origin ($\Im z \equiv r = 0$). The analytic function

$$z(w) = Le^{\pi w/L} \equiv Le^{\pi(u+iv)/L} \tag{5.128}$$

maps the upper half-plane into the strip $0 \leq v \leq L$, which corresponds to a system of a finite length L, with the positive real ray mapped on the lower, and the negative real ray - on the upper boundary of the strip. Let's take $\tau_1, \tau_2 > 0$. Then the boundary condition at $r = L$ will be always "A". The Green's function (5.127) is transformed according to (5.124):

$$\langle AA|\mathcal{O}(u_1)\mathcal{O}^\dagger u_2)|AA\rangle = \left(\frac{dz}{dw(u_1)}\right)^x \left(\frac{dz}{dw(u_2)}\right)^x \left(\frac{1}{z(u_1)-z(u_2)}\right)^{2x}$$

$$= \left(\frac{\pi/2L}{\sinh[(\pi/2L)(u_1-u_2)]}\right)^{2x} \underset{(u_2-u_1)\gg L}{\longrightarrow} \left(\frac{\pi}{L}\right)^{2x} e^{-\frac{\pi x(u_2-u_1)}{L}},$$

$$(5.129)$$

where $|AA\rangle$ is the ground state of the system of length L with the boundary conditions "A" (i.e., with zero phase shifts) at both ends. On the other hand, by directly inserting the closure relation $\mathcal{I} = \sum |n\rangle\langle n|$ in the expression $\langle AA|\mathcal{O}(u_1)\mathcal{O}^\dagger(u_2)|AA\rangle$, we find

$$\langle AA|\mathcal{O}(u_1)\mathcal{O}^\dagger(u_2)|AA\rangle = \sum |\langle AA|\mathcal{O}|AB,n\rangle|^2 e^{-(E_{AB}^n - E_{AA}^0)(u_2-u_1)}. \quad (5.130)$$

Here n labels all energy eigenstates in the system of length L with the boundary conditions "B" at $r = 0$ and "A" at $r = L$.

In the limit $(u_2 - u_1) \gg L$ the leading exponent in (5.130) should coincide with (5.129). This exponent corresponds to the lowest-energy state of the system with the phase shift, which is usually the ground state energy of the system with the boundary conditions "A" and "B", i.e.

$$\left(\frac{\pi}{L}\right)^{2x} e^{-\frac{\pi x(u_2-u_1)}{L}} \sim |\langle AA|\mathcal{O}|AB,0\rangle|^2 e^{-(E_{AB}^0 - E_{AA}^0)(u_2-u_1)}. \quad (5.131)$$

Therefore the conformal dimension of the boundary condition change operator can be obtained from the shift of the ground state energy due to change in the boundary conditions (Affleck and Ludwig [7]):

$$x = \frac{L}{\pi}(E_{AB}^0 - E_{AA}^0). \quad (5.132)$$

The matrix element $\langle AA|\mathcal{O}|AB,0\rangle$ is the overlap between the ground states of the system with and without the scattering potential at the origin. Thus, the power x in (5.131) is the Anderson orthogonality exponent of Eq. (5.11), and the relation (5.132) provides a convenient way of computing it directly from the energy spectrum of the system.

5.3.2.1 Ground State Energy in the Presence of Scattering Potential

In order to find the ground state energy shift due to scattering potential, let us recall Eq. (5.20) for the asymptotic form of the wave function:

$$\Psi_{\tilde{k}_n}(x_j) \sim \sin\left[\tilde{k}_n x_j + \delta(\tilde{k}_n)\right], \quad j \gg 1. \quad (5.133)$$

Here $\tilde{k}_n$ is the wave vector shifted from its value k_n in the absence of scattering potential, $x_j = ja$, and a is the lattice constant. The ground state energy is then

$$E_0 = \sum_{n=1}^{N} \epsilon(\tilde{k}_n) \tag{5.134}$$

(recall that in Sect. 5.1.2 we were imposing free boundary conditions on the 1D chain; therefore $k_n = \pi n / L$, $n > 0$). Taking in (5.133) $x_j = L$ and demanding, in order to satify the boundary condition, that

$$\tilde{k}_n L + \delta(\tilde{k}_n) = k_n L,$$

we obtain

$$\tilde{k}_n = k_n - \frac{\delta(\tilde{k}_n)}{L} \approx k_n - \frac{\delta(k_n)}{L} + \frac{\delta(k_n)\delta'(k_n)}{L^2} \equiv K(k_n), \tag{5.135}$$

where $\delta'(k) = d\delta/dk$.

The energy (5.134) can be evaluated using the Euler-MacLaurin formula:

$$\sum_{n=1}^{N} F\left(n - \frac{1}{2}\right) = \int_0^N dx\, F(x) - \frac{1}{24}(F'(N) - F'(0)) + O(F''). \tag{5.136}$$

Setting

$$F\left(n - \frac{1}{2}\right) = \epsilon(K(k_n)), \tag{5.137}$$

we get $F(N) = \epsilon(K(k_F))$, and

$$E_0 = \int_0^N dn\, \epsilon\left[K\left(\frac{\pi\left(n + \frac{1}{2}\right)}{L}\right)\right] - \frac{\pi v_F}{24L}, \tag{5.138}$$

where now $v_F = \epsilon'(k_F)$ (while still $\hbar = 1$), and we have kept only terms $O(1/L)$. Changing the integration variable to $\xi = \pi(n + 1/2)/L$, using (5.135) and again neglecting corrections of order $1/L^2$, we find

$$E_0 = L \int_0^{k_F} \frac{d\xi}{\pi} \epsilon\left[K(\xi)\right] - \frac{\pi v_F}{24L}$$

$$= L \int_0^{k_F} \frac{d\xi}{\pi} \left[\epsilon(\xi) - \frac{\epsilon'(\xi)\delta(\xi)}{L} + \frac{\epsilon''(\xi)\delta^2(\xi)}{2L^2} + \frac{\epsilon'(\xi)\delta'(\xi)\delta(\xi)}{L^2}\right] - \frac{\pi v_F}{24L}. \tag{5.139}$$

Integrating by parts and using $\epsilon'(0) = 0$, we eventually obtain the desired result:

$$E_0 = L \int_0^{k_F} \frac{d\xi}{\pi} \epsilon(\xi) - \frac{1}{\pi} \int_{\epsilon(0)}^{\epsilon(k_F)} d\epsilon \delta(\epsilon) + \frac{\pi v_F}{L}\left[\frac{1}{2}\left(\frac{\delta(k_F)}{\pi}\right)^2 - \frac{1}{24}\right] + O(\frac{1}{L^2}).$$

(5.140)

It can be directly checked using the tight-binding model of Sec. 5.1.2 with free boundary conditions. There $\epsilon(k) = -2T\cos k$, $v_F = 2T\sin k_F$, and the ground state energy is given exactly by a geometric series,

$$E_0 = T - \frac{T\sin k_F}{\sin(\pi/2l)} = \frac{l}{\pi}v_F + T - \frac{\pi v_F}{24l} + O(\frac{1}{l^2}).$$

(5.141)

If add to the ground state n extra electrons directly above the Fermi level, the energy of the system becomes

$$E_n = E_0 + \sum_{m=1}^{n} \epsilon\left(k_F - \frac{\delta(k_F)}{L} + \frac{\pi(m-1/2)}{L}\right)$$

$$= E_0 + n\epsilon(k_F) + \frac{\pi v_F}{L}\sum_{m=1}^{n}\left(m - \frac{1}{2} - \frac{\delta(k_F)}{\pi}\right) + O(\frac{1}{L^2}).$$

(5.142)

Then, instead of (5.140), we find [12]

$$E_n = L \int_0^{k_F} \frac{d\xi}{\pi} \epsilon(\xi) - \frac{1}{\pi} \int_{\epsilon(0)}^{\epsilon(k_F)} d\epsilon \delta(\epsilon) + n\epsilon(k_F)$$

$$+ \frac{\pi v_F}{L}\left[\frac{1}{2}\left(n - \frac{\delta(k_F)}{\pi}\right)^2 - \frac{1}{24}\right] + O(\frac{1}{L^2}).$$

(5.143)

5.3.2.2 Anderson Orthogonality Exponent

Returning to Eq. (5.132) for the Anderson exponent, we substitute there the $O(1/L)$-correction to the ground state energy from (5.140) and immediately find, that we have successfully rederived Eq. (5.12):

$$x = \frac{1}{2}\left(\frac{\delta(k_F)}{\pi}\right)^2.$$

Moreover, we are now equipped to find out what happens, if the core hole potential creates a bound state. Then the Green's function (5.130) becomes the sum of two terms, corresponding to the bound state being empty or filled [12]:

$$\langle AA|O(u_1)O^\dagger(u_2)|AA\rangle = \sum |\langle AA|O|AB, n, e\rangle|^2 e^{-(E^n_{AB,e}-E^0_{AA})(u_2-u_1)}$$
$$+ \sum |\langle AA|O|AB, n, f\rangle|^2 e^{-(E^n_{AB,f}-E^0_{AA})(u_2-u_1)}.$$

$$(5.144)$$

They give rise to two peaks in the absorption rate, separated by the bound state energy $|\epsilon_B|$. The processes of core hole creation with or without filling the bound state can be in the long time limit considered independently, as due to separate operators. In case of filled bound state the result will be the same as Eq. (5.12): the wave function of the bound state is exponentially decaying and cannot influence the $O(1/L)$-terms in the ground state energy. Therefore

$$x_f = \frac{1}{2}\left(\frac{\delta(k_F)}{\pi}\right)^2.$$

$$(5.145)$$

If the bound state is empty, then the corresponding operator not only changes the boundary condition, but also creates an extra electron above the Fermi level. Using (5.143) with $n = 1$, we finally reproduce, using quite a different approach, the result of [9, 10]:

$$x_e = \frac{1}{2}\left(1 - \frac{\delta(k_F)}{\pi}\right)^2.$$

$$(5.146)$$

We have reached the goal in a circuitous way, but learned some useful techniques and did not need the assumption $\delta \ll \pi$!

5.4 Problems

- *Problem 1*

Verify that Eq. (5.87) is equivalent to Eqs. (5.62, 5.63). When integrating, make use of $\rho(x + L) = \rho(x)$.

- *Problem 2*

Obtain the explicit expression (5.114) for the thermal Green's function of free fermions in the Tomonaga-Luttinger model. Taking the integral over k, consider it as a complex variable and use the method of contour integration. Take into account that the integrand has infinitely many simple poles on the imaginary axis, and close the integration contour in either upper or lower half-plane of complex wave vector k, depending on the sign of x.

- *Problem 3*

Using conformal mapping, obtain the finite-temperature Green's function of free one-dimensional fermions (5.115) from the zero-temperature one (5.113).

References

Books and Reviews

1. Di Francesco, P., Mathieu, P., Sénéchal, D.: Conformal Field Theory, Springer, GTCP, New York (1997) (A fundamental textbook on conformal field theory in 1+1 and higher dimensions.)
2. Eggert, S.: One-dimensional quantum wires: A pedestrian approach to bosonization. In: Kuk Y. et al. (eds.) Theoretical Survey of One Dimensional Wire Systems, Sowha Publishing, Seoul (2007); arXiv:0708.0003 (chapter 2)
3. Mahan, G.D.: Many-Particle Physics, 2nd edn. Plenum Press, New York (1990)
4. Nagaosa, N.: Quantum Field Theory in Strongly Correlated Electronic Systems. Springer, TMP, Berlin-Heidelberg (2010)
5. Tsvelik, A.M.: Quantum Field Theory in Condensed Matter Physics. Cambridge University Press (1995)
6. von Delft, J., Schoeller, H.: Bosonization for Beginners - Refermionization for Experts, Ann. Phys. 4, 225 (1998) (A very detailed tutorial, where bosonization is introduced constructively and different approaches are compared.)

Articles

7. Affleck, I., Ludwig, A.W.W.: J. Phys. A· Math. Gen. 27, 5375 (1993)
8. Cardy, J.L.: Nucl. Phys. B 324, 581 (1989)
9. Combescoot, M., Nozières, P.: J. Physique 32, 913 (1971)
10. Hopfield, J.J.: Comment. Solid State Phys. 11, 40 (1969)
11. Nozières, P., De Dominicis, C.T.: Phys. Rev. 178, 178 (1969)
12. Zagoskin, A.M., Affleck, I.: J. Phys. A: Math. Gen. 30, 5743 (1997)
13. Anderson, P.W., Phys. Rev. Lett. 18, 1049 (1967)
14. Thouless, D.J.: Quantum mechanics of many-body systems. Academic Press, New York (1972)
15. Mattis, D.C., Lieb, E.H., J. Math. Phys. N.Y. 6, 304 (1965)

Appendix A
Friedel Oscillations

In the static limit the polarization operator, which describes screening of the Coulomb potential by the Fermi gas, is given by Eqs. (2.71, 2.72):

$$\Pi_0(p) = -\frac{m p_F}{2\pi^2}\left(1 + \frac{p_F^2 - p^2/4}{p_F p}\ln\left|\frac{p_F + p/2}{p_F - p/2}\right|\right).$$

This formula was obtained in the random phase approximation at zero temperature. As mentioned before, the second term in the parentheses is non analytic at $p = 2p_F$. Indeed, though

$$\lim_{p\to 2p_F}\frac{p_F^2 - p^2/4}{p_F p}\ln\left(\frac{p_F + p/2}{p_F - p/2}\right) \equiv \lim_{p\to 2p_F} g(p) = 0, \qquad (A.1)$$

all the derivatives of $g(p)$ diverge at this point. The screened Coulomb potential in the coordinate representation is then

$$
\begin{aligned}
U_{\text{eff}}(r) &= \int \frac{d^3 p}{(2\pi)^3}\frac{e^{i\mathbf{pr}}4\pi e^2/p^2}{1 - (4\pi e^2/p^2)\Pi_0(p)}\\
&= \frac{e^2}{\pi}\int_0^\infty dp\, p^2 \int_{-1}^1 d(\cos\theta)\frac{e^{ipr\cos\theta}}{p^2 + (1/2)q_{TF}^2(1 + g(p))}\\
&= \frac{2e^2}{\pi r}\int_0^\infty dp\frac{p\sin(pr)}{p^2 + (1/2)q_{TF}^2(1 + g(p))}\\
&= \frac{e^2}{\pi r}\Im\int_{-\infty}^\infty dp\frac{p e^{ipr}}{p^2 + (1/2)q_{TF}^2(1 + g(p))}. \qquad (A.2)
\end{aligned}
$$

Here we took into account that $g(p)$ and $p\sin(pr)$ are even functions of p and extended the integration to all of the real axis. Now we can, as usual, close the integration contour in the upper halfplane of the complex variable p and reduce

A. Zagoskin, *Quantum Theory of Many-Body Systems*,
Graduate Texts in Physics, DOI: 10.1007/978-3-319-07049-0,
© Springer International Publishing Switzerland 2014

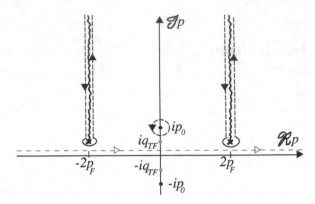

Fig. A.1 Analytic structure of the integrand in (A.2) after regularization, and the initial (**a**) and deformed (**b**) integration contours

the integral to the contributions from the singularities of the integrand. Replacing $g(p)$ in this expression with its limiting value $g(0) = 1$ we would indeed obtain the exponential screening $\sim\exp(-q_{TF}r)$ due to the simple poles at $p = \pm iq_{TF}$.

Taking into account the actual expression for $g(p)$ drastically changes the outcome. It is straightforward to see that the poles will be slightly shifted along the imaginary axis,

$$p = \pm ip_0 \approx \frac{\pm iq_{TF}}{\sqrt{1 - \frac{1}{12}\left(\frac{q_{TF}}{p_F}\right)^2}}.$$

But the main difference comes from the logarithm having singularities - branch points $p = \pm 2p_F$.

Following [3], let us regularize the logarithm:

$$\ln\left|\frac{p_F + p/2}{p_F - p/2}\right| = \frac{1}{2}\ln\left[\frac{|p_F + p/2|^2}{|p_F - p/2|^2}\right] = \lim_{\epsilon \to 0}\ln\left[\frac{(p + 2p_F)^2 + \epsilon^2}{(p - 2p_F)^2 + \epsilon^2}\right].$$

This way we shift the singularities to the points $p = \pm 2p_F \pm i\epsilon$, away from the real axis (see Fig. A.1). The branch cuts of the logarithm in the upper halfplane are chosen along the rays $\pm 2p_F + i\epsilon + is$, where $0 \le s < \infty$. Since

$$\frac{1}{2}\ln(z^2 + \epsilon^2) = \frac{1}{2}[\ln(z - i\epsilon) + \ln(z + i\epsilon)], \quad z = p \pm 2p_F,$$

it is clear that going full circle around a branch point one adds to the logarithmic term an extra $\pm \pi i$.

Now we can deform the integration contour to run mostly along the infinitely large cemicircle in the upper complex half-plane. Its contribution to the integral will be suppressed due to the exponential factor $\exp(ipr) \to 0$. The surviving terms come from the integrations along the branch cuts and around the pole at ip_0 (Fig. A.1). The latter, together with the prefactor in (A.2), will yield the exponentially screened potential,

$$\sim \frac{e^2}{r} e^{-p_0 r},$$

but as we shall see, at $r \to \infty$ it becomes irrelevant compared to the slower-decaying contributions from the former.

Consider the integral around the left branch cut. Taking $\epsilon \to 0$, we find (the prime indicates that we first integrate along the left, and then along the right bank of the cut):

$$I_{-2p_F} = \frac{e^2}{\pi r} \Im \left\{ \left[\int_{-2p_F+i\infty}^{-2p_F} + \int_{-2p_F}^{-2p_F+i\infty} \right]' dp \frac{p e^{ipr}}{p^2 + (q_{TF}^2/2)(1 + g(p))} \right\}$$

$$= \frac{e^2}{\pi r} \Im \left\{ \int_{-2p_F}^{-2p_F+i\infty} dp \Delta \left[\frac{p e^{ipr}}{p^2 + (q_{TF}^2/2)(1 + g(p))} \right] \right\}.$$

In the last expression $\Delta[...]$ is the difference between the right and left banks of the branch cut, which is solely due to the multivaluedness of the logarithm. Writing now $p = -2p_F + is$ and replacing the slowly varying terms under the integral with their values on the real axis (which is justified in the limit $r \to \infty$ by the exponent, $\exp(ipr) = \exp(-2ip_F r)\exp(-sr)$), rapidly decaying away from the real axis, we can write

$$I_{-2p_F} \approx \frac{e^2}{r} \frac{(1/2)q_{TF}^2}{(4p_F^2 + (1/2)q_{TF}^2)^2} \Im \left\{ e^{-2ip_F r} \left[\frac{i}{r^2} \int_0^\infty du\, u\, e^{-u} + \frac{1}{r^3} \int_0^\infty du\, u^2\, e^{-u} \right] \right\}$$

$$\approx \frac{e^2}{r^3} \frac{(1/2)q_{TF}^2}{(4p_F^2 + (1/2)q_{TF}^2)^2} \cos(2p_F r). \tag{A.3}$$

We have kept the slowest-decaying term. The integral over the branch cut at $p = 2p_F$ yields the same expression, and therefore, as promised, at large distances the screened potential behaves as

$$U_{\text{eff}}(r) \sim \frac{e^2}{r^3} \cos(2p_F r). \tag{A.4}$$

The non-analytic behaviour of the polarization operator at $p = \pm 2p_f$ is due to the sharp edges of the Fermi distribution; the momentum transmission $\pm 2p_F$ clearly corresponds to the transitions between the opposite points of the Fermi surface. Therefore one expects that when the Fermi step is smeared by finite temperature or interactions (like in a superconductor), an exponential screening should be restored. This is indeed the case (see [3], p.180). At a finite temperature T in a normal system, or at zero temperature in a superconductor with the superconducting gap Δ, Eq. (A.4) acquires an exponentially decaying factor, respectively

$$\exp[-2\pi r(mk_B T/\hbar^2 p_F)] \quad \text{or} \quad \exp[-r p_F(\Delta/\epsilon_F)].$$

You should not be too surprized realizing that the exponents, up to a factor of order unity, are simply the ratios of the distance r to the normal metal (resp. superconducting) coherence length,

$$\frac{\hbar v_F}{k_B T} \quad \text{and} \quad \frac{2\hbar v_F}{\pi \Delta},$$

which provide the natural length scales for these systems.

Appendix B
Landauer Formalism for Hybrid Normal-Superconducting Structures

B.1 The Landauer–Lambert formula

An important generalization of Landauer's formula (3.176) was made by Lambert [7]. He considered a situation in which besides elastic scatterers, the system contains a superconducting "island" (Fig. B.1). We address here the simplest, single-channel, case. That is, in the spirit of Landauer's approach, the equilibrium electronic reservoirs are connected to the "scattering part" of the system by perfect one-dimensional leads A, A'. The former can be considered as a "black box" containing, along with normal conductors and scatterers, a superconductor in some way connected to the rest of the system.

Sweeping all the details of this inner structure under the rug, we can describe it by a 4×4 matrix $\hat{\mathbf{P}}$ that gives us *probabilities* for a quasiparticle injected in a lead from the respective reservoir to be scattered to this or the other lead:

$$\hat{\mathbf{P}} = \begin{pmatrix} R_{ee} & R_{eh} & T_{ee'} & T_{eh'} \\ R_{he} & R_{hh} & T_{he'} & T_{hh'} \\ T_{e'e} & T_{e'h} & R_{e'e'} & R_{e'h'} \\ T_{h'e} & T_{h'h} & R_{h'e'} & R_{h'h'} \end{pmatrix}. \tag{B.1}$$

The matrix has a rich structure (a purely normal system would be described simply by transmission, T, and reflection, R, coefficients, related by the unitarity condition $T + R = 1$), because in the presence of a superconductor, quasiparticles can switch between particle and hole branches of the spectrum due to Andreev reflections (as in Fig. 4.16). For example, $R_{ee}(R_{eh})$ is the probability for an electron in the left lead to be reflected as an electron (hole) to the same lead, while $T_{e'e}(T_{h'e})$ gives the probability of its normal (Andreev) transmission to the other lead, etc. (see Fig. B.3). In other words, we have added to the system an off-diagonal scattering potential. The probability flux conservation (unitarity) requires that the elements in any row or column of $\hat{\mathbf{P}}$ add up to unity:

A. Zagoskin, *Quantum Theory of Many-Body Systems*,
Graduate Texts in Physics, DOI: 10.1007/978-3-319-07049-0,
© Springer International Publishing Switzerland 2014

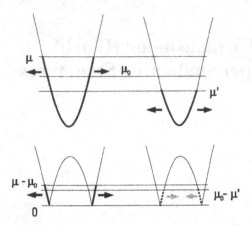

Fig. B.1 Landauer conductance in a normal-superconducting system

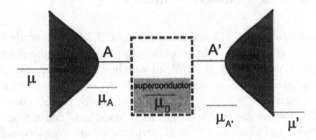

Fig. B.2 Measuring quasiparticle energies from zero or from μ_0: two equivalent pictures. Occupied states are shown by *solid lines* for quasielectrons, *dotted lines* for quasiholes

$$\sum_j P_{ij} = 1; \quad \sum_i P_{ij} = 1. \tag{B.2}$$

Let us denote the chemical potential of the superconductor by μ_0. If we apply a small bias $eV = \mu - \mu'$ between the normal reservoirs, evidently $\mu > \mu_0 > \mu'$. In the presence of a superconductor, it is expedient to use the "folded" dispersion law (Fig. B.1), which we used when considering Andreev reflections in Sect. 4.5, and to measure the quasiparticle energies from μ_0. Then we see that at zero temperature the left reservoir injects into the system quasielectrons with energies in the interval $[0, \ \mu - \mu_0]$; the right reservoir injects *quasiholes*, with energies within $[0, \ \mu_0 - \mu']$ (Fig. B.2).

To calculate the *two-probe* conductance of the system, we now find the current in, e.g., lead A and divide it by $(\mu - \mu')$:

$$G = \frac{I}{V}.$$

Fig. B.3 Matrix $\hat{P}$ and the physical sense of its elements

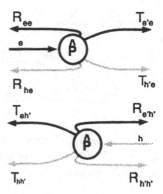

The current is

$$= ev_F \cdot \frac{2}{hv_F}[(\mu - \mu_0)(1 - R_{ee} + R_{he}) + (\mu_0 - \mu')(T_{hh'} - T_{eh'})]. \tag{B.3}$$

Here $2/(hv_F)$ is the one-dimensional density of states per velocity direction, and the meaning of the terms in brackets is self-evident.

Now we must somehow get rid of μ_0. This can be done if we impose the condition that there be no net electric current in or out of the superconductor. This will be so, e.g., if the superconductor is finite: otherwise it would accumulate electric charge until its field stops the further charge transfer. This gives us an extra equation necessary to exclude μ_0 from the answer.

The total current from the left reservoir flowing into the system is carried by quasielectrons and equals

$$\delta i = \frac{2e}{h}(\mu - \mu_0)(1 - R_{ee} + R_{he} - T_{e'e} + T_{h'e}) = \frac{4e}{h}(\mu - \mu_0)(R_{he} + T_{h'e}) \tag{B.4}$$

(we have used $1 = R_{ee} + R_{he} + T_{e'e} + T_{h'e}$). The current from the right reservoir is carried by quasiholes (thus the minus sign):

$$\delta i' = -\frac{4e}{h}(\mu_0 - \mu')(R_{h'e'} + T_{eh'}). \tag{B.5}$$

From the "no-charging" condition $\delta i + \delta i' = 0$ we find

$$\mu - \mu_0 = (\mu - \mu')\frac{R_{e'h'} + T_{eh'}}{R_{e'h'} + T_{eh'} + R_{he} + T_{h'e}}, \tag{B.6}$$

$$\mu - \mu_0 = (\mu - \mu')\frac{R_{he} + T_{h'e}}{R_{e'h'} + T_{eh'} + R_{he} + T_{h'e}},$$

and

$$G = \frac{Ie}{\mu - \mu'}$$

$$= \frac{2e^2}{h} \frac{(R_{e'h'} + T_{eh'})(1 - R_{ee} + R_{he}) + (R_{he} + T_{h'e})(T_{hh'} - T_{eh'})}{R_{e'h'} + T_{eh'} + R_{he} + T_{h'e}}$$

$$= \frac{2e^2}{h} \frac{(R_{e'h'} + T_{hh'})(R_{he} + T_{h'e}) + (R_{e'h'} + T_{eh'})(T_{e'e} + R_{he})}{R_{e'h'} + T_{eh'} + R_{he} + T_{h'e}}. \tag{B.7}$$

If there is particle–hole symmetry ($R_{hh} = R_{ee} = R_N$, $R_{eh} = R_{he} = R_A$, $T_{he'} = T_{eh'} = T'_A$ etc.), N, A denoting normal and Andreev processes, then the above formula reduces to

$$G = \frac{2e^2}{h} \frac{(R'_A + T'_A)(R_A + T_N) + (R_A + T_A)(R'_A + T'_N)}{R'_A + T'_A + R_A + T_A}. \tag{B.8}$$

Finally, if the system is spatially symmetric (that is, the difference between primed and nonprimed coefficients disappears), we see that simply

$$G = \frac{2e^2}{h}(T_N + R_A). \tag{B.9}$$

This is an intuitively clear result: In addition to the normal transmission channel, $\frac{2e^2}{h} T_N$, which we had in the normal case, another conductivity channel opens due to Andreev reflections.

This is only one of Landauer-type formulas that describe conductivity of normal-superconducting mesoscopic systems. We could, e.g., calculate the *four-probe* conductance,

$$\tilde{G} = \frac{Ie}{\mu_A - \mu_{A'}},$$

where μ_A, $\mu_{A'}$ are the chemical potentials in the leads. Evidently, $\mu > \mu_A \geq \mu_0 \geq \mu_{A'} > \mu'$, and therefore $\tilde{G} > G$. For example, if there are *no* scatterers in the system, then $\mu_A = \mu'_A$, and $\tilde{G}$ becomes infinite. On the other hand, $G = 2e^2/h$ stays finite, being an inverse of what is an analogue to Sharvin resistance of a clean point contact.

The chemical potentials of the wires are determined by the charge densities brought there by the currents, that is (we must take into account both directions of velocity, hence the factor of 2):

$$2 \times \frac{2}{hv_F}(\mu_A - \mu_0) = \frac{2}{hv_F}[(\mu - \mu_0)(1 - R_{he} + R_{ee}) + (\mu_0 - \mu')(T_{eh'} - T_{hh'})],$$

$$2 \times \frac{2}{hv_F}(\mu_{A'} - \mu_0) = \frac{2}{hv_F}[(\mu_0 - \mu_{A'})(-1 - R_{h'h'} + R_{e'h'}) \tag{B.10}$$

$$+ (\mu - \mu_0)(T_{e'e} - T_{h'e})].$$

Finding from here $\mu_A - \mu_{A'}$ and eliminating μ_0 as before, we obtain the expression

$$\tilde{G} = \frac{2e^2}{h} \frac{(R_{e'h'} + T_{eh'})(R_{he} + T_{e'e}) + (R_{he} + T_{h'e})(R_{e'h'} + T_{hh'})}{(R_{e'h'} + T_{eh'})(R_{ee} + T_{h'e}) + (R_{he} + T_{h'e})(R_{h'h'} + T_{eh'})}, \tag{B.11}$$

which in the case of particle–hole and spatial symmetry simplifies to

$$\tilde{G} = \frac{2e^2}{h} \frac{R_A + T_N}{R_N + T_A}. \tag{B.12}$$

By the way, if there is no superconductor in the system, then $R_A = 0$, $T_A = 0$, $T_N + R_N = 1$, and

$$\tilde{G} = \frac{2e^2}{h} \frac{T_N}{R_N} = \frac{2e^2}{h} \frac{T_N}{1 - T_N}. \tag{B.13}$$

This is the original Landauer formula for the four-point conductance; due to $(1 - T_N)$ in the denominator, it indeed diverges in the limit of ideal transparency of the barrier, $T_N \to 1$.

Many more theoretical and experimental results in this very dynamic field are discussed in [1, 2, 4].

B.2 Giant Conductance Oscillations in Ballistic Andreev Interferometers

As an example, we will consider an *Andreev interferometer*, that is, a mesoscopic device, where the "black box" of Fig. B.1 contains two or more separate NS interfaces with different superconducting phases. Since as we know, Andreev reflection coefficients are phase sensitive, the resulting conductance between normal reservoirs may depend on the (controllable) superconducting phase difference between these interfaces.

A simple version of this device is shown in Fig. B.4: a ballistic Andreev interferometer. It is essentially a clean SNS junction, its normal part being a wire AD with $N_\perp$ transverse modes, to which normal electronic reservoirs are only weakly linked in points B and C. Those points are the only places where normal scattering takes place: The quasiparticles fly through the wire ballistically, and reflections at the NS interfaces are purely of Andreev type. To further simplify the situation, we assume that normal scattering does not mix different transverse modes. Thus the wire reduces to a stack of independent one-dimensional "wires," each with its own effective longitudinal Fermi velocity $v_{F,v}^{\parallel} = \sqrt{v_F^2 - (v_{F,v}^{\perp})^2}$ ($v_{F,v}^{\perp}$ being determined by the transverse quantization conditions).

The longitudinal motion of the electrons in a vth transverse mode is quantized, giving rise to a set of Andreev levels (see 4.175)

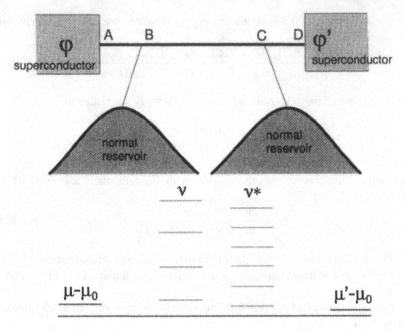

Fig. B.4 Landauer conductance of a ballistic Andreev interferometer. *Below* Andreev levels in the vth and v^*th transverse modes (see text)

$$E^{\pm}_{v,n} = \frac{\hbar v^{\|}_{F,v}}{2L}((2n+1)\pi \mp \Delta\varphi) \tag{B.14}$$

controlled by the superconducting phase difference $\Delta\varphi = \varphi - \varphi'$. The normal conductance of the system is due to the current-carrying states at the Fermi energy (that is, the levels with $E \approx 0$, the Fermi level being our reference point). Therefore, if the Andreev level coincides with the Fermi level, $E = 0$, we should expect a resonant peak in the conductance of the order of maximal quantum conductance $2e^2/h$.

Because longitudinal velocities in different modes differ, typically Andreev levels are nondegenerate (see Fig. B.4). But it is clear from (B.14) that the condition $E^{\pm}_{v,n} = 0$,

$$\Delta\varphi = \pm(2n+1)\pi, \tag{B.15}$$

is independent on $v^{\|}_{F,v}$! When it is satisfied, not one, but $N_{\perp}$ Andreev levels (one in each of $N_{\perp}$ transverse modes) are simultaneously aligned with the Fermi energy, thus producing a giant conductance peak with amplitude $N_{\perp}2e^2/h$. The width of the peak is of the order of the single-electron transparency of the barriers at B, C.

There is one more interesting detail. The very Andreev levels that are responsible for the normal conductance are responsible for the Josephson current that flows between the superconductors when $\Delta\varphi \neq 2\pi n$. When the Andreev level coincides with the Fermi level, the Josephson current abruptly changes sign, while normal conductance peaks. One could therefore infer that there is a relation between I_J and G in this system such as $G(\varphi) \approx -\partial I_J(\Delta\varphi)/\partial\varphi$.

The quantitative consideration of the problem [6] is based on the Landauer-Lambert formalism of the previous paragraph. (The presence of the Josephson current is consistent with our assumption that there is no net current flow in or out of the superconducting part of the system.) We assume that the electron–hole symmetry holds and that the normal part of the system is spatially symmetric. Then the conductance can be expressed as

$$G = \frac{2e^2}{h} 2 \int_0^\infty d\xi (T_N(\xi) + R_A(\xi)) \left(-\frac{\partial n_F(\xi)}{\partial \xi} \right) + \eta, \qquad (B.16)$$

where $T_N(\xi)(R_A(\xi))$ is the probability for normal transmission (Andreev reflection) of an electron incident from the left normal reservoir with energy ξ, $n_F(\xi)$ is the Fermi distribution function, and the energy ξ is measured from the Fermi level. (This is an evident generalization of our formula (B.9) to finite temperatures.) The additional term η in (B.16) reflects the fact that (B.9) requires spatial symmetry of the system, which would also include $\varphi = \varphi'$. In the presence of finite $\Delta\varphi$ we should use formula (B.8) instead, but due to the fact that this correction term is a rapidly oscillating function of the electron momentum ($\eta \sim \exp 2ik_F L$), we can neglect it if we are not interested in investigating the fine structure of conductance peaks.

The scattering coefficients in (B.16) can be found by solving the Bogoliubov-de Gennes equations in the normal part of the system. To do this, we must somehow describe scattering of electrons and holes at the junctions B and C. This is conveniently done by introducing (after [5]) identical, real scattering matrices

$$\mathbf{S} = \begin{pmatrix} -\epsilon/2 & 1-\epsilon/2 & \sqrt{\epsilon} \\ 1-\epsilon/2 & -\epsilon/2 & \sqrt{\epsilon} \\ \sqrt{\epsilon} & \sqrt{\epsilon} & -1+\epsilon \end{pmatrix}. \qquad (B.17)$$

Here $\epsilon \ll 1$ parametrizes the weak coupling of the system to normal reservoirs. For example, a quasiparticle reflected from the left superconductor has probability $|1 - \epsilon/2|^2 \approx 1$ to pass through B unhindered, while the probability of it being diverted to the normal reservoir is $|\sqrt{\epsilon}|^2$ and of being reflected back only $|-\epsilon/2|^2$, while a quasiparticle incident from the left normal reservoir is reflected back with probability $|-1+\epsilon|^2 \approx 1$.

Neglecting all quickly oscillating terms and all terms of order higher than ϵ^2, we finally obtain

$$T_N(\xi) \approx R_A(\xi) \approx \sum_{\sigma=\pm 1} \frac{\frac{1}{2}\epsilon^2}{1 + 2\epsilon^2 + \cos(\Delta\varphi + \sigma \frac{2L}{\hbar v_F^\parallel} \xi)}. \qquad (B.18)$$

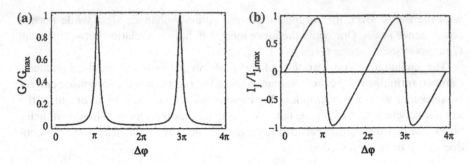

Fig. B.5 Phase dependence of the normal conductance (**a**) and Josephson current (**b**) in a ballistic Andreev interferometer at zero temperature and $\epsilon = 0.1$

Resonance is achieved at energies of Andreev levels (B.14), in agreement with our qualitative reasoning.

At zero temperature the resonant conductance depends only on $T_N(0)$, $R_A(0)$. Since the contribution to the resonant conductance of each transverse mode is exactly the same, the total resonant conductance of the system (within accuracy of ϵ^2) is (Fig. B.5a)

$$G(\Delta\varphi) = N_\perp \frac{2e^2}{h} \frac{2\epsilon^2}{1 + 2\epsilon^2 + \cos\Delta\varphi}. \tag{B.19}$$

We have described the method of calculation of the Josephson current in this system in Sect. 4.5.4. The only difference is that Andreev levels are broadened not due to impurity scattering, but due to ϵ-proportional "leakage" into the normal reservoirs, and we take $T = 0$. As a result,

$$I_J^{(\epsilon)}(\Delta\varphi) = N_\perp \frac{2e\overline{v}_F^\parallel}{\pi L} \sum_{n=1}^{\infty} \frac{(-1)^{n+1} e^{-2|n|\epsilon} \sin n\Delta\varphi}{n}, \tag{B.20}$$

where $\overline{v}_F^\parallel = N_\perp^{-1} \sum_{\nu=1}^{N_\perp} v_{F,\nu}^\parallel$ (Fig. B.5b).

Comparing (B.20) and (B.19), we see that the normal conductance and Josephson current in this system are indeed related by

$$G(\Delta\varphi) = \epsilon \left(-\frac{eL}{\hbar\overline{v}_F^\parallel} \frac{dI_J^{(\epsilon)}}{d\Delta\varphi} + \frac{2e^2}{h} N_\perp \right) \tag{B.21}$$

(within the accuracy of ϵ^2, that is, neglecting the details of the conductance peak structure on a finer scale).

References

Books and reviews

1. Beenakker, C.W.J.: Quantum transport in semiconductor-superconductor microjunctions. In: Akkermans, E. Montanbaux, G. Pichard, J.L. (eds.). Mesoscopic Quantum Physics North-Holland, Amsterdam (1994)
2. Beenakker, C.W.J.: Random-matrix theory of quantum transport. Rev. Mod. Phys. **69**, 731 (1997)
3. Fetter, A.L., Walecka, J.D.: Quantum Theory of Many-Particle Systems. McGraw-Hill, San Francisco (1971)
4. Lambert, C.J., Raimondi, R.: Phase-coherent transport in hybrid superconducting nanostructures. J. Phys. Cond. Matter 10, 901 (1998)

Articles

5. Büttiker, M., Imry, Y.: Phys. Rev. A **30**, 1982 (1984)
6. Kadigrobov, A., Zagoskin, A., Shekhter, R.I., Jonson, M.: Phys. Rev. B **52**, R8662 (1995)
7. Lambert, C.J.: J. Phys.: Cond. Matter 3, 6579 (1991)

Index

A. Zagoskin, *Quantum Theory of Many-Body Systems*,
Graduate Texts in Physics, DOI: 10.1007/978-3-319-07049-0,
© Springer International Publishing Switzerland 2014

Printed in the United States
By Bookmasters